EXTRASOLAR PLANETS

This volume presents the lectures from the sixteenth Canary Islands Winter School, which was dedicated to extrasolar planets. Research into extrasolar planets is one of the most exciting fields of astrophysics, and the past decade has seen research leap from speculations on the existence of planets orbiting other stars to the discovery of over 200 planets to date.

The book covers a wide range of issues involved in extrasolar planet research, from the state-of-the-art observational techniques used to detect extrasolar planets, to the characterizations of these planets, and the techniques used in the remote detection of life. It also presents insights we can gain from our own Solar System, and how we can apply them to the research of planets in other stellar systems.

The contributors, all of high standing in the field, provide a balanced and varied introduction to extrasolar planets for research astronomers and graduate students, with the aim of bridging theoretical developments and observational advances.

Intended for students, researchers, lecturers and scientifically minded amateur astronomers, this book provides a suitable introduction to the field, and can form the basis for a specialist course in extrasolar planets.

Canary Islands Winter School of Astrophysics

Volume XVI

Editor in Chief
F. Sánchez, *Instituto de Astrofísica de Canarias*

Previous volumes in this series

I. Solar Physics
II. Physical and Observational Cosmology
III. Star Formation in Stellar Systems
IV. Infrared Astronomy
V. The Formation of Galaxies
VI. The Structure of the Sun
VII. Instrumentation for Large Telescopes: a Course for Astronomers
VIII. Stellar Astrophysics for the Local Group: a First Step to the Universe
IX. Astrophysics with Large Databases in the Internet Age
X. Globular Clusters
XI. Galaxies at High Redshift
XII. Astrophysical Spectropolarimetry
XIII. Cosmochemistry: the Melting Pot of Elements
XIV. Dark Matter and Dark Energy in the Universe
XV. Payload and Mission Definition in Space Sciences

Participants of the XVI Canary Islands Winter School, in front of the Congress Center in Puerto de la Cruz, Tenerife.

Lecturers and scientific organizers of the Winter School, in front of Mt. Teide. Back row, from left to right: Franck Selsis, Juan Antonio Belmonte, Tim Brown, Stephane Udry, Laurance Doyle and Hans Deeg. In the front: Agustín Sánchez-Lavega and Günther Wuchterl. Not present are James Kasting, Rafael Rebolo and Garik Israelian.

EXTRASOLAR PLANETS
XVI Canary Islands Winter School of Astrophysics

Edited by

HANS DEEG, JUAN ANTONIO BELMONTE,
and ANTONIO APARICIO

Instituto de Astrofísica de Canarias, Tenerife

CAMBRIDGE UNIVERSITY PRESS
Cambridge, New York, Melbourne, Madrid, Cape Town, Singapore,
São Paulo, Delhi, Dubai, Tokyo, Mexico City

Cambridge University Press
The Edinburgh Building, Cambridge CB2 8RU, UK

Published in the United States of America by Cambridge University Press, New York

www.cambridge.org
Information on this title: www.cambridge.org/9780521155601

© Cambridge University Press 2008

This publication is in copyright. Subject to statutory exception
and to the provisions of relevant collective licensing agreements,
no reproduction of any part may take place without the written
permission of Cambridge University Press.

First published 2008
First paperback edition 2010

A catalogue record for this publication is available from the British Library

ISBN 978-0-521-86808-2 Hardback
ISBN 978-0-521-15560-1 Paperback

Cambridge University Press has no responsibility for the persistence or
accuracy of URLs for external or third-party internet websites referred to in
this publication, and does not guarantee that any content on such websites is,
or will remain, accurate or appropriate.

Contents

List of contributors		*page* viii
List of participants		ix
Preface		xi
Acknowledgements		xiv
1	Overview of extrasolar planet detection methods *Laurance R. Doyle*	1
2	Statistical properties of exoplanets *Stéphane Udry*	24
3	Characterizing extrasolar planets *Timothy M. Brown*	65
4	From clouds to planet systems: formation and evolution of stars and planets *Günther Wuchterl*	89
5	Abundances in stars with planetary systems *Garik Israelian*	150
6	Brown dwarfs: the bridge between stars and planets *Rafael Rebolo*	162
7	The perspective: a panorama of the Solar System *Agustín Sánchez-Lavega*	178
8	Habitable planets around the Sun and other stars *James F. Kasting*	217
9	Biomarkers of extrasolar planets and their observability *Franck Selsis, Jimmy Paillet and France Allard*	245

Contributors

FRANCE ALLARD, CRAL-ENS Lyon, France

TIMOTHY M. BROWN, High Altitude Observatory, National Center for Atmospheric Research, Boulder, CO, USA

LAURANCE R. DOYLE, SETI Institute, Mountain View, CA, USA

GARIK ISRAELIAN, Instituto de Astrofísica de Canarias, La Laguna, Spain

JAMES F. KASTING, Department of Geosciences, Penn State University, University Park, PA, USA

JIMMY PAILLET, CRAL-ENS Lyon, France

RAFAEL REBOLO, Instituto de Astrofísica de Canarias, La Laguna, Spain

AGUSTÍN SÁNCHEZ-LAVEGA, Dpto. Física Aplicada, Escuela Superior de Ingenieros, Universidad del Pais Vasco, Bilbao, Spain

FRANCK SELSIS, CRAL-ENS Lyon, France

STÉPHANE UDRY, Observatoire de Genève, Switzerland

GÜNTHER WUCHTERL, Astrophysikalisches Institut Jena, Germany

Participants

Participants (students and lecturers) of the XVI Canary Islands Winter School, Puerto de la Cruz, Tenerife, 22 November to 3 December 2004

Aigrain, Suzanne	University of Cambridge, United Kingdom
Antichi, Jacopo	University of Padova, Italy
Antonova, Antoaneta Emilova	Armagh Observatory, United Kingdom
Artigue Carro, Fernanda	Instituto de Astrofísica de Canarias, Spain
Benavidez, Paula Gabriela	Universidad de Alicante, Spain
Bendjoya, Philippe Maurice	University of Nice Sophia Antipolis (UNSA), France
Belmonte, Juan Antonio	Instituto de Astrofísica de Canarias, Spain
Berezhnoy, Alexey	Sternberg Astronomical Institute, Russia
Berton, Alessandro	Ruprecht Karls Universität Heidelberg, Germany
Bianco, Federica	University of Pennsylvania, USA
Bienenda, Wolfgang Hermann	EADS, Germany
Bihain, Gabriel	Instituto de Astrofísica de Canarias, Spain
Bounama, Christine	Potsdam Institute for Climate Impact Research, Germany
Brown, Tim	High Altitude Observatory, USA
Caballero Hernández, Jose Antonio	Instituto de Astrofísica de Canarias, Spain
Carmona González, Andrés	European Southern Observatory (Garching), Germany
Cassan, Arnaud Jean Marie Joseph	Institut d'Astrophysique de Paris, France
Chavero, Carolina Andrea	Universidad Nacional de Córdoba, Argentina
Crespo Chacón, Inés	Universidad Complutense de Madrid, Spain
Deeg, Hans	Instituto de Astrofísica de Canarias, Spain
Díaz, Rodrigo Fernando	Universidad de Buenos Aires, Argentina
Domínguez Cerdeña, Carolina	Instituto de Astrofísica de Canarias, Spain
Doyle, Laurance	SETI Institute, USA
Ecuvillon, Alexandra	Instituto de Astrofísica de Canarias, Spain
Enoch, Rebecca Louise	Open University, United Kingdom
Esposito, Massimiliano	Universitá degli Studi di Salerno, Italy
Fortier, Andrea	Universidad Nacional de La Plata, Argentina
Galland, Franck Louis	Universite Joseph Fourier, France
Gómez Martín, Cynthia	University of Florida, USA
González García, Beatriz Maria	Laboratorio de Astrofísica Espacial, Spain
Grenman, Tiia	Lulea University of Technology, Sweden
Hekker, Saskia	University of Leiden, Netherlands
Hernán Obispo, Maria Magdalena	Universidad Complutense de Madrid, Spain
Hernández Alarcón, Jesús Omar	Universidad de Los Andes, Venezuela
Hogan, Emma	University of Leicester, United Kingdom
Hole, Karen Tabetha	University of Wisconsin – Madison, USA
Israelian, Garik	Instituto de Astrofísica de Canarias, Spain
Joergens, Viki	University of Leiden, Netherlands
Joos, Franco	ETH Zurich, Switzerland
Jorge Roque, Silvia Cristina	Universidade de Coimbra, Portugal

Kasting, Jim F.	Penn State University, USA
Krumpe, Mirko	University of Potsdam, Germany
Kubas, Daniel	University of Potsdam, Germany
Madsen, Felipe Ramos Hald	Instituto Nacional de Pesquisas Espaciais, Brazil
Mahadevan, Suvrath	Pennsylvania State University, USA
Mandell, Avram Max	Pennsylvania State University, USA
Martín Fleitas, Juan Manuel	GRANTECAN, Spain
Melendez Moreno, Jorge Luis	California Institute of Technology, USA
Montalto, Marco	University of Padova, Italy
Montañés Rodríguez, Maria del Pilar	New Jersey Institute of Technology, USA
Montojo Salazar, Francisco Javier	Real Observatorio de la Armada, Spain
Mordasini, Christoph Andrea	Universität Bern, Switzerland
Näränen, Jyri Antero	University of Helsinki, Finland
Navarro Góngora, Diana Alicia	Universidad Nacional Autónoma de México, Mexico
Ocaña, Breezy	Instituto de Astrofísica de Canarias, Spain
O'Donovan, Francis Thomas	California Institute of Technology, USA
Pallé Bago, Enric	New Jersey Institute of Technology, USA
Platt, Elizabeth Jacqueline	University of London, United Kingdom
Pribulla, Theodor	Astronomical Institute, Slovakia
Quanz, Sascha Patrick	Max-Planck Institut für Astronomie, Germany
Rebolo Lopéz, Rafael	Instituto de Astrofísica de Canarias, Spain
Regandell, Samuel Gustav	University of Uppsala, Sweden
Sánchez-Lavega, Agustín	Universidad del País Vasco, Spain
Selsis, Franck	CRAL-ENS Lyon, Spain
Sosa Ibarra, Nancy	Universidad de La República, Uruguay
Stamatellos, Dimitris	University of Cardiff, United Kingdom
Tamuz, Omer	Tel Aviv University, Israel
Tingley, Brandon West	Australian National University, Australia
Udry, Stéphane	Observatoire de Genève, Switzerland
Viironen, Kerttu	University of Oslo, Norway
Vince, Oliver	University of Belgrade, Serbia
Voss, Holger	DLR (German Aerospace Center) Berlin-Adlershof, Germany
Weldrake, David Thomas Frederick	Australian National University, Australia
Wuchterl, Günther	Astrophysikalisches Institut, Jena Germany
Yakut, Kadri	University of Ege, Turkey
Zhou, Liyong	Nanjing University, China P.R.

Preface

Contemplating the existence and character of 'other worlds' has a long history, giving rise to an ample body of philosophical and artistic works. But only in 1995 could we begin to put these musings on a scientific basis, with the detection of the first extrasolar planet by Michel Mayor and collaborators at Geneva Observatory. Since that time, the field of extrasolar planets (exoplanets for short) has undergone extremely rapid development and has delivered some of the most exciting results in astronomy. Research today on exoplanets has established itself as a major branch of current astronomy. The growing importance of this field can be shown from the rising number of publications in the field. Starting with a few scattered papers over ten years ago, currently about 2% of all of the papers published in astronomy deal with extrasolar planets. Similarly, the number of projects searching for extrasolar planets has risen from five in 1995 to over 70 at present. Training in exoplanets may therefore be considered very valuable for young researchers. Due to the novelty of the subject, new research groups are frequently still being formed, giving excellent opportunities for participation by qualified personnel.

With exoplanetary science essentially starting in 1995 and with its very rapid development in the following years, this topic has hardly found its way into the astronomy/astrophysics curricula taught at universities. There are still relatively few lecturers familiar with the topic. The exceptions are those departments where active exoplanets research is being pursued; in such cases it is typically taught in optional advanced courses. Coinciding with this lack of curricular diffusion is a lack of monographs suitable for university courses. This book is intended to remedy both of these shortcomings; we hope that it may serve as a useful basis for intermediate- to advanced-level university courses.

The milestone of 200 known extrasolar planets has been passed, and over twenty systems of two or more planets orbiting the same star are known. Scientific work on extrasolar planets, however, begins only with their detection. For most planets, their *characterization* is still limited to basic physical parameters such as period and distance to the central star, and to certain further parameters, such as the planet's mass and estimates of its approximate surface temperature. Only for very few extrasolar planets is significantly more known; the first ingredients of an atmosphere were recently detected for one of them. It will be through these more detailed characterizations that the upcoming observing projects will have the greatest impact. The employment of a wider variety of detection methods (such as transit detection, precision astrometry and interferometry; see the figure on the next page) will give us a wider range of knowledge on these planets in the coming years. Also, the launch of the first space based missions dedicated to exoplanets will lead to a further enlargement in the parameter space of detectable planets, the most desired being the detection of small Earthlike planets. One of the most ambitious goals will be the detection of biological signals on exoplanets. Though very difficult, this goal is already being contemplated in the design of the most advanced space missions that may launch within about 15–20 years.

It is the great interest in communicating exoplanetary research findings to the general public, and the potential for further important discoveries (most notably, potentially inhabitable Earthlike planets) that has convinced all major space agencies to dedicate missions to the detection and characterization of exoplanets. The coming decade will therefore see a series of launches. The first will be the Franco-European *COROT* mission, which will be the first experiment to test for the presence of massive terrestrial planets. This will be followed by NASA's *Kepler*, which will be the first mission to look for the presence of true Earthlike planets. The most ambitious projects are *Darwin* (ESA)

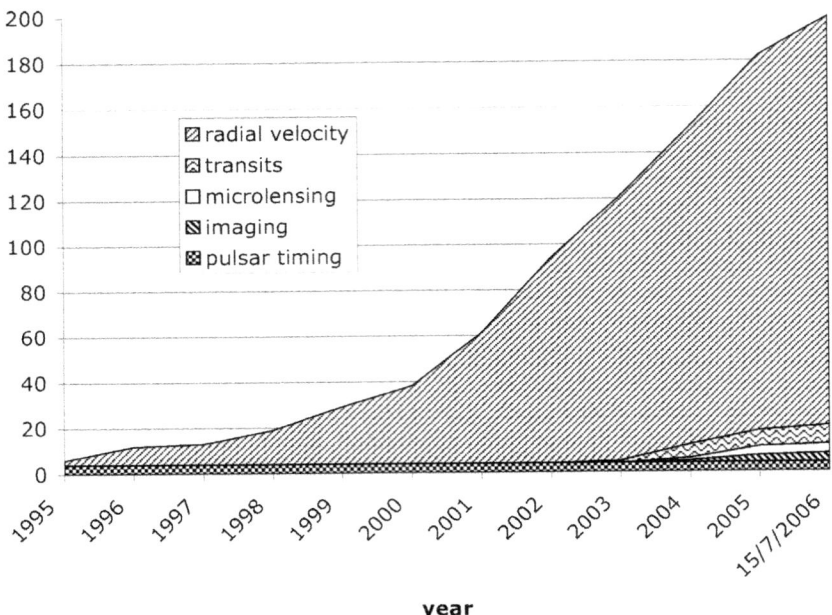

Cumulative number of exoplanets, with the method of their detection. Before 1995, only four planets around pulsars, found by timing, were known. Radial velocity detection was the only successful method until 2003, and today maintains a clear dominance on detection rates. Since then, several other methods have had their first successes, allowing a more varied characterization of the detected planets. At the time of writing, there were 179 radial velocity, eight transit, four microlensing and four imaging detections. Together with the pulsar planets, this gives a total of 199 extrasolar planets. (Numbers based on The Extrasolar Planets Encyclopaedia, www.exoplanet.eu.)

and *TPF* (NASA), planned around the year 2020. These missions will attempt the direct detection of Earthlike planets around nearby stars by coronagraphy and interferometry, and perform a fairly detailed analysis of their atmospheres, with the major goal of probing for the presence of biomarkers. Since the first exoplanet was discovered rather recently, a layperson might expect that only very specialized equipment and large telescopes can provide important results in this field. However, small telescopes are also playing an important role, as has been shown by the detection of transiting planets of the stars HD 209458b and TrES-1 with the 10 cm STARE telescope. Currently, an ample variety of small telescopes for similar detections are being constructed or are already operational. These are mainly aimed at the detection of relatively large planets in nearby stellar systems. These planets are still important discoveries, since they allow the most detailed studies with current observing techniques, employing large telescope. Their characterization is also an important driver for the development of future extremely large telescopes, or for the employment of telescopes at very special sites like Antarctica's Dome C.

While current exoplanet science is certainly being driven by observations, a number of theoretical interpretations have undergone a great refinement since the first planet discoveries. These theories are fundamental to our understanding of these objects. They are also needed to formulate the questions that may be resolved by the next generation of observing projects, where they may be drivers for their design. Hence, observers need to have a theoretical understanding in order to be able to define observing projects that are able to advance theory. This book addresses this duality between observation and theory. Its principal contents are an observational part dealing with planet detection methods and giving a description of the current state of knowledge from observations. This is followed by a theoretical part on the formation and evolution of planets, with a section devoted to habitability and biomarkers.

Once the first Earthlike planets have been discovered, we expect that this field will become a melting pot for activities of astronomers, paleontologists, geologists and biologists alike. Surely, the subject of extrasolar planets will undergo an exciting development, of which we are currently witnessing only the beginnings.

The Editors
La Laguna, Tenerife

Acknowledgements

The organizers of the XVI Canary Islands Winter School would like to express their sincere gratitude first and foremost to the lecturers, for making it a great scientific and scholarly event. Preparing the lectures, attending the school and writing the chapters for this book has been a major, but we hope rewarding, commitment in their very busy agendas. We specially acknowledge Tim Brown who, despite a strong familiar concern, was able to make his way to Tenerife, and to Agustín Sánchez-Lavega who, despite a major personal event, had the time to offer us a fascinating perspective of our Solar System, including Pluto's very recent loss of planetary status. Terry Mahoney, of the IAC's Language Correction Service, had a major part in the creation of this book, revising minutely and with enormous patience all submitted manuscripts, from native and non-native English speakers equally, and this book would not have been possible without him.

The soul of the school has been without any doubt our secretary Nieves Villoslada. Without her help, diligence and permanent availability the school would have not worked as perfectly as it did. Her colleague Lourdes González was also very helpful on many relevant occasions. Jesús Burgos of the OTRI at IAC, one of the most efficient persons we know of, provided invaluable help with all the issues concerning the preparation of applications needed to receive sufficient funding. The school's poster was prepared by Ramón Castro and offers a vision of extrasolar planet research that may entice young scholars to enter this field. For Carmen del Puerto and her team at the Gabinete de Dirección of the IAC the Winter Schools are a time of great pressure, since they have to prepare a special issue of the IAC *Noticias* Newsletter for the end of the school, a task they accomplished excellently. The dedication and enthusiasm of all these people is an essential ingredient to the Canary Islands Winter School Programme, and is warm-heartedly acknowledged.

We greatly acknowledge the financial assistance from the Spanish Ministerio de Educación y Ciencia and from the Cabildo de Tenerife who kindly provided the excellent facilities of the Congress Palace of Puerto de la Cruz where the event took place. The Ayuntamiento of Puerto de la Cruz generously offered all participants a very nice banquet at the restaurant Casa Régulo. This was one of the most intimate moments of the school when, after two weeks of apparent seriousness, a group of the students started performing short sketches, pulling the legs of both organizers and lecturers. Many thanks for such a delightful and unforgettable moment! Last but not least, we would like to acknowledge all the participants of the school, both lecturers, students and supporting personnel when they, even under exceptional circumstances like the visit at IAC of the Spanish Crown Prince Don Felipe and his wife, provided the finest ambience one could imagine for such an event.

1. Overview of extrasolar planet detection methods

LAURANCE R. DOYLE

In this chapter we will describe in a general manner each planet detection method and examine the fundamental astrophysical parameters each technique measures as well as its present measurement limitations for the detection of inner giant planets, jovian outer planets, and Earthlike planets. We then outline several secondary detection methods that may be instituted in the near future with increased detection sensitivity. We then discuss the ranges of each detection method and sketch several cases in which additional parameters may be derived through the acquisition of data from several methods combined. In the final section we discuss habitable zones around M-dwarf systems as potential near-term targets for the detection of life-supporting planets.

1.1. Introduction

In the following sections an overview of the main methods of extrasolar planet detection is presented. This is not a historical review – an excellent review, for example, can be found in Perryman (2000) and the 469 references therein. It is also not an up-to-date listing of extrasolar planet detections or candidates; these can be found at the comprehensive site of the *Extrasolar Planets Encyclopedia* by J. Schneider (www.obspm.fr/encycl/encycl.html). In this chapter we do, however, describe in a general manner each detection method and examine the general astrophysical parameters each technique measures as well as its present measurement limitations. We mention some secondary detection methods that may find application in the near future and what additional parameters may be derived through the acquisition of data from several methods combined. We finally discuss M-dwarf star habitable zones, as these are likely to be the near-term targets for the detection of exobiology on extrasolar planets. This chapter is aimed, in explanatory detail, at the interested college student level.

We note that the detection parameters for the pulsar timing, radial velocity, astrometric imaging, reflected light and eclipsing binary timing methods depend, at any given time, on the orbital phase, $\varphi(t)$, of the extrasolar planet, which is a function of the geometry involved in that detection method. However, detectability depends on the maximum signal produced for a given method, and it is this that we formulate in the equations below. However, we shall point out at which phases this maximum occurs. In keeping with eclipsing binary protocol, the planetary orbital phase $\varphi(t) = 0$ degrees will be when the (darker) planet is in inferior conjunction, that is when it is closest to the observer.

1.2. Pulsar timing

Unexpectedly, the first planetary-mass objects detected around another star were closer to terrestrial-mass than to jovian-mass. The parent star was the pulsar PSR B1257+12, 500 parsecs distant, and the two planetary objects detected around it are a 2.8 Earth (projected) mass ($M_\oplus$) body with a period of 98.22 days and a 3.4 $M_\oplus$ body with a period

Extrasolar Planets, eds. Hans Deeg, Juan Antonio Belmonte and Antonio Aparicio.
Published by Cambridge University Press.
© Cambridge University Press 2007.

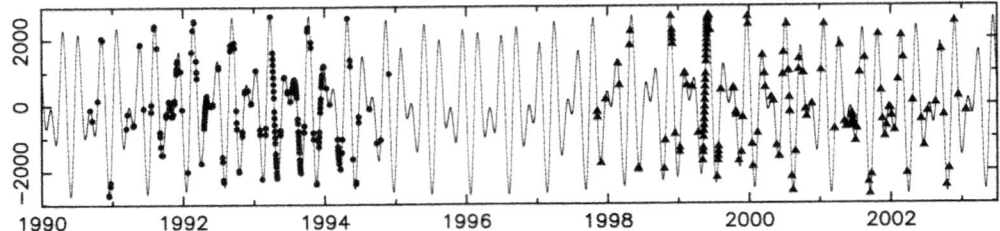

FIGURE 1.1. Time of arrival residuals (in microseconds) of 430 MHz signals from the 6.2-millisecond pulsar PSR B1257 + 12 (from Konacki and Wolszczan 2003) showing that the residuals are dominated by the Keplerian orbits of two planets of actual mass 4.3 (B) and 3.9 (C) Earth-masses (of three in the system, the third being very close to the pulsar). These planets are nearly coplanar (around 50 degrees orbital inclination) and are in actual 3:2 orbital resonance with each other (66.5 and 98.2-day periods; planet A having a period of 25.3 days).

of 66.54 days (Wolszczan 1994; Wolszczan and Frail 1992). The precise radio pulse rates of pulsars (seconds to milliseconds) and their stability as timing 'clocks' (variations in pulse timing on the order of only about a trillionth of a second per year) allow variations in the position of the pulsar to be measured precisely. The variation in timing can occur due to a positional shift in the pulsar around the pulsar–planet barycentre. If such a second mass (planet) is in orbit around the pulsar, the two bodies will orbit around a mutual barycentre, each distance from the barycentre being determined directly by their mass-ratios, where M_* and a_* are the mass and distance (semi-major axis) from the barycentre to the centre of the pulsar and M_p and a_p are the mass and distance from the barycentre to the planet. The motion of the pulsar around the barycentre causes the addition of (or subtraction of) the light travel time across this distance, which will result in a delay (or early arrival) of the periodic variations in the timing of the pulsar pulses. For a planet in a circular orbit, the maximum amplitude of the delay time will be:

$$\tau = \sin i \left(\frac{a_p}{c}\right)\left(\frac{M_p}{M_*}\right), \qquad (1.1)$$

where i is the inclination of the planet's orbit ($i = 90°$ being edge-on), and c is the speed of light. The pulses will be 'on time' at phases $\varphi(t) = 90$ and 270 degrees, late by an amount τ at $\varphi(t) = 0$ degrees, and early by an amount τ at $\varphi(t) = 180$ degrees orbital phase angle, where zero degrees phase is when the planet is closest to the observer (i.e. inferior conjunction). Note that the sine function in Eq. (1.1) is not negative because it is the pulsar signals that are being measured directly, and the pulsar is at the opposite orbital phase from the planet. Thus, via the foreshortened light travel time across the stellar-barycentre distance, the pulsar timing method can measure the projected planet-to-star mass ratio, the true orbital period of the planet (or planets), and its orbital eccentricity (if the orbit is not circular). General relativistic precessional phase drifts may allow further constraints on the pulsar mass, but only for closer-in planets over longer observing times (see Figure 1.1).

If we define a typical close-in extrasolar giant planet (CEGP) as a 3 jovian-mass planet with a circular orbital semi-major axis (i.e. orbital radius) of 0.05 AU (astronomical unit), and a 'Jupiter' and an 'Earth' as planets with the mass and orbital location (distance from their star) of Jupiter and Earth in the Solar System, respectively, then the half-amplitude timing offsets for such planets around a typical pulsar (assuming the pulsar to be 1.35 solar masses) would be $\tau = 140$ milliseconds (ms) for a CEGP, 1.65 seconds for a 'Jupiter', and 3 ms for an 'Earth'. That is, these will be the expected maximum delays in the pulse arrival times at a planetary orbital phase $\varphi(t) = 0$ degrees.

1.3. Periodic radial velocity variations

The radial velocity or 'Doppler shift' method has been the most successful extrasolar planet detection method to date, detecting the vast majority of planets as of this writing. The first extrasolar planets around solar-type stars were discovered in this way (Mayor and Queloz 1995; see also Marcy and Butler 1998 and reference therein). Radial velocity variations again cause a wobble in the parent star, but the stellar light flux is generally very constant, so that timing of variations cannot be used to detect this stellar offset around the star–planet barycentre. However, very high precision spectral line measurements (one part in a hundred millionth of a spectral line width) can be performed by superimposing a comparison spectrum with many lines (like an iodine cell in the light path at the observatory) on to the stellar spectrum for a precise measurement of periodic movement in the star's spectral lines.

The stellar spectral lines will move periodically redward or blueward due to the Doppler shift by $\Delta\lambda/\lambda = v/c$, caused by the periodic motion, with a maximum velocity v of the star about the star–planet barycentre. Again, the spectral line variations only measure the component of the motion directly towards or away from the observer, and hence the mass of the body (planet) causing the reflex motion of the star is a minimum mass measurement for the planet, $M_\mathrm{p} \sin i$. The maximum amplitude of this periodic radial velocity variation is given by:

$$K = \left(\frac{2\pi G}{P}\right)^{1/3} \frac{M_\mathrm{p} \sin i}{(M_\mathrm{p} + M_*)^{2/3}} \frac{1}{(1-e^2)^{1/2}}, \tag{1.2}$$

where P is the planetary orbital period, e is the planetary orbital eccentricity, and G is the gravitational constant. K, P and e can be derived from several measurements of the Doppler shift during a planet's orbit. The maxima in K will occur at planetary orbital phases of $\varphi(t) = 90$ degrees (blueshift of stellar spectra lines) and $\varphi(t) = 270$ degrees (redshift of stellar spectral lines). With knowledge of M_* from stellar classification (typically based on low-resolution spectra), the term $M_\mathrm{p} \sin i$ can then be derived. Kepler's third law, $(P/\mathrm{yr}) = (a_\mathrm{p}/1\mathrm{AU})^{3/2}(M_*/M_\odot)^{-1/2}$, where $M_\odot$ is one solar mass, allows then also a derivation of the semi-major axis of the planet.

The precision possible for this detection method is about 1 m/s, this limit imposed by intrinsic stellar surface fluctuations – i.e., variations present in even the most stable solar-type stars (see Figure 1.2). For a CEGP the radial velocity amplitude will be about 56 m/s, for a 'Jupiter' about 13 m/s, and for an 'Earth' about 0.1 m/s. Thus this method may not be expected to detect Earthlike planets around solar-like stars but can, however, detect any jovian-mass bodies within a star's circumstellar habitable zone (CHZ).[1] Hypothetical Earth-sized moons around such bodies have been suggested as being of interest to exobiologists. The detection of Jupiterlike planets are of interest both because of their comparability with our own Solar System as well as the ability of jovian-type planets to remove cometary debris, serving as a possibly necessary 'shield' for any biosystems developing on the inner terrestrial planets of the star system. This method is also, at

---

[1] The circumstellar habitable zone (CHZ) is defined here as the distance regime around a star where liquid water can persist on the surface of a sufficiently large planet. For a discussion of the CHZ see Chapter 8.

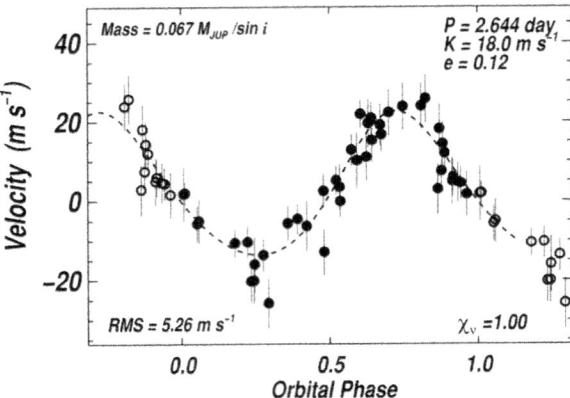

FIGURE 1.2. A Neptune-mass planet orbiting the nearby 0.41 solar-mass M-dwarf GJ 436 (from Butler et al. 2004). The 18.1 metre/second variation in the spectral lines of the star with a period of 2.644 days is caused by a planet with a projected mass of about 1.2 Neptune-masses.

present, limited to detection of planets around fairly stable, single star systems as the measurement of these radial velocity variations demands such high spectral line precision measurements.

1.4. Gravitational microlensing

Due to general relativistic effects of bending spacetime, a star moving very close to alignment with a background star will bend – that is, focus – the light of the background star, causing a temporary increase in the combined brightness of the stars by amplifying the light from the background star. The phenomenon, first observed with galaxies, is known as gravitational lensing, A perfect stellar alignment will cause symmetric images around the lensing star; this is known as the 'Einstein ring' (or sometimes an 'Einstein cross'). The Einstein ring radius is given by:

$$R_{\rm E} = \left[\frac{4GM_{*\rm L}}{c^2}\frac{(D_{\rm S} - D_{\rm L})D_{\rm L}}{D_{\rm S}}\right]^{1/2}, \quad (1.3)$$

where $M_{*\rm L}$ is the mass of the lensing star, $D_{\rm L}$ is the distance to the lensing star and $D_{\rm S}$ is the distance to the source star. The angle on the sky of the Einstein radius (the Einstein angle) is then given as: $\theta_{\rm E} = R_{\rm E}/D_{\rm L}$. The microlensing magnification, which varies with time, is given by:

$$Q(t) = \frac{u^2(t) + 2}{u(t)[u^2(t) + 4]^{1/2}}, \quad (1.4)$$

where $u(t)$ is the projected distance between the image of the lensing star and the source star in units of the Einstein radius (Perryman 2000). We can see that for an exact alignment the magnification would become infinite, theoretically. If a planet is in orbit around the lensing star, then observable deviations from the amplification pattern given by Eq. (1.4) may occur, which are caused by a planet-mass distorting the stellar gravitational field.

The probability of alignment among two stars is, even in the Galactic Centre, only about one in 10^6, but once a star is aligned with another star the probability that a planet may also cause an amplification that exceeds 5% of the brightness of the star's amplification itself becomes about one in five (Schneider et al. 1999). For this superposition

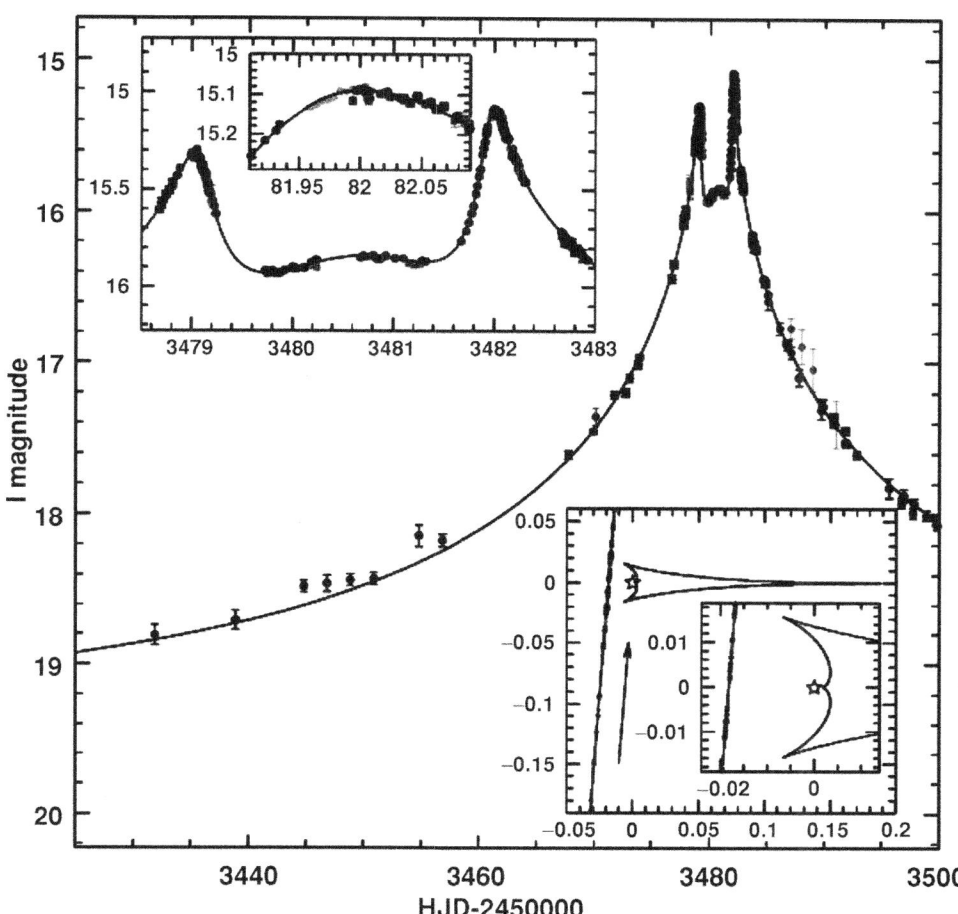

FIGURE 1.3. The light curve of OGLE-2005-BLG-071 (adapted from Udalski *et al.* 2005), showing a binary peak, indicative of the binary nature of this microlensing event, where both a star and a planet move in front of a background star. A magnification of the three peaks (the middle one with a low amplitude) is shown in the inset in the upper left. The bottom right inset shows the caustic surfaces (closed-curve regions of very high magnification) consistent with a binary lens mass ratio of $q = 0.0071$. From an analysis of parallax effects in the wings of the microlensing event, the lensing stellar mass is constrained to be between 0.5 and 0.08 solar masses, at a distance between 1.5 and 5 kiloparsecs, giving a planetary mass from 0.05 to 4 Jupiter masses.

of a brightening due to a planet on top of that due to the amplified star, the term M_{*L} becomes the mass of the planet, M_p in Eq. (1.3).

The duration of a microlensing event is given by:

$$t_E = \sqrt{4GM_p dV/c^2}, \tag{1.5}$$

where d is the distance to the lensing star in parsecs and V is the orbital velocity. If the distance d to the lensing star–planet system is assumed to be 5 kiloparsecs, then for a CEGP the brightening will be about four magnitudes and last for about five days. For a 'Jupiter' the brightening will be about three magnitudes and last about three days. And for an 'Earth' the brightening will be about one magnitude and last about four hours (see

Figure 1.3). Thus we see that these events are quite detectable. The main difficulty of this approach is that the detection of these events requires a large number of stars to be observed that are also pretty far away (a few kiloparsecs), and finally that the detection event cannot be expected to repeat. However, as pointed out in Udalski *et al.* (2005) planetary parameters (such as orbital period) may be determined that are longer than the duration of the observations (note no orbital phase dependence in Eqs. (1.3), (1.4), or (1.5)). In general, however, the planet-to-star mass ratio is what is determined, and since target systems are generally so far away, precision spectra determining the stellar mass are more difficult to obtain than for nearby systems. However, this is at present the planet detection method that works over the longest distance and might give insights into the planetary distribution in quite different stellar populations – for example disc compared to bulge galactic populations.

1.5. Astrometry

Astrometry is perhaps the oldest method to search for extrasolar planets, several having been reported in the mid-twentieth century. This method measures a periodic variation in the position of the star on the 'plane of the sky', subtracting out the star's apparent motion due to the yearly parallax motion and the projection of its real proper motion through space. The motion of a star around the star–planet barycentre thus describes an elliptical motion with semi-major axis (in arcseconds) of:

$$\alpha = \frac{M_\mathrm{p}}{M_*} \frac{a_\mathrm{p}}{d}, \qquad (1.6)$$

where a_p is in AU (astronomical units), and d, the distance to the stellar system, is given in parsecs. For a circular planetary orbit the inclination of the planetary orbit from the observer's line of sight is just $\sin i = b/\alpha$, where b is the semi-minor axis measured. The maximum semi-major axis position of the star occurs at elongation when the planetary orbital phases are $\varphi(t) = 90$ and 270 degrees. Here at maximum elongation $\varphi(t) = 1$. This technique measures the motion of the photometric centroid position of the star in images taken over at least a large fraction of a planet's orbit. It is complementary, for example, to the radial velocity detection method in that it is most sensitive to long-period (large semi-major axis) planets, while the radial velocity method is most sensitive to inner, short-period planets with higher velocity variations. As for the factor d (as indicated in Eq. (1.6)), astrometric detection of extrasolar planets is very sensitive to the distance to the system, and at present is limited to somewhat nearby stellar systems (see Figure 1.4).

For the full amplitude of a CEGP the astrometric offset on a solar-type star at a distance of 5 pc would be about 0.03 milliarcseconds. For a 'Jupiter' it would be about 1 milliarcsecond, and for an 'Earth' the offset would be about 0.6 microarcseconds. This technique can be extended to search for extrasolar planets around radio-emitting stars using very long baseline radio interferometry (see Perryman 2000). Upcoming wide field searches for transiting planets (for example, the NASA *Kepler* mission) may also allow astrometric searches for planets to take place using the same photometric data, since the pointing precision as well as the photometric centroiding of star images should be near the 1 milliarcsecond precision required for astrometry. Near-term spacecraft missions such as *SIM (Space Interferometry Mission)* will be specifically designed to optimize astrometric measurements both for stellar parallax determinations and the detection of extrasolar planets in the solar neighbourhood astrometrically. *SIM* should be able to detect nearby extrasolar planets while mapping exact distances to stars by using interferometry to

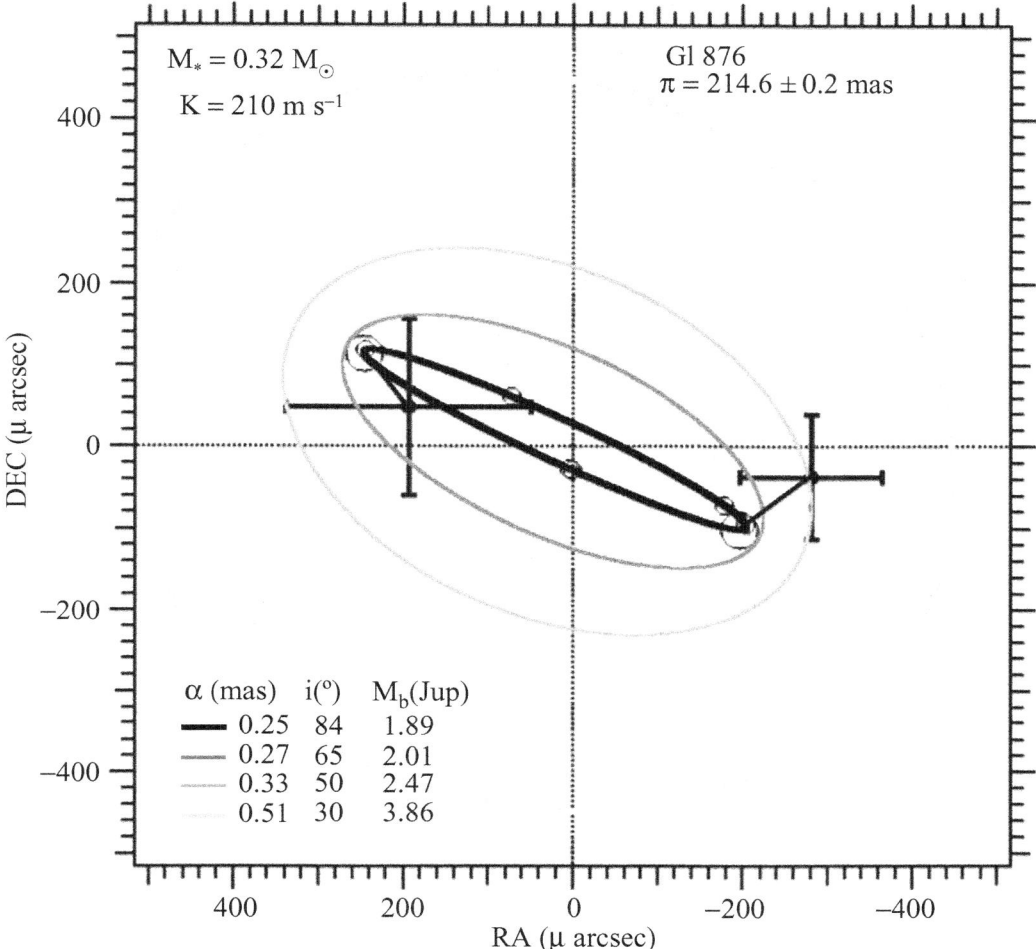

FIGURE 1.4. Astrometric deviations on the plane of the sky measured by the *Hubble Space Telescope* Fine Guidance Sensor in fringe-tracking mode (from Benedict *et al.* 2002). The best fits to the astrometric variations of 0.25 milliarcseconds (the inner ellipsoid, with measurement errors indicated by the crosses) of the star Gliese 876 (with a parallactic mass of about 0.32 solar masses) gives a planetary orbital inclination of about 84 degrees, and thus a planet mass of about 1.89 jovian masses.

accurately measure astrometric wobbles of stars, caused by orbiting planets, to about one microarcsecond in angular resolution.

1.6. Imaging

Direct imaging of an extrasolar planet at visible wavelengths depends on the reflected light from the star that the planet produces which, in turn, depends on its distance from the star, the planet's size, and the nature of its atmosphere (i.e. the product of the geometric albedo, A, and particle phase function $p(\varphi(t))$, which is a measure of the light-scattering nature of the particles in the planetary atmosphere, such as a Lambertian function or, more likely, a steeper function of viewing angle). The ratio of brightness of

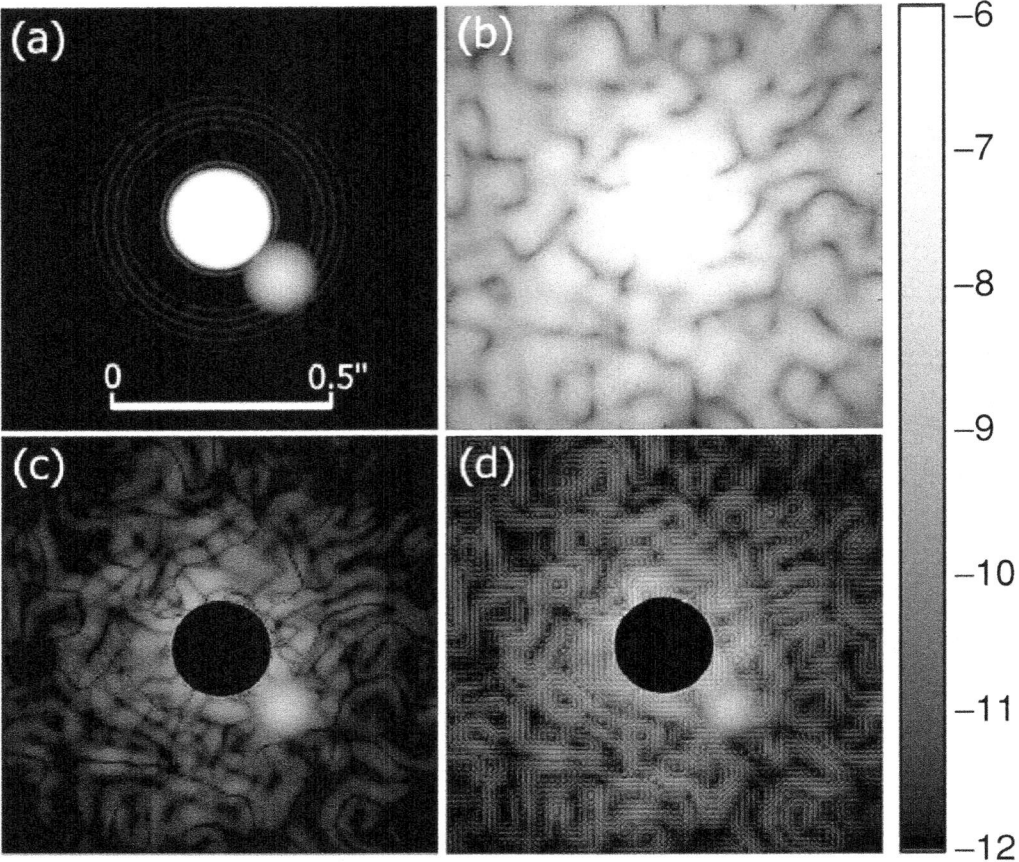

FIGURE 1.5. (a) A star that is ten-thousand times brighter than a companion planet at 0.2 arcsecond separation within the halo of the star (faint rings) with ideal apodization (from Codona and Angel 2004). (b) The system with low-order residuals in the nulling process. (c) Additional destructive interference using an anti-halo technique. (d) Same process as (c) but with spatial light modulation.

the planet to the star is the important factor as the planet, even at a reasonably large angular distance, a_p/d, from the star, will be 'lost' in the brightness of the diffraction rings of the star as imaged by a telescope (Jupiter viewed from Alpha Centauri would be about 4 arcseconds in angular distance away from the Sun, but typical angular distances of exoplanets are much smaller). The brightness ratio of planet-to-star is:

$$L_p/L_* = Ap(\varphi(t)) \left(\frac{R_p}{a_p}\right)^2, \qquad (1.7)$$

where one would want to image a planet at maximum distance from the parent star. This would occur at phases of $\varphi(t) = 90$ and 270 degrees. The maximum separation of the planet and its parent star (see Figure 1.5) is simply (a/d). For $d = 5$ parsecs, at visible wavelengths L_p/L_* for a CEGP would be about 16 magnitudes (with a star–planet separation of 0.01 arcseconds), for a 'Jupiter' about 23 magnitudes (separation of about 1 arcsecond), and for an 'Earth' about 25 magnitudes (separation of about 0.2 arcseconds). The contrast between star and planet may improve as one moves into the infrared, if the planet is a warm or hot one. In these cases the planet not only reflects the stellar light but also emits significant thermal IR radiation. The scattered light from the star can, however, be blocked somewhat by a coronagraphic telescope, and this is the approach being taken by, for example, the first part of the NASA *Terrestrial Planet Finder* (*TPF-C*) mission. The second *TPF* mission (*TPF-I*) is planning to use nulling interferometry, as does the European *Darwin* mission. In that approach the telescope (or telescopes) can be designed such that the light paths from the star will destructively interfere, thus largely cancelling out the stellar flux. The planet, being in a different angular locale, will not undergo destructive interference, and the system may even be designed to allow a planet in a definite position (in the circumstellar habitable zone, for example) to constructively interfere. Significant improvement in ground-based imaging may also be realized with the further development of new techniques. One of several interesting examples is 'dark speckle' imaging in which atmospheric scintillation may cause random destructive interference around a star, allowing the planet image to emerge momentarily, and such images – taken very rapidly – may be summed to give a planetary image (Labeyrie 1995).

1.7. Radio flux

Jovian extrasolar planets with sufficient magnetic field (i.e. rotation rates and metallic cores to produce a dynamo) can emit significant flux at radio wavelengths. The flux produced by a planet can be characterized by:

$$F_\mathrm{p}(\nu) \left(\frac{a\nu}{cd}\right), \quad (1.8)$$

where ν is the radio frequency being observed, $F_\mathrm{p}(\nu)$ is the radio flux from the planet, and d is the distance to the star system. Jupiter, for example, would produce a flux density of about 0.3 μJy (micro-Janskys) at a distance of 4 parsecs at a wavelength of 1 mm (synchrotron radiation; Jones 1994). A CEGP at this distance may be expected to produce less than about 0.03 μJy of flux due to being tidally locked in rotation with periods more on the order of several days, thus decreasing the dynamo effect. An Earthlike planet may also be expected to produce a similar strength signature to a CEGP (see Figure 1.6).

For interferometric detection techniques, the flux ratio between that of the planet and the star, $F_*(\nu, t)$, at a given detection frequency and time is the important limiting criterion:

$$F\left(\frac{\mathrm{planet}}{\mathrm{star}}\right) = \left(\frac{F_\mathrm{p}(\nu,t)}{F_*(\nu,t)}\right), \quad (1.9)$$

where the time dependence refers to both the radio fluctuations in the star and planet, and also to the planet's orbital position for maximum angular separation from the star (i.e. elongation at orbital phases of $\varphi(t) = 90$ and 270 degrees for inclinations not too close to zero degrees).

At 10 MHz this ratio, for a jovian planet around a solar-type star, is about 4×10^{-3} during active phase, but could be as high as 4 during a typical quiet starspot cycle – that is, the planet is brighter than the star at this time! However, interstellar electrons add substantial noise to the detection of flux at this wavelength over any appreciable distance. This is even more the case at lower frequencies. However, at a frequency of about 100 kHz the flux ratio of a jovian-type planet to its star could be as high as 100 during active stellar starspot phases, and as high as 2000 at the quiescent phase of the stellar activity cycle. The proposed square kilometre array (SKA) may have some possibilities of detecting the nearest jovian-type planets in this way. The SKA would be an array of

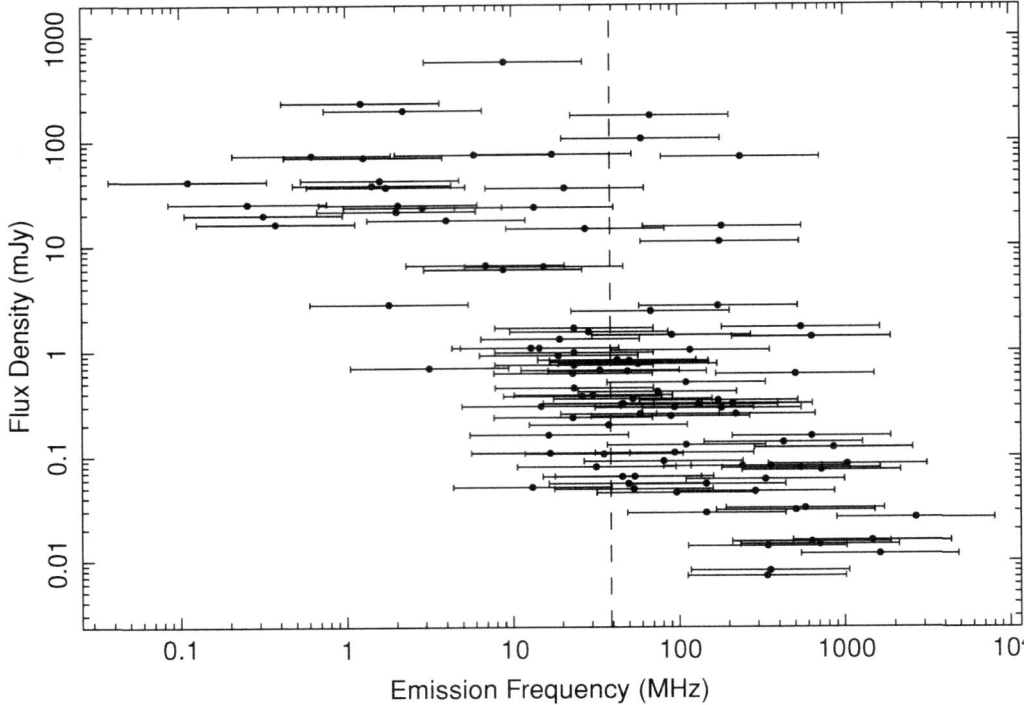

FIGURE 1.6. Theoretical 'burst' flux densities for 106 extrasolar planets (from Lazio et al. 2004). The range of expected frequencies is given with their expected flux densities (in milli-Janskys).

centimetre-to-metre wavelength radio telescopes making up a total collecting area of one million square metres with 50% of the collecting area within five square kilometres (for sensitivity), 25% within 150 kilometres (intermediate resolution), and the remaining 25% of the array out to as far as 3000 kilometres for very high angular resolution.

Current radio technology on Earth emits at narrow-band (less than 1 Hertz wide) microwave frequencies in the range 1–10 GHz, and such signals can be many millions of times more powerful than the natural flux from the Sun. This is the basis for the radio searches for extraterrestrial intelligence (SETI) projects (Tarter, 2001). Having but one example of such technological development, constraints on the expected success of detecting such signals are few. One can nevertheless state that at present such SETI projects remain the most unambigous way proposed to detect exobiology since no source in interstellar space is known to produce such narrow-band radio signals (OH masers having a bandpass of several hundred hertz, for example).

1.8. Transit photometry

Although there is a significant coverage in this volume on extrasolar planet detection by transit photometry (see Chapter 3), it is centred on the detection of transits across single stars. Thus, in this section we shall add a discussion of transits across close double star systems – in particular, eclipsing binaries.

1.8.1. Single-star transits

Photometric transit detection of extrasolar planets actually detects the shadow of the planet as it crosses in front of the stellar disc. The probability of such an alignment of the planet's orbital plane with the observer's line of sight increases linearly with decreasing distance; an Earthlike planet has a probability of alignment of about 0.5%, while the shortest period CEGPs can have a probability of alignment of more like 10%.

The brightness, or luminosity, change, ΔL, produced by the geometric blocking of light by the planet as it orbits (transits) in front of the star is proportional to the areas of the planetary and stellar discs; that is, the differential luminosity will be:

$$\frac{\Delta L}{L_0} \propto \left(\frac{R_p}{R_*}\right)^2, \quad (1.10)$$

where R_p is the radius of the planet, R_* is the radius of the star and L_0 is the system's total off-transit brightness. The proportion sign is used to point out that the limb-darkening of the star will cause a slightly different profile and total (colour-dependent) depth depending on the inclination from an edge-on orbit. Estimating the duration, T, of such a transit event for circular planetary orbits of period P gives:

$$T \propto \left(\frac{P}{\pi}\right)\left(\frac{R_*}{a}\right). \quad (1.11a)$$

The full equation for the duration of a single star planetary transit (after Tingley and Sackett 2005) is given by:

$$T = 2Z(R_* + R_p)\frac{\sqrt{1+e^2}}{1+e\cos\varphi(t)}\left(\frac{P}{2\pi GM_{\text{tot}}}\right)^{-3}, \quad (1.11b)$$

where Z incorporates geometrical effects of the orbital inclination (limb darkening, etc.), e is the orbital eccentricity and M_{tot} is the mass of the star and planet. The photometric change in brightness due to a jovian-type planet (Jupiter in size and at 5.2 AU from a solar-type star) is about 1% and lasts for about 30 hours. The photometric change due to a CEGP may be expected to be about 1–2% and last 2–4 hours. And, for an Earthlike planet, the drop in brightness would be about 0.1% around a solar-type star and last for about 12 hours (see Chapter 3).

At present, this method allows the smallest planets to be detected because it does not rely on the planet-to-star mass ratio but instead measures the ratio of planetary to stellar disc areas, thus cutting out one dimension of scaling. It may also be noted that for a given detection limit (photometric precision of $\Delta L/L \cong 0.1\%$ seems to be a usual constraint on ground-based observations), smaller stars may allow the detection of smaller planets. Thus, the closest star to us, the M5.5-dwarf Proxima Centauri, has such a small stellar disc that a 0.4% deep transit would be produced by an Earth-sized planet and therefore might be detectable from the ground.

Perhaps the largest confusion for transit detections is caused by grazing eclipsing binaries (and by eclipsing binaries that are in the background of the sample star for observatonal systems with low angular resolution). At marginal photometric precision, the low amplitude eclipses generated by either of these binary configurations can masquerade as a large planet transit – the more gradual slope of ingress and egress of a central planetary transit being produced by stellar limb-darkening effects having a signal somewhat comparable to the geometric area of the stellar disc gradually being blocked by grazing stellar eclipses. Distinguishing these effects can, however, be addressed with higher precision photometric measurements (e.g. Seager and Mallen-Ornelas 2003; Tingley and Sackett 2005), and by the inclusion of colour terms (i.e., filters), for many cases of binaries with unequal components (see Chapter 3).

Secondary effects may allow one to distinguish stellar characteristics (e.g. grazing eclipsing binaries or ellipsoidal stars) from planetary transits. For grazing eclipsing binary

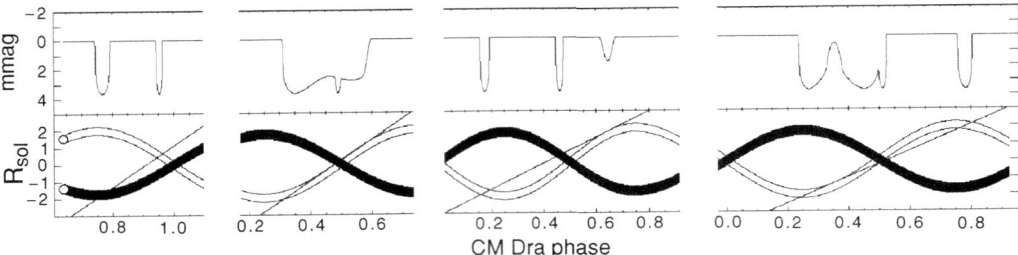

FIGURE 1.7. Quasi-periodic shapes (top row) of theoretical planetary transits across the eclipsing binary CM Draconis in millimagnitudes (from Deeg et al. 1998), along with (bottom row) binary component's phases plotted against orbital elongation of a transiting planet (in solar radii). The two left panels are for a planet with a period of 9 days, and only differ by the phase of the binary during the planetary transit. The two right panels are similar, but for a planet with a 36 day period. While the quasi-periodic nature of these types of transits makes them difficult to recognize, it also removes many ambiguities.

systems, one may be able to detect the signature of ellipsoidal variations characteristic of contact systems, for example (Drake 2003). Also, multi-spectral characteristics (two or more filter observations) should be able to distinguish between double stars and single-star transit events at photometric precisions (in most cases) already achievable (Jha et al. 2000; Seager and Mallen-Ornelas 2003; Tingley and Sackett 2005). This is because stellar limb-darkening should produce a more extreme variation due to a transit in a blue than in a red filter – i.e. a central planetary transit blocks more blue light as it passes over the central regions of a star while an edge-grazing binary would not show such an effect, the edges of the star being redder than the centre. The effect may produce as much as a 30% difference in the area blocked during a given obscuration event (Tingley, personal communication).

1.8.2. *Eclipsing binary transits*

Planets orbiting around both components of a close eclipsing binary may produce light variations that are predictable, but not periodic. This is because rather than the planet moving across a single stellar disc, the planet transits across the stellar discs of both binary components, while the stars move even more rapidly around each other 'behind' the planet. The stars have different velocities relative to the planet, and the exact shape and duration of the planet transits depend on the orbital phase of the binary components (see Figure 1.7). This produces a quasi-periodic drop in brightness with limb darkening of the stellar discs adding further complexities. Individual transit features will also vary; transits produced by the star moving in the same direction as the planet will have a longer duration than transits produced by the other star moving in the opposite direction. For the detached M4-dwarf system CM Draconis (Deeg et al. 1998; Doyle et al. 2000) it was found that transits as short as 20 minutes could occur, with typical transit events lasting about one hour. Transits during stellar elongation could be somewhat longer as the stars are moving away from or towards the observer rather than across the field of view, and transits during stellar eclipses could last many more hours if the planet crosses the stellar discs in such a way as to move from one star right across the other sequentially.

Depending on the distance of the planet from the binary, multiple transits can occur per planetary pass. Since a minimum of three transits is usually required for a positive detection (two for the period and then at least one additional predicted event), eclipsing binary transits have the advantage of obtaining four transits in only two planetary orbital periods. An Earthlike planet around a single solar-type star would clearly take a minimum of 3 years to confirm a detection. However, while a planet within the CHZ around an eclipsing binary consisting of two solar-type stars would have a period of about 1.18 years (the CHZ is farther out by $\sqrt{2}a_\mathrm{p}$, but the period is then shortened by $1/\sqrt{2M_*}$) one could then expect at least four transit features within less than half the time required for confirmation around single stars. The first potentially Earthlike planets might therefore well be found around such a close binary system.

Eclipsing binary transits also cannot generally be confused with grazing eclipsing binaries as the dynamics and timing of the quasi-periodic signals will be distinct for stellar radii compared with planetary radii. Finally, transits across eclipsing binaries can occur occasionally for non-planar planetary orbits due to precession of the nodes across the observer's line-of-sight (Schneider and Doyle 1995). A rough estimate of this precession rate (by assuming a toroidal stellar mass distribution is 'seen' on average by the planet) can be given as:

$$p \cong \frac{16}{3} P_\mathrm{p} \left(\frac{a_\mathrm{p}}{s_*}\right)^2, \qquad (1.12)$$

where s_* is the separation of the binary stellar pair. Since stable planetary orbits must be at least about three times the stellar separation, it is clear that precession times will be rather long. The NASA *Kepler* mission is a wide-field space telescope that will observe one or two crowded stellar fields (in the constellation Cygnus) for four to six years, with unprecedented 0.003% photometric precision on over 100 000 stars. Over the extended six-year lifetime of this mission, only about a dozen such binary systems can be expected to produce transits that 'disappear' for a season, and then start to recur after several years due to this precession effect.

Finally, it may be pointed out that planetary transits detected during mutual stellar eclipses will produce up to twice the signal (in terms of relative brightness variation $\Delta L/L_0$) of transits across a single star, since the stellar disc area during eclipse minima may be as small as half that when both stars are visible (for two equal-sized stars with a binary inclination of 90 degrees). Thus, it will be important, when detecting transits across eclipsing binary stars, to perform accurate modelling of stellar eclipse lightcurves in order to distinguish the much smaller superimposed variations due to planetary transits. It may be that such transits already exist in the high precision photometric data of eclipsing binary star studies to date, but have not been recognized due to their quasi-periodic nature, appearing as outlying non-periodic points due to, for example, observational noise.

1.9. Phase-reflection variations

Similarly to the Moon, planets present brightness variations with changing orbital phase due to reflection angle with their star(s). This was recognized as possibly measureable for CEGPs soon after their discovery (Doyle 1996; Borucki *et al.* 1996) but was not fully developed until recently because the characteristics of CEGPs were not yet widely modelled (Seager *et al.* 2000; Jenkins and Doyle 2003). This periodic variation in apparent luminosity can be formulated as:

$$\Delta l = A\Phi(\varphi(t)) \left(\frac{R_\mathrm{p}}{a}\right)^2, \qquad (1.13)$$

where A is the geometric albedo of the planet, and $\Phi(\varphi)$ is the planetary atmospheric phase function dependent on the orbital phase $\varphi(t)$. For a planetary albedo of about 1.0, and a Lambertian atmospheric phase function (which are upper limits; see Jenkins and Doyle 2003 for a discussion of realistic values), the amplitude of the brightness variations for a CEGP will be about 3×10^{-5}, for a jovian-type planet about 2×10^{-9}, and for an Earthlike planet also about 2×10^{-9}. Thus, even for best estimates of spacecraft photometric precision, only CEGPs will be detectable using this method. If the planetary orbital plane is not close enough to 90 degrees to produce obscuration (i.e. an occultation) of the planet, then one may expect that at orbital phases of $\varphi(t) = 180$ degrees the planet will be brightest (i.e. maximum value of the atmospheric phase function, or 'full'), and at an orbital phase of $\varphi(t) = 0$ degrees the planet will be dimmest (i.e. minimum value of the atmospheric phase function, or 'new').

This detection method does not, of course, rely upon the planetary orbital inclination being close to 90 degrees. However, detectability does depend on planetary orbital inclination as well as stellar activity as correlated with starspot activity. Thus, it is more difficult to detect giant inner planets around fast rotators; fortunately, the main targets of planet searches are late-type slow stellar rotators. At the photometric precision achievable by the NASA Kepler mission, and the statistics of the frequency of occurrence of CEGP, that mission should be able to add over one thousand additional giant planet discoveries using this detection method on the same photometric data used for planetary transit detections.

An advantage of this method is also that detailed studies of the planetary atmospheric particle size may be undertaken. The periodic nature of the brightness variations nevertheless depends on the steepness of their rise and fall in amplitude relative to the phase function of the planet, which is in turn dictated by either the atmospheric particle size distribution or the roughness of the planetary surface. In the case of CEGPs modelled to date, atmospheres of highly refractive elements are implied. In the transition from Rayleigh scattering to the Mie particle scattering realm (where the atmospheric particle circumference is approaching that of the wavelength of the light being used for the observation) and beyond, the forward diffraction lobe of the particle can create a rather steep phase function, with much of the light being reflected back during a full phase only at viewing angles very close to full phase. Thus, while the amplitude of the phase variation of the reflected light is dictated by the planet's albedo, a shallow phase function (Rayleigh scattering) will spread this light variation out more evenly in a sinusoidal variation, while a steep phase function will be indicative of larger particles in the planetary atmosphere.

A steep phase function will thus produce more Fourier components in the power spectrum than a shallow (more precisely sinusoidal) phase function (Jenkins and Doyle 2003). Thus, the number of components in the power spectrum of the photometric observations can be indicative of the particle size distribution in the CEGP's atmosphere, with the total amplitude of the variations giving a strong indication directly of the albedo of the planet (see Figure 1.8).

1.10. Eclipsing binary minima timing

Photometric data obtained during planetary transit searches that include eclipsing binaries can also be used in a complementary way to detect circumbinary jovian-like planets whose orbits aren't required to be coplanar along the line of sight – that is, non-transiting planets (Deeg et al. 2000, Doyle et al. 1998). The eclipse minima themselves can act as a clock (similar to the pulsar timing method) to mark variations in the displacement

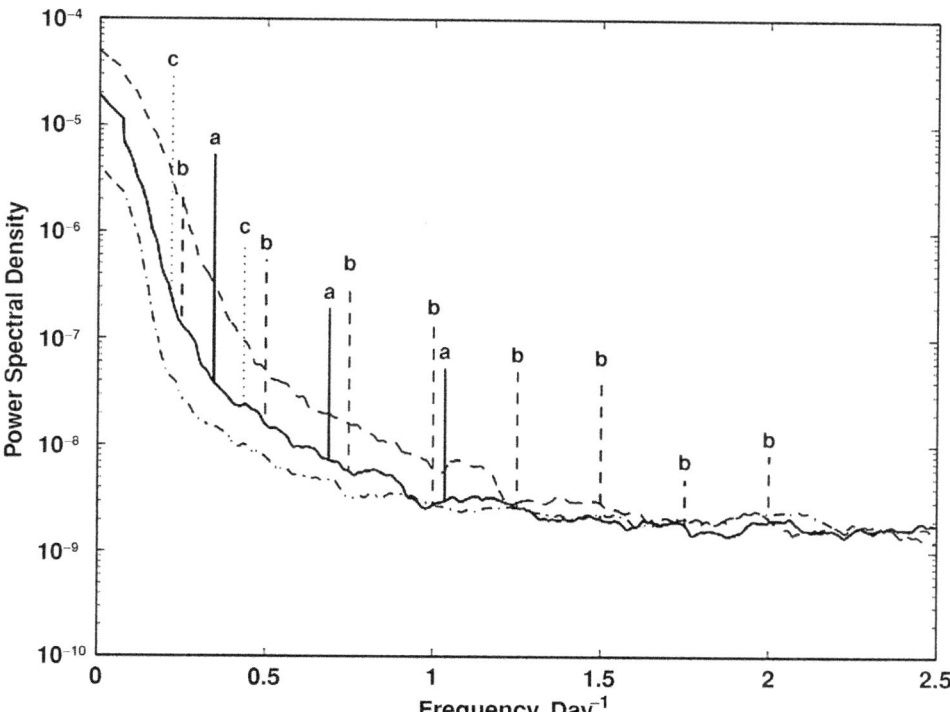

FIGURE 1.8. Power spectrum (power spectral density) of light variations due to three types of giant close-in extrasolar planets (CEGPs) with different atmospheric-particle phase functions against the background of noise from the Sun (taken from high precision visual-wavelength spacecraft data), detectable by the reflected light phase variation method (see Jenkins and Doyle 2003). The planet size in each case was taken to be 1.2 jovian radii. The vertical lines are actually spikes in the power spectral density seen at very low resolution (so appear as lines rather than histogram-like rectangles). The spikes labelled with 'a' are for a planet with a four-day orbital period with an atmosphere of one-micron-sized particles. The spikes labelled 'b' are for a 2.9-day orbital period planet with an atmosphere made up of 0.1-micron-sized particles. And the spikes labelled 'c' are for a four-day orbital period planet consisting of atmospheric particles well modelled as a Lambert sphere with an albedo of 0.67. There are fewer components in the power spectrum of model planet 'c', as a Lambert sphere produces light variations with orbits that are closer to a perfect sinusoid. Models 'a' and especially 'b' represent larger deviations from a perfect sinusoidal variation and so have more components in their power spectra. Thus the number of components in the power spectra of reflected-light-variation-detected extrasolar planets can be indicative of the planet's atmospheric particle size distribution.

of the binary pair across a binary–giant planet barycentre. If the binary is offset by a giant planet in circumbinary orbit, the time delay (or early arrival) of the eclipse minima must indicate this event. The half-amplitude of this offset time is given (for circular orbits, and binary components of equal mass M_*) by:

$$\tau = \sin i \left(\frac{a_\mathrm{p}}{c}\right) \left(\frac{M_\mathrm{p}}{2M_*}\right). \tag{1.14}$$

In this case, similar to the case with pulsar planet detection (or any detection method that measures a 'time stamp'), the maximum delay in the eclipse minimum occurs, for

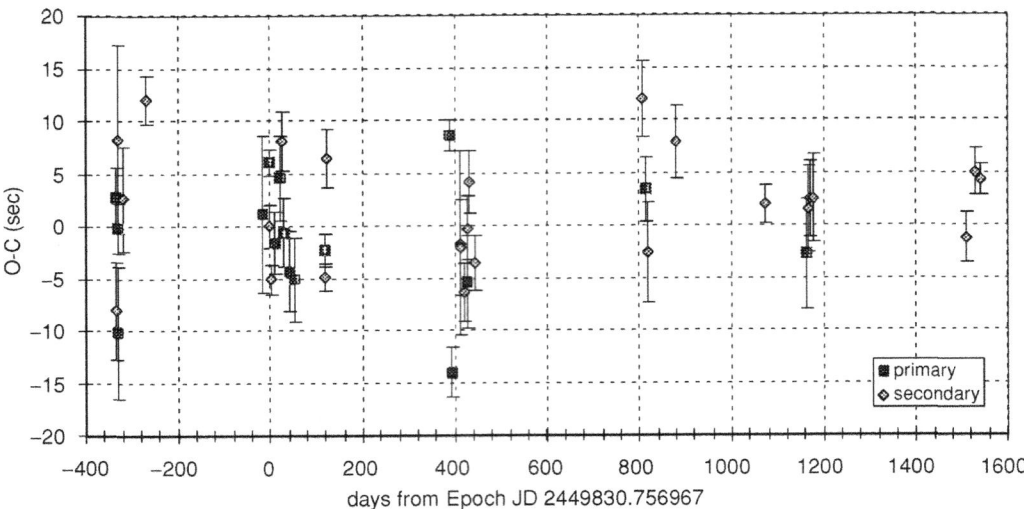

FIGURE 1.9. Observed minus the calculated (O-C) times of primary or secondary eclipse minima during six years of observations of the CM Draconis system (from Deeg et al. 2000). A power-spectrum of these (O-C) times indicate a candidate for a circumbinary planet of several jovian masses offsetting the binary star system periodically by about a few light-seconds around the barycentre.

circular orbits, when the planetary orbital phase $\varphi(t) = 180$ degrees. The eclipse minima will occcur 'on time' at phases $\varphi(t) = 90$ and 270 degrees, and be late by an amount τ at the planetary orbital phase of $\varphi(t) = 0$ degrees (see Doyle et al. 1998 for a short review of other effects that can produce changes in the binary–stellar eclipse period – most of which are, however, non-periodic). This approach to detecting extrasolar outer planets is most sensitive to long orbital period planets since these will produce the largest timing offsets, τ. So, while one must observe stellar eclipses over more than half a planetary orbital period, this method may also perhaps be the least observationally and photometrically demanding. This is because one can record several eclipses per season knowing when eclipse events are going to occur ahead of time. Also, eclipse minima may be quite precisely determined without prohibitive levels of photometric precision being required for planets at distances of several AU. For circumbinary planets within distances of an AU or less, sub-second precisions may be required (depending on the mass of the binary), and extremely good photometric precision (for example, by spacecraft missions) will be required in these cases (see Doyle and Deeg 2003).

For the detection of planets orbiting around two solar-type stars, a CEGP (as defined above) would produce a total timing variation of about 0.2 s. A jovian-mass planet in Jupiter orbit would have a variation of about 3 s, and an Earthlike planet a variation of about 0.004 s – per eclipse minimum measured. The precision will increase directly with the photometric precision, and inversely with the number of eclipse minima measured (Doyle and Deeg 2003). But this method clearly favours the detection of several jovian-mass planets at one AU or farther from the binary pair. However, again, detection is not generally dependent on planetary orbital inclination, and periodic variations in the binary eclipse minima epoch (although not in overall period) *must* occur if sufficiently massive planets are present. Thus non-detections are also meaningful. This detection method has already been applied to the M4-dwarf eclipsing binary CM Draconis (see Figure 1.9).

1.11. Secondary effects in extrasolar planet detection methods

What might be termed more 'exotic' extrasolar planet techniques are those that are secondary effects at the present state of detectability – those that may detect moons, rings, subtle changes in the atmosphere of the planet, and so on. One of these has already been demonstrated in the detection of sodium during a transit (see other reviews in this volume). We mention here just a few.

One may ask, for example, if transits of white dwarf stars can be distinguished from transits of planets of a similar size (Earth-sized planets). White dwarf transits may be detected photometrically but they will not cause a drop in brightness as in the usual transit events. Rather, because of gravitational microlensing, the transit of a white dwarf will produce an achromatic brightness increase that more than offsets the light blocked by the actual geometric area of the white dwarf itself (Sahu and Gilliland 2003). For an 0.6 solar mass white dwarf transiting a solar-type star, for example, the photometric drop due to the transit will be exactly cancelled when the white dwarf is coplanar with a semi-major axis of 0.0463 AU. At an orbital distance of 1 AU, however, the brightening due to the gravitational lens focusing of the white dwarf will be about 24 times the transit depth produced by its geometric area blocking the solar-sized disc.

Regarding planets around white dwarf systems, a 2.5 jovian-mass planet (the oldest planet known) has been found in a circumbinary orbit around a white dwarf–pulsar binary (B1620–26) in the globular cluster M4 (NGC 6121). This planet was detected by the pulsar timing method mentioned above. However, another suggestion for the detection of what might be called a 'former planet' has been put forth by Li et al. (1998). In this scenario a planet's outer layers are ablated during the star's red giant stage, leaving only a metallic core. As this core orbits through the magnetic field of the white dwarf, an electric current is generated that heats the outer regions of the white dwarf at its poles. Ohmic dissipation will cause the planet-core to spiral inward, but the heating of the poles of the white dwarf should be detectable as H_α emission. Li et al. suggest that the peculiar white dwarf GD 356 could be such a system.

Another secondary effect that may be detectable is the oblateness of the transiting planet itself. The detection of this variation in the ingress and egress of the transit event – a measure actually of the planet's rotation rate – however, requires precisions of a few times 10^{-5}, which should be achieved by spacecraft missions such as the NASA *Kepler* mission. Lensing by a giant planet's atmosphere may also be detectable at this precision (Hui and Seager 2002; Seager and Hui 2002). Variations of a planet's brightness or spectrum in rhythm with its rotational period, indicative of continental, atmospheric, or even plant-induced variations remain something for future planned missions (Ford et al. 2001 as an early example), and this is addressed in other chapters in this volume.

As outlined above, transiting planets may be expected to have generally short orbital periods as the probability of alignment goes down with increasing orbital semi-major axis. However, a change in timing of the transits themselves can be indicative of additional planets in the system due to their gravitational interaction with the transiting inner planet (Holman and Murray 2005). Such variations of the terrestrial planets produced by the outer giant planets have already been well studied for our Solar System, producing variations of hundreds of seconds over a few years.

Moons of transiting planets may also be detected by a slight increase in brightness during a transit (the moon transiting the transiting planet and thereby not blocking out light for a time; Sartoretti and Schneider 1999). Such slight increases in brightness have been observed during transits of HD209458 but attributed rather to starspots being transited by the planet rather than a moon in orbit around such short-period planets as HD209458b where moon-orbits can apparently not be expected to be stable (Barnes and O'Brien 2002). Also, moons may cause step-shaped transits – if the transit of the moon is detectable – since the moon and the planet won't ingress or egress at the same time (Sartoretti and Schneider 1999).

Another way to detect moons around extrasolar planets is by timing of the transit event itself (Doyle and Deeg 2003, for example), but the periodic transit events must be distinguished from eccentricity, apsidal motion, further orbiting bodies (as mentioned previously) and other possible effects that could cause transit ingress/egress changes. The Earth's Moon viewed from another star system – if leading when crossing the solar disc during the onset of a transit of the Sun – would, for example, delay the actual Earth's ingress by about 158 seconds.

In summary, secondary effects to the actual planet detection may continue to be of significant interest as detection precision increases. Planetary rings, aurorae, light echoes from flares (visible stellar flares echoed then in increased radio activity from a giant planet's magnetosphere), nulling interferometric spectroscopy, white dwarf pulsation timing, and density wave trails in circumstellar discs should all be detectable within the next decade.

1.12. Ranges and combinations of detection methods

As pointed out, the detection techniques that can be used in the solar neighbourhood are imaging and astrometry techniques, as they rely on the resolution of the angular separation of the star from the planet. By 'solar neighbourhood' here we mean within 20 parsecs or so of the Sun for present methods. At present, this group also includes direct radio detection of jovian-type extrasolar planets due to noise limits at radio frequencies. Detection of extrasolar planets at short range – out to several hundred parsecs – includes the radial velocity detection method since high-precision spectra require lots of photons and hence bright stars. Larger optical telescopes (30 and 100 metre apertures) will allow the range of this technique to be extended in the near future.

Stellar neighbourhood-to-intermediate-distance methods (to nearby spiral arms at one kiloparsec or so for near-term spacecraft missions) include the phase variation reflection method, which requires very high precision photometry just for the detection of giant inner extrasolar planets. Detection methods for distant stars (to the galactic bulge at 2 kiloparsecs or more) include the gravitational microlensing method which requires a large distance to intercept light from a lensing star, and also very crowded stellar fields in order to detect any events. However, the events are expected to be quite bright (one-magnitude increases being typical, as mentioned above).

The transit method works for both single and double stars and will work from solar-neighbourhood to distant ranges, although larger telescopes (i.e. greater than 1 metre) may be required for sufficient photon statistics on galactic bulge stars. Finally, eclipsing binary timing is also effective from solar-neighbourhood to distant-field ranges, requiring varying levels of photometric precision – depending on the period of the planet – and, in some cases, significantly less observing time for the detection (or ruling out the presence) of circumbinary giant planets. Given the state of the art today, for repeatable follow-up observations, the transit method is the only technique that may detect Earth-sized extrasolar planets around solar-type stars. However, since transits are a differential technique (the ratio of planet-to-star area is measured), ground-based photometric precision would allow, for example, the detection of an Earth-sized planet around stars as small as Proxima Centauri, as previously mentioned.

Multiple methods of extrasolar planet detection can be combined to produce additional parameters. To date, the most useful mixture of detection methods has been combining radial velocity measurements with transit measurements. In case of HD 209458b, which was discovered by radial velocity variations but then shown to also transit its parent star, the transit resolved the 'projected' $M_p \sin i$ mass measured by radial velocity variations, since $\sin i \cong 1.0$ in the cases where a transit can actually occur. In addition, the transit method gives the ratio of the disc-area of the planet to the star. Knowing the stellar radius then gives the planet's volume, and thereby – combined with the planet's actual mass – the planetary density, which in the case of HD 209458b could then be tested against thermal models (see reviews in this volume).

The gravitational lens method, while restricted to systems far enough away to detect the lensing effect, nevertheless does measure the distance to the star system with a planet that is lensing the background star, as well as the planetary mass and orbital velocity. There can be, however, ambiguities in the model fits to the brightness variations of such microlensing events, such as cases where a wide and a narrow binary configuration can both provide solutions (e.g. Kubas *et al.* 2005). In most cases this can be resolved, but in the case where the stellar system might also show a transit, the transit will certainly resolve any close/wide binary ambiguity. With a constraint on distance from gravitational lensing detection, the stellar magnitude may be accurately determined, and this allows a determination of the true star size and therefore the true planet size, again, for example, giving the planet's density.

To give another example, if a planet that has been detected using reflected phase variations is also found to transit its parent star – and this should occur between 14% and 40% of the time (Jenkins and Doyle 2003) – then an occultation will also occur. This means that the planet's contribution to the total brightness of the system – at an orbital phase of $\varphi(t) = 180$ degrees when it is actually at maximum brightness (in 'full' phase) – will 'turn off'. Thus, knowing the size of the planet from the transit, and consequently the geometric ('full' phase) albedo contribution, one may then isolate the planet's atmospheric phase function with much higher precision.

If a planetary system has been detected through its parent star's radial velocity variations, then the true mass of the planet can be determined by astrometry, as was done in the case of the planet of Gliese 876 (Benedict *et al.* 2002), where changes in the optical centroid position of the point-spread function of the star were determined using the fine guidance sensor on the *Hubble Space Telescope*.

Astrometric planet detections give the position angle of a planet–star system but with an ambiguity of 180 degrees in its orientation; they also indicate the inclination of the system. If the planet also transits, inclination is independently determined but with an ambiguity across which of the stellar hemispheres (north or south) the planet is transiting; this is identical to an ambiguity in the position angle of 180 degrees. In addition to the usual yearly parallactic shift due to the Earth's yearly orbit, the motion of our Solar System relative to the observed planetary system's galactic position will also cause a variation of the perceived inclination of the planetary system, and hence in the duration of the transits. Depending on whether the transit duration increases or decreases, the ambiguity in the position angle may then be resolved.

Finally, as another example, if a circumbinary planet is detected with eclipse minima timing, and the system is near enough to allow direct imaging interferometry, then the location of the planet during maximum elongation can be calculated, because then the planet will be farthest in angular extent from the star (i.e. at orbital phases of $\varphi(t) = 90$ and 270 degrees). Also, one might want to image the planet during the stellar eclipse to also cut down the contribution of stellar flux (by half, in the case of equal stellar component sizes, temperatures, and binary orbital inclination of about 90 degrees).

Many other interesting combinations are possible and will no doubt be tried in the near future.

1.13. Future mission near-term targets: M-dwarf star systems

Future extrasolar planet missions (after *Kepler*, to be launched in 2008) include mostly interferometric missions such as *SIM (Space Interferometer Mission)*, and *TPF (Terrestrial Planet Finder)/Darwin*, as mentioned. Nulling interferometry will probably be the methodology to isolate the light from terrestrial-type extrasolar planets in order to attempt to detect exobiological signatures in their atmospheres. (A coronagraphic *TPF* is also being planned but will block out the inner CHZ of M-stars.) While these missions are also reviewed in more detail in this volume, it may be noted that, since the present range of these mission targets is at about 15 parsecs, perhaps as many as 80% of the nearby target stars for these missions will be red dwarfs – i.e. the cool M-spectral type stars ranging from about 0.5 to 0.08 solar masses and, on average, about 1% the brightness of the Sun. It has been known for some time that, since the CHZ distance increases with the square-root of the brightness, but tidal forces decrease with the cube of the distance, virtually all M-dwarf stars will tidally lock into synchronous rotation any planets orbiting within their CHZ.

It was assumed for decades, then, that planets with the right size and composition to be habitable would, due to synchronous rotation around M-dwarfs, lose their atmospheres – boiling off on the stellar side and condensing out on the dark side. However, with initial work by Haberle *et al.* (1996) it was shown that 0.1 bar of carbon dioxide added to an Earthlike atmosphere would lower the thermal gradient between the light and dark sides of such a tidally-locked planet to the point where liquid water might persist on the surface for long periods of time.

More detailed modelling (Joshi *et al.* 1997; Heath *et al.* 1999; Joshi 2003) determined that the sub-stellar point (the point on the planet with the star at its zenith) would have massive rains, but that surface winds would nevertheless not exceed about 15 m/s, so that hurricane-force winds would not be expected as a result of the temperature differences between light and dark sides of the atmosphere. A hydrological cycle would exist, and help to mitigate the temperature effects of both flare and starspot activity. Certain kinds of bacteria – known on Earth to be able to photosynthesize at wavelengths as red as 0.9 microns (e.g. *clorobium*), could thrive on such a planet. However, these bacteria do not produce oxygen as a byproduct, and thus an atmospheric ozone layer would likely not form on such an M-star planet. However, given the flux distribution of typical M-stars, sufficient ultraviolet light that may damage complex plant biota is not produced except in infrequent flare activity. In other words, the photodissociation flux, with wavelengths shortward of about 0.26 microns, is not produced in any great quantity in late M-dwarf photospheres. Thus there is no runaway greenhouse effect possible, in the classical sense (photodissociation of the water molecule). Joshi *et al.* (1997) hypothesized that the inner boundary of the M-dwarf CHZ might then be able to withstand up to three times the stellar flux compared with the CHZ defined by the runaway greenhouse effect conditions, i.e. the CHZ may be 70% closer to the star.

Present three-dimensional atmospheric models (Joshi 2003) predict a dark-side equatorial ice cap, temperatures well within habitable (liquid water) ranges around the terminator zones extending out as far as half the planet on the day side, and ground fog throughout most of the planet. However, some problems remain to be solved if, for example, the *TPF/Darwin* mission is to target primarily M-dwarf stars in looking for evidence of disequilibria in planetary atmospheres caused by exobiological processes.

M-dwarfs evolve at a much slower rate than solar-type stars and, therefore, are believed to remain at least ten times as long on the main-sequence as the Sun. Although the CHZ of M-dwarfs may remain stable for perhaps a hundred billion years, there are three aspects that may turn out to be of concern regarding a planet being able to remain within this CHZ over a significant fraction of such timescales. One concern is M-dwarf mass loss; rates up to 10^{-10} solar masses per year may have been detected in some systems (see references in Doyle *et al.* 1996). Significant reduction of the M-dwarf stellar mass over a time period of billions of years (assuming the process could be sustained) would cause the CHZ to move inward (towards the star) dramatically – the stellar luminosity function is $L_* \propto M_*^{4.75}$ for solar-type stars but is possibly only about $L_* \propto M_*^{2.5}$ for M-dwarf stars. Assuming this latter mass–luminosity relationship, an Earthlike planet within the CHZ would be expected to migrate outward with a rate proportional to the square root of the mass decrease (assuming angular momentum conservation – e.g. the planet was not subject to significant drag by the mass outflow material). As an example, at a mass-loss rate of 10^{-12} solar masses a year for an M4-dwarf (0.25 solar masses), over ten billion years, about 10% of the stellar mass could be lost. The luminosity would decrease by about 77% (assuming the second relationship above with a mass–luminosity exponent of 2.5), the CHZ would then have migrated inward by about 12%, and the planet's orbit would have migrated outward by about 20%. Thus we see that this may be a significant consideration in planning for searches for exobiological processes on planets around such stars.

Another consideration is the rate of tidal dissipation of rotational angular momentum on an Earthlike planet, the rate of which can be estimated by:

$$T = 0.027 \left(\frac{P_i t}{Q}\right)^{1/6} M_*^{1/3}, \tag{1.15}$$

where P_i is the initial rotation rate of the planet relative to the central star, and Q is the tidal dissipation function (for present Earth with shallow seas $Q = 13$, but more likely $Q = 100$ for deeper oceans). For a typical M4-dwarf mass, and a planet within the CHZ, Eq. (1.15) gives a time less than 10^5 years for tidal locking (largely independent of initial rotation period). Thus, tidal locking of planets within the CHZ of M-dwarfs can be thought to essentially take place almost immediately in such systems. A tidally locked planet nevertheless rotates (with its orbital period), and frictional dissipation caused by any shallow seas within the hydrological cycle, as well as frictional dissipation due to internal tectonic activity, may produce a significant rate of angular momentum dissipation over geologic time. Heath *et al.* (1999) have shown that such planets indeed may be expected to have hydrological cycles so that such tidal friction of water with the planet should be taking place. However, the planet will not be able to slow down in rotation so that dissipation of angular momentum must take place rather in a shortening of the orbital semi-major axis. Depending on condition on any given planet, an Earthlike planet initially within the M-dwarf CHZ may migrate inwards, in this case, out of the CHZ on timescales where the evolution of biota might be affected. In addition, if the star is rotating faster than the planet's orbit, it could migrate inwards due to the tidal bulge produced by the planet on the star. The angular momentum of a sphere with a mass-distribution like the Earth (the moment of inertia adjusted by multiplying a uniform sphere by 83%) is about:

$$\Lambda_r = 0.33 M_p R_p^2 \omega, \tag{1.16}$$

where ω is the angular velocity. The Earth is presently decreasing in angular velocity by about -5.5×10^{-22} radians/second due to shallow seas (mostly the Bering Sea). However, a tidally locked 'Earth' cannot lose angular momentum through rotational slowdown, and might be expected to, again, shorten its orbital semi-major axis. If we take, as an order-of-magnitude estimate of the rate of rotational slowdown given above over, say five billion years, we have an angular momentum loss of about 7×10^{40} g cm^2/five billion years. The orbital angular momentum of such an 'Earth' within the CHZ of an M4-dwarf would be about $\Lambda_o = M_p a_p^2 \omega_o = 4.7 \times 10^{46}$ g cm^2/s (assuming a semi-major axis of about 0.1 AU). If we set the rate of rotational angular momentum loss equal to the rate of orbital angular momentum loss over the same time period, then we obtain a decease in the planetary orbital semi-major axis of about 4.8×10^{11} cm $\cong 0.03$ AU, a significant decrease to about two-thirds of the planet's original distance from the star with an increase in insolation of a factor of 50%. However, if the planet started in the outer CHZ it could still be well within the CHZ at the end of this orbital evolutionary process.

A third concern about the habitability of M-star planets is that tidal locking implies a low rotation rate of the planet, thus also reducing the planetary magnetic field by perhaps as much as an order of magnitude relative to a freely rotating planet. If the magnetic field is reduced substantially, the main problem for maintaining the planet may be that the atmosphere could be sputtered away on the order of a gigayear (Backman, personal communication). All three of these aspects of maintenance of the habitability of M-dwarf planets will need to be examined in detail as the precise targets are identified. However, it is clear that the habitability of M-dwarf planets cannot a priori be assumed to be as stable as the CHZ of the M-dwarf star itself.

Acknowledgements

The author greatly thanks Drs Hans-Jörg Deeg and Juan Antonio Belmonte for the kind invitation and honour to teach at the IAC Winter School as well as their patience in getting this chapter. I would also like to thank IAC faculty and staff in general for an outstanding meeting, their wonderful hospitality, and also thank the excellent students for their interesting questions and enthusiasm. I also thank Drs Martin Heath and Manoj Joshi for the in-depth discussions of M-dwarf circumstellar habitable zones through the years.

REFERENCES

BARNES, J. W. & O'BRIEN, D. P., 2002, *ApJ* **575**, 1087.
BENEDICT, G. F., MCARTHUR, B. E., FORVEILLE, T., et al., 2002, *ApJ* **581**, L115.
BORUCKI, W., JENKINS, J., KOCH, D. & DOYLE, L. R., 1996, *Bull. Am. Astron. Soc.* **28**, 100.
BUTLER, R. P., VOGT, S. S., MARCY, G. W., et al., 2004, *ApJ* **617**, 580.
CODONA, J. L. & ANGEL, R., 2004, *ApJ* **604**, L117.
DEEG, H. J., DOYLE, L. R., KOZHEVNIKOV, V. P., et al., 1998, *A&A* **338**, 479.
DEEG, H. J., DOYLE, L. R., KOZHEVNIKOV, V. P., BLUE, J. E., MARTIN, E. L. & SCHNEIDER, J., 2000, *A&A* **358**, L5.
DOYLE, L. R., 1996, SETI Institute white paper presented to the FRESIP (now Kepler) team, March 1996.
DOYLE, L. R. & DEEG, H. J., 2003, in *Bioastronomy 2002: Life Among the Stars*, Proc. of IAU Symposium 213, eds. R. Norris & F. Stootman, San Francisco: Astronomical Society of the Pacific, p. 80.

Doyle, L. R., Heather, N. C., Vikramsingh, R. & Whitmire, D. P., 1996, in *Circumstellar Habitable Zones: Proceedings of the First International Conference*, ed. L. R. Doyle, Menlo Park, CA: Travis House Publications, 157.

Doyle, L. R., Deeg, H. J., Jenkins, J. M., et al., 1998, in *Brown Dwarfs and Extrasolar Planets*, eds. R. Rebolo, E. L. Martin & M. R. Z. Osorio, ASP Conference Series 134, 224.

Doyle, L. R., Deeg, H. J., Kozhevnikov, V. P., et al., 2000, *ApJ* **535**, 338.

Drake, A. J., 2003, *ApJ* **589**, 1020.

Ford, E. B., Seager, S. & Turner, E. L., 2001, *Nature* **412**, 805.

Haberle, R. M., McKay, C. P., Tyler, D. & Reynolds, R. T., 1996, in *Circumstellar Habitable Zones*, ed. L. R. Doyle, Menlo Park, CA, Travis House Publications, 29.

Heath, M., Doyle, L. R., Joshi, M. M. & Haberle, R., 1999, *Origins of Life* **29**, 405.

Holman, M. J. & Murray, N. W., 2005, *Nature* **307**, 1288.

Hui, L. & Seager, S., 2002, *ApJ* **572**, 540.

Jenkins, J. M. & Doyle, L. R., 2003, *ApJ* **595**, 429.

Jha, S., Charbonneau, D., Garnavich, P. M., Sullivan, D. J., Sullivan, T., Brown, T. M. & Tonry, J. L., 2000, *ApJ* **540**, L45.

Jones, D. L., 1994, in *Planetary Systems: Formation, Evolution, and Detection*, eds. B. F. Burke, J. H. Rahe & E. E. Roettger, Dordrecht, Kluwer Academic Publishing, p. 391.

Joshi, M. M., 2003, *Astrobiology J.* **3** 415.

Joshi, M., Haberle, R. M. & Reynolds, R. T., 1997, *Icarus* **129**, 450.

Konacki, M. & Wolszczan, A., 2003, *ApJ* **591**, L147.

Kubas, D., Cassan, A., Beaulieu, J. P., et al., 2005, *A&A* **435**, 941.

Lazio, T., Joseph, W., Farrell, M., et al., 2004, *ApJ* **612**, 511.

Labeyrie, A., 1995, *A&A* **298**, 544.

Li, J., Ferrario, L. & Whikramasinghe, D., 1998, *ApJ* **503**, L151.

Marcy, G. W. & Butler, R. P., 1998, *Ann. Rev. A&A* **36**, 57.

Mayor, M. & Queloz, D., 1995, *Nature* **378**, 355.

Perryman, M. A. C., 2000, *Rep. Progr. Phys.* **63**, No. 8, 1209.

Sahu, K. C. & Gilliland, R. L., 2003, *ApJ* **584**, 1042.

Sartoretti, P. & Schneider, S., 1999, *A&A Supp.* **134**, 553.

Schneider, J. & Doyle, L. R., 1995, *Earth, Moon, and Planets* **71**, 153.

Schneider, P., Ehlers, J. & Falco, E. E., 1999, *Gravitational Lenses*, Berlin: Springer-Verlag.

Seager, S. & Hui, L., 2002, *ApJ* **574**, 1004.

Seager, S. & Mallen-Ornelas, G., 2003, *ApJ* **585**, 1038.

Seager, S., Whitney, B. A. & Sasselov, D. D., 2000, *ApJ* **540**, 504.

Tarter, J., 2001, *Ann. Rev. A&A*, **39**, 511.

Tingley, B. & Sackett, P. D., 2005, *ApJ* **627**, 1011.

Udalski, A., Jaroszyáski, M., Paczyaski, B., et al., 2005, *ApJ* **628**, L109.

Wolszczan, A., 1994, *Science* **264**, 538.

Wolszczan, A. & Frail, D. A., 1992, *Nature* **355**, 145.

2. Statistical properties of exoplanets

STÉPHANE UDRY

Since the detection a decade ago of the planetary companion of 51 Peg, more than 200 extrasolar planets have been unveiled by radial-velocity measurements. They present a wide variety of characteristics such as large masses with small orbital separations, high eccentricities, period resonances in multi-planet systems, etc. Meaningful features of the statistical distributions of the orbital parameters or parent stellar properties have emerged. We discuss them in the context of the constraints they provide for planet-formation models and in comparison to Neptune-mass planets in short-period orbits recently detected by radial-velocity surveys, thanks to new instrumental developments and adequate observing strategy. We expect continued improvement in velocity precision and anticipate the detection of Neptune-mass planets in longer-period orbits and even lower-mass planets in short-period orbits, giving us new information on the mass distribution function of exoplanets. Finally, the role of radial-velocity follow-up measurements of transit candidates is emphasized.

2.1. Motivation and context

The hypothesis of the formation of planets in our Solar System from a *solar nebula*, in a flattened gaseous disc in differential rotation, is more than two centuries old. This approach was first proposed by Kant around 1755 and then developed by Laplace (1796). The idea came in a natural way from the observation of the planet configuration in our Solar System: they turn in the same direction, on quasi-circular orbits, in a quasi-common plane. Although only qualitative, the ideas developed at that time are very similar to fundamental precepts accepted today: local instability of a gaseous cloud, collapse in a rotating disc, privileged location of planet formation. The contemporary aspect of the formation of the Sun and its cortege of planets is supported by independent dating of the Sun (astrophysics), the Earth (geology), the Moon and meteorites (rock analysis and crater dating). Stars and planets are then formed in a common global process. Planets are natural by-products of stellar formation and are thus expected to be very common in the Universe.

These theoretical indications of the widespread existence of planets around stars were first supported by high-quality observations with a new generation of instruments capable of high-resolution imaging like the *HST*, the Keck telescopes and the VLT. In particular, infrared (IR) images of star forming regions revealed the presence of protoplanetary discs around a large number of very young forming stars (McCaughrean *et al.* 2000; Beckwith 2000). These results confirmed the already existing indirect indications of the presence of dust and gaseous discs coming from IR and UV excesses in the light of young HL Tauri (Beckwith & Sargent 1993, 1996). Traces of planetary formation were also observed in the form of dust and debris discs around stars in regions corresponding to our Kuiper belt (Lagrange *et al.* 2000).

Before 1995, the Solar System was the only known example of a planetary system in orbit around a sun-like star, and the question of its uniqueness was more a philosophical than a scientific matter. The discovery of an extrasolar planet orbiting another

Extrasolar Planets, eds. Hans Deeg, Juan Antonio Belmonte and Antonio Aparicio.
Published by Cambridge University Press.
© Cambridge University Press 2007.

solar-type star, 51 Peg (Mayor and Queloz 1995), has changed this fact and opened the road to a steadily increasing suite of exoplanet detections. In the following years, we learned first that planets are very numerous and most importantly that the planetary formation process may produce very different configurations: masses considerably larger than the one of Jupiter, planets moving on highly eccentric orbits, planets very close to their parent stars, planets in resonant multi-planet systems, planets orbiting the components of stellar binaries, etc. Understanding the reasons for such wide variations in outcome remains a central issue in planet formation theory. The role of observation is then to provide constraints that will help theoreticians to build models able to reproduce the large variety of properties observed for extrasolar planets. With the number of known exoplanets now surpassing 200,[1] statistical trends appear in the distribution of planetary orbital elements or primary-star properties. They should help us to constrain the planet-formation models, as those distributions are supposed to retain fossil traces from the processes active during the formation and/or evolution stages of these systems.

We will try here to do a census of the main statistical results obtained from spectroscopic observations since the discovery ten years ago of 51 Peg b. Section 2.2 will first recall important notions related to the radial-velocity detection method (the most efficient today), and to its limitations. Then, in addition to the orbital properties described in Sections 2.3 and 2.5, and the primary-star characteristics discussed in Section 2.6, we will discuss major changes in our view of the future of radial-velocity measurements in the exoplanet domain that have emerged during the past two years, namely (i) the role played by follow-up radial-velocity measurements in confirming and helping the characterization of planetary objects among the many candidates detected by photometric-transit programmes (Section 2.7) and (ii) the development of specially designed high-resolution spectrographs achieving precisions for radial velocities below the 1 ms^{-1} limit that are allowing us to detect planets in the Uranus/Neptune mass range (Section 2.4). This opens the way to still better precision and to the detection of Earth-type planets with radial-velocity measurements (Section 2.8).

2.2. Radial-velocity planet-search programmes

A census of the different planet-detection methods is given in Chapter 1. I will just recall here some aspects related to the radial-velocity technique because this method is by far the most efficient to date in terms of planet detections and is thus of prime importance to understanding how it works and what its limitations are for a correct interpretation of the results obtained.

2.2.1. Measuring high-precision radial velocities

The radial-velocity detection method is based on the measure of the variation of the velocity of a star in its orbit around the centre of mass of the planet–star system. From Newton's equations and Kepler's second law describing two-body motion, the semi-amplitude of the radial-velocity variation of the primary star ($K_\star$) relates to the mass of the two components in the following way:

$$\frac{(m_{\rm pl} \sin i)^3}{(m_\star + m_{\rm pl})^2} = \frac{P}{2\pi G} K_\star^3 (1 - e^2)^{3/2},$$

---

[1] A comprehensive list of known planets, with references to related papers, is maintained by the *Extrasolar Planets Encyclopaedia* – www.exoplanet.eu. The two most prolific teams of exoplanet searchers also propose their own sites: Geneva: www.exoplanets.eu; Berkeley: exoplanets.org/.

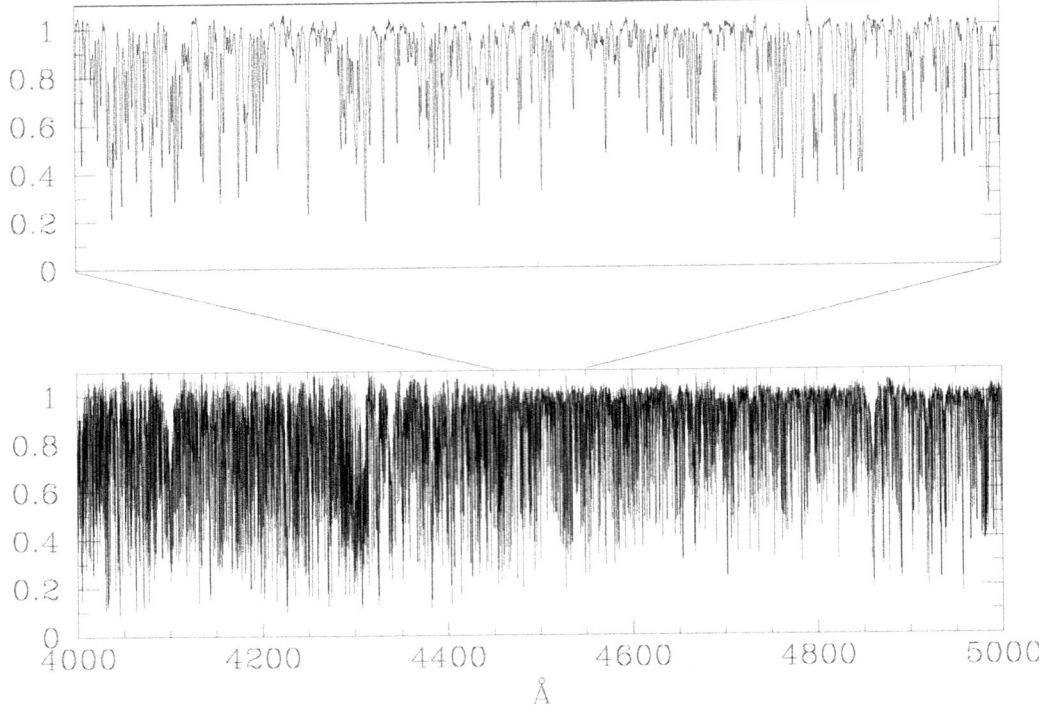

FIGURE 2.1. Blue part of an ELODIE spectrum of 51 Peg (∼30% and 3% of the spectral window), illustrating the large number of lines available in solar-type stars and used for the cross-correlation in the simultaneous-thorium approach.

with P and e the orbital period and eccentricity and i the inclination angle between the sky and orbital planes. The left term of the equation is called the *mass function*. In the case of a star–planet system ($m_{\rm pl} \ll m_\star$), the *minimum mass* of the planetary companion is, to a first approximation, a simple function of the primary star mass and orbital parameters:

$$m_{\rm pl} \sin i \simeq \left(\frac{P}{2\pi G}\right)^{1/3} K_\star m_\star^{2/3} \sqrt{1-e^2}.$$

To fix orders of magnitudes, in the case of a circular orbit around a solar-mass star, we can write (in standard units):

$$\begin{array}{cccc} m_{\rm pl} \sin i & \simeq & 3.5 \cdot 10^{-2} & K_\star & P^{1/3}. \\ {[M_{\rm Jup}]} & & & [{\rm ms^{-1}}] & [{\rm yr}] \end{array}$$

The influence of Jupiter on the Sun is about 12 ms^{-1} (when $\sin i = 1$). The Saturn effect is ∼4 ms^{-1}, similar to the influence of a 15 $M_\oplus$ planet on a 10 day period orbit.

The actual estimate of the radial velocity is obtained by the measurement of the Doppler shift of the star spectrum compared to its position at rest. For radial-velocity changes of the order of 10 ms^{-1}, this Doppler shift is very small, of the order of 1/1000 of a typical spectral line width, i.e. ∼10^{-4} Å. To reach such precision, high-resolution, high-S/N spectra are required, as well as a multi-line approach of the Doppler shift estimate, such as cross-correlation or spectrum-fitting techniques. The radial-velocity method is thus especially efficient for 'cold' stars with a large number of spectral lines (Figure 2.1),

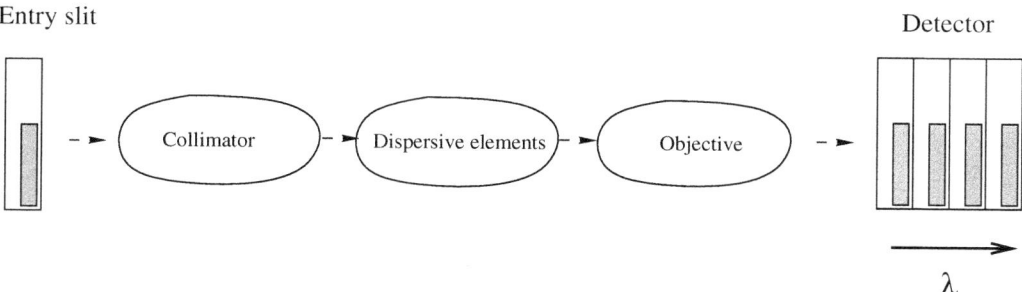

FIGURE 2.2. Cartoon representation of the effect on the CCD of a shift in the star photocentre on the slit entrance of the spectrograph.

not too much widened by stellar rotation and intrinsically stable in radial velocity (i.e. non-pulsating). The most favourable targets are thus dwarf stars with spectral type later than ∼F5 (often called *solar-type stars*, for reasons of simplicity).

2.2.2. *Wavelength calibration and long-term stability*

To maintain a long-term high radial-velocity precision (required for planet detection and orbital characterization), two major problems have to be taken care of.

(1) The position of the photocentre of the star at the spectrograph entrance has to be very stable from one observation to the next. A small shift in this position is directly reported as a small physical displacement of the position of the spectrum on the CCD, seen as an *equivalent Doppler shift* (Figure 2.2). This shift can be very large in terms of radial velocity. Typically, for a slit spectrograph with a good guiding system, the precision achieved on the position of the star photocentre is of the order of 1/20 of the slit width. In velocity units, this corresponds to several tens or even hundreds of ms^{-1}, depending on the spectrograph resolution. The effect is minimized with fibre-fed spectrographs. A scrambling device is often added along the path to azimuthally average the light and have an illumination of the spectrograph entrance *as uniform as possible*.

(2) The wavelength calibration defining the radial-velocity reference (*zero point*) has also to be stable. Each radial-velocity measurement is in practice a differential measurement between the velocity of the star and the zero-point reference. Variations in the reference velocity will thus directly translate into undesired variations of the measured star velocity. The precision of the zero point definition sets the ultimate precision you can reach with your instrument, on top of other sources of errors (systematics, photon noise).

The radial-velocity variations to be measured are very tiny and thus require a specifically designed instrumentation to be accurately estimated. Two competing techniques based on different radial-velocity references have proven to be very efficient, reaching a precision of ≪ 10 ms^{-1} on a timescale of years.

The 'simultaneous-thorium' technique

The *simultaneous-thorium* technique is applied with fibre-fed echelle spectrographs with two fibres, one fibre for the star and the other illuminated by a reference lamp. The light paths from the two fibres are kept close to each other in the spectrograph and the spectral orders are recorded in an alternating way on the CCD (the orders of the lamp spectrum are located in the inter-orders of the stellar spectrum; Figure 2.4).

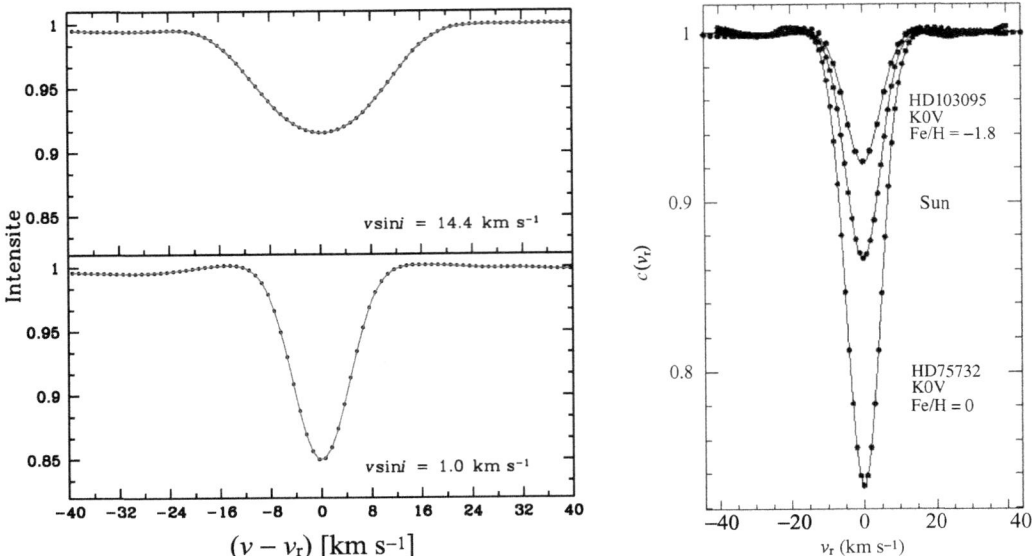

FIGURE 2.3. Left: cross-correlation functions (CCFs) of CORALIE spectra for two stars with different rotational velocities. The effect of the rotational broadening is seen directly on the CCF. Right: CCFs for two similar stars of different metallicities. The Sun's CCF is shown for comparison.

Usually a thorium–argon (ThAr) lamp is used. Thorium has numerous emission lines in the visible that allow for (i) precise wavelength calibration and (ii) tracking (with the second fibre) of the spectrograph drifts due to local astroclimatic changes,[2] and correct for them supposing that the effects are the same on the two fibres (close optical paths). To ensure sufficient stability of the instrument, the spectrograph has to be kept in a temperature- and pressure-controlled environment.

Both stellar and reference velocities are calculated by numerical cross-correlation of the stellar and lamp spectra with corresponding templates (Baranne *et al.* 1996). The template can be a high S/N spectrum of the star (lamp) itself or a binary transmission function with 'holes' (transmission = 100%) at the positions of the lines to be used in the cross-correlation. The simultaneous-thorium technique is used with the ELODIE spectrograph in Haute-Provence (France), with the CORALIE spectrograph on the 1.2-m Swiss telescope and the HARPS spectrograph on the 3.6-m ESO telescope, both at La Silla (Chile). Typical precisions are from 5–7 ms^{-1} (CORALIE, ELODIE) to below 1 ms^{-1} for HARPS (see later in this section and Section 2.4).

The cross-correlation function (CCF) obtained corresponds in essence to an average of the actual stellar lines selected in the template. It carries a rich astrophysical content. Every physical phenomenon that changes the characteristics of these lines will affect the CCF in the same way. For example, for solar-type stars, calibration of the width of the CCF provides a good estimate of the stellar rotational velocity (Figure 2.3, left). The surface (or equivalent width) of the CCF is in the same way a good metallicity indicator (Figure 2.3, right) as most lines of solar-type stars in the template used are iron-related lines. Santos *et al.* (2002) derived such correlations for the CORALIE targets.

---

[2] Temperature or pressure changes in the spectrograph induce variations of the refraction index and thus of the position of the spectrum on the CCD, and of the measured radial velocity.

Then, with a single observation, in addition to the radial velocity, you get 'for free' estimates of the $v \sin i$ and [Fe/H] of the star.

The 'iodine cell' technique

Contrarily to the ThAr technique in which the stability of the radial-velocity reference relies on the spectrograph's stability, with the iodine cell method the reference is 'stuck' to the stellar spectrum itself. Before entering the spectrograph, the light from the star is sent through a cell filled with iodine gas of fixed pressure and temperature. The known absorption lines of the iodine superimpose themselves on to the stellar spectrum and thus directly provide a precise velocity calibration. The Doppler shift is then estimated by a global fit to the observed spectrum of a standard composite spectrum built from high S/N reference spectra of the star and the iodine (Butler *et al.* 1996).

The technique is interesting because, thanks to the iodine spectrum, it allows us to follow temporal changes of the *point spread function* (PSF) of the instrument. It is also weakly dependent on spectrograph variations as the light paths are strictly the same for the star and the calibration. The method is easily implemented on existing spectrographs. Most of the large telescopes operating with a high-resolution echelle spectrograph (Keck, HET, Magellan, VLT, etc.) are equipped with such a device. The typical precision obtained for radial-velocity measurements is generally better than ~ 5 ms^{-1} (~ 2–3 ms^{-1} in the best cases).

The method, however, is expensive in its need of photons because (i) the fit procedure requires high S/N; (ii) the iodine lines are in absorption; (iii) the 'small' size of the wavelength window with iodine lines (5000–6000 Å) limits the portion of the stellar spectrum usable for the radial-velocity estimate. In total, for a similar precision, the iodine cell technique requires about six to ten times more photons than the simultaneous-thorium technique. On the same facility and in the same integration time, the simultaneous-thorium method will thus allow us to measure stars more than two magnitudes fainter than stars observable with the iodine technique. This has implications for the definition of observing programmes and in particular for the radial-velocity follow-up of planetary transit candidates around faint stars (Section 2.7).

HARPS: below 1 ms^{-1}

The simultaneous-thorium technique has been pushed to new limits with the development of the HARPS spectrograph mounted on the ESO 3.6 m telescope at La Silla. HARPS is a cross-dispersed echelle spectrograph illuminated by two fibres, from the telescope and the calibration lamps. An echelle spectrum of 72 orders is recorded on the CCD for each fibre (Figure 2.4). The spectral domain covered ranges from 380 nm to 690 nm. The resolution of the spectrograph is given by the fibre diameter and attains a value of about $R = 115'000$. At this resolution each spectral element is still sampled by 3.5 CCD pixels (Pepe *et al.* 2002).

HARPS is an 'ordinary' spectrograph with outstanding efficiency and spectral resolution. The main characteristic of HARPS is its extraordinary stability. The ThAr-reference technique is able to measure and correct for the tiniest instrumental drifts; nevertheless, lots of efforts have been invested in making the spectrograph intrinsically stable in order to avoid as far as possible any kind of second-order instrumental errors. As a result, the spectrograph is operated in a vacuum, since ambient pressure variations would have produced huge drifts (typically 100 ms^{-1} per mbar). The operational pressure is always kept below 0.01 mbar such that the drifts will never exceed the equivalent of 1 ms^{-1} per day. Not only the pressure but also the temperature is sharply controlled. Because of the huge thermal inertia of the vacuum vessel and the excellent thermal insulation between

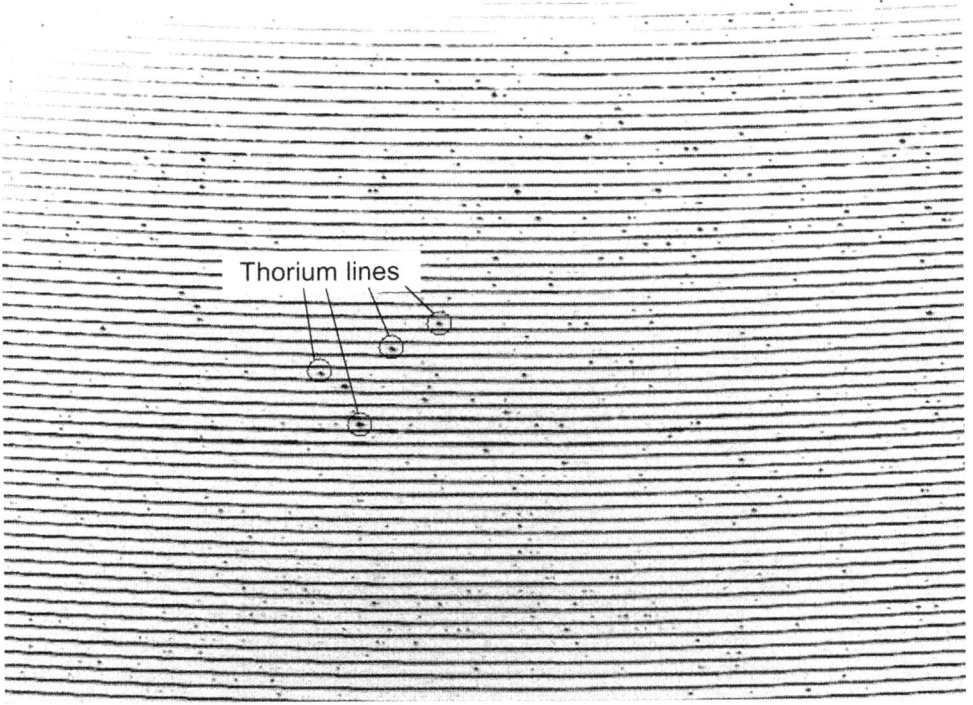

FIGURE 2.4. Top: view of the HARPS-spectrograph grating inside the vacuum tank. Bottom: raw CCD image obtained with HARPS, with the alternate orders of a star and the lamp spectra.

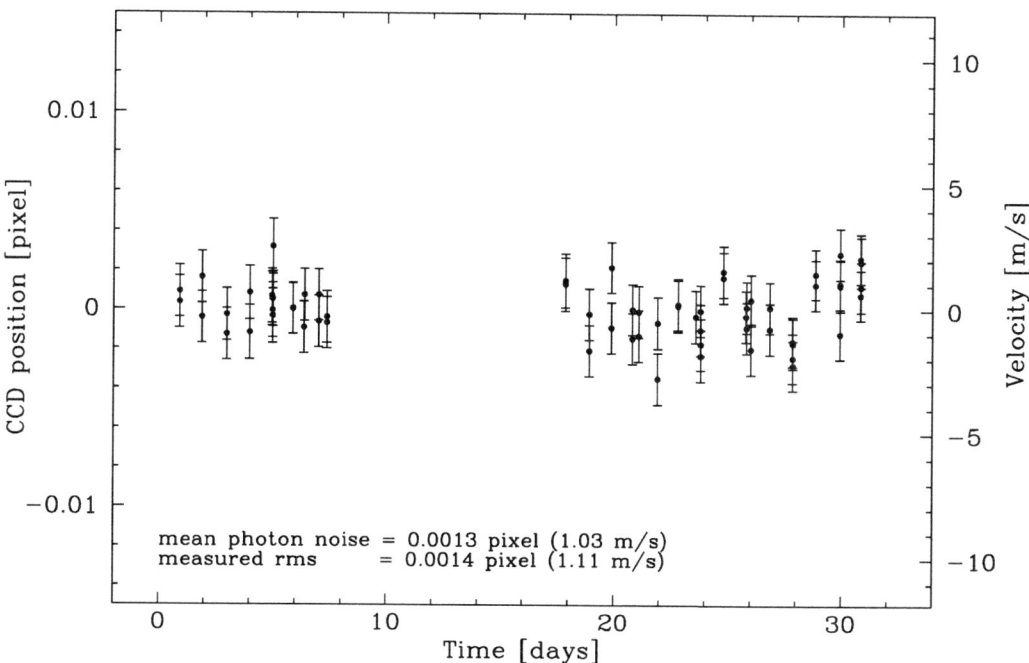

FIGURE 2.5. Absolute position of a thorium line over more than one month, without new wavelength calibration or instrumental changes. This illustrates the extraordinary stability of the spectrograph operated in vacuum.

the spectrograph and the vessel provided by the vacuum, the short-term temperature stability actually obtained is even better. Over one day we have measured variations of the order of 0.001 K rms (0.1 ms^{-1}). With such stability, around the 1 ms^{-1} precision level, it is not even necessary any longer to worry about the spectrograph drift during one night and use the second-fibre tracking.

An illustration of the spectrograph stability achieved is shown in Figure 2.5, displaying the position on the CCD of a single (strong but non-saturated) emission line of the thorium spectrum, over more than one month, without intermediate wavelength calibration or any instrument changes. This extraordinary stability reflects itself directly in the stability of the radial velocity measurements with a precision well below 1 ms^{-1} (Section 2.4).

2.2.3. *Intrinsic stellar limitations on radial-velocity searches*

So far, the discovery of extrasolar giant planets has essentially relied on the detection of radial-velocity variations of solar-like stars. Since radial-velocity variations can also be induced by motions of the photosphere due to pulsation and/or stellar activity-related variations (like rotation of starspots, Figure 2.6) or convective inhomogeneities and their temporal evolution, it is very important to be able to distinguish between them.

A quantification of those effects is possible by comparing, in the large planet-search surveys, the weighted radial-velocity dispersion (corrected for the mean internal error and orbital contribution from the known planets) with stellar characteristics such as spectral type, rotation and magnetic activity (Saar *et al.* 1998; Santos *et al.* 2001a). The amplitude of the radial-velocity variations associated with intrinsic phenomena may

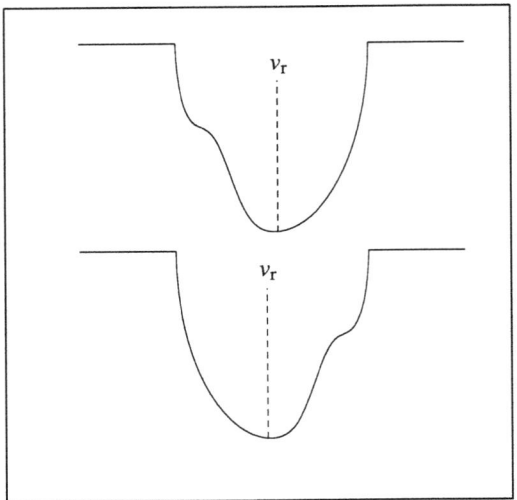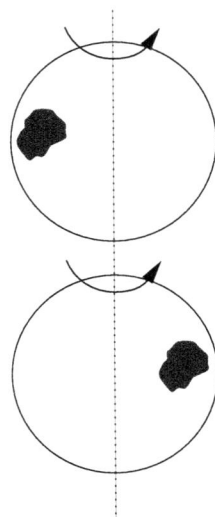

FIGURE 2.6. A spot on the surface of a rotating star produces changes in the shape of the spectral lines (or CCF), thus inducing a subsequent temporal variation of the measured radial velocities (centre of gravity of the lines/CCF) that may mimic a planet effect with a period equal to the stellar rotation period on a timescale corresponding to the lifetime of the spot.

reach a few tens of m s^{-1} and can possibly inhibit (when non-coherent) or confuse planet detection (in a few rotational or pulsation periods).

Stellar activity

Stellar activity is the empirical impetus driving models of dynamo activity. In magnetic dynamo models, stellar activity is produced by the interaction of magnetic fields in or immediately below the subsurface convection zone with the motions of rotation and convection (e.g. Noyes *et al.* 1984, Baliunas *et al.* 1995). Saar *et al.* (1998), Santos *et al.* (2001a) and Wright (2004) have shown a clear relation between stellar activity[3] and the level of non-orbital radial-velocity variations (Figure 2.7). The same trend is observed with $v \sin i$, the projected rotational velocity. This is expected because of the coupling of high activity and rapid rotation in the magnetic dynamo model and, on the other hand, because of the line broadening due to stellar rotation that degrades the radial-velocity precision.

In order to avoid spurious velocity variations due to activity and false planetary detections associated with periodic radial-velocity changes induced by spots over a rotational period, a careful pre-selection of the star sample is mandatory when setting up a planet-search programme. In addition to evolved stars known for their frequent intrinsic radial-velocity variations (pulsation or jitter), young rapidly rotating active dwarfs should be left over. At the precision of the ELODIE and CORALIE results, an empirical limit at $v \sin i = 5$–6 kms^{-1} seems adequate. When available, an activity estimator ($R'_{\rm HK}$) rather than projected rotational velocity (confused by inclination effects) should be used. For large samples, the a priori stellar rotation and activity characteristics are not always

---

[3] A 'standard' activity estimator is given by $R'_{\rm HK}$, the fractional flux of the chromospheric reemission feature in the core of the Ca II H and K lines (Figure 2.7) corrected for the photospheric flux (Baliunas *et al.* 1995).

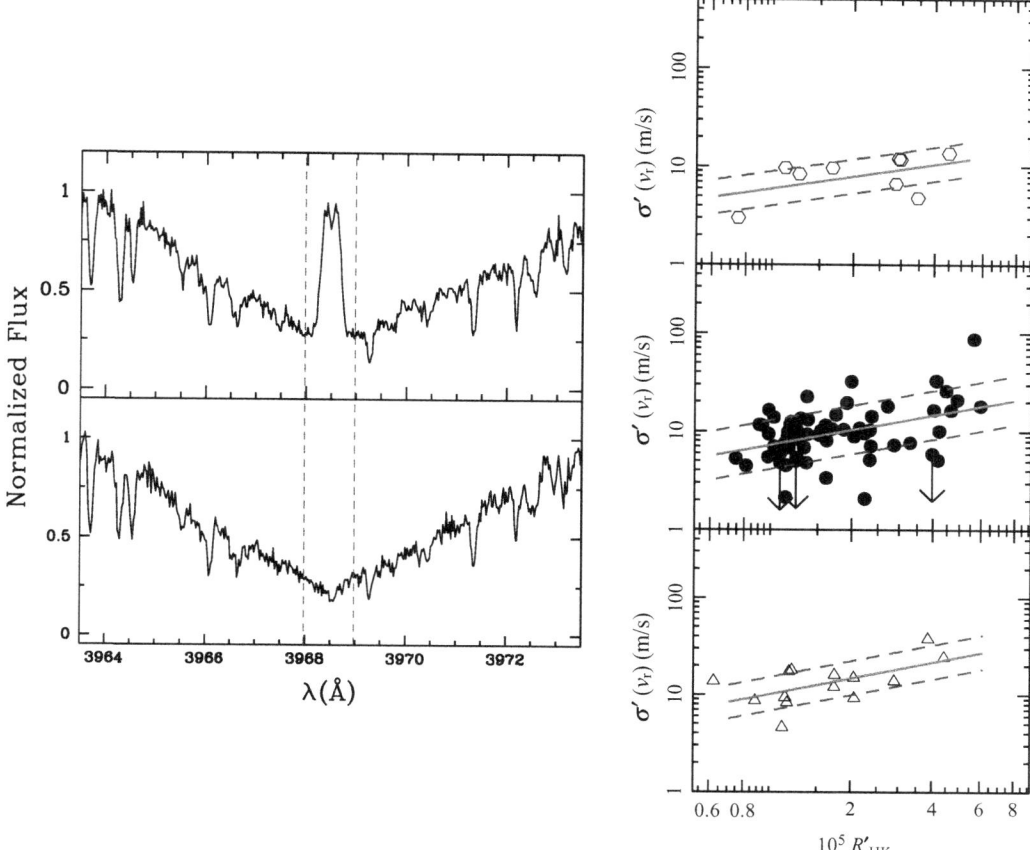

FIGURE 2.7. Left: CaII H absorption line measured with ELODIE. Upper panel: for a chromospherically active star (HD 166435) for which a stellar activity-related emission line is superimposed on the core of the line. Lower panel: for a 'quiet' star showing no additional special feature. Right: Measured radial-velocity dispersion vs stellar activity for a subsample of CORALIE F- (bottom), G- (middle) and K-dwarf (top) stars.

available. However, already after the first measurement, an a posteriori check may be done. In particular, in the case of periodic radial-velocity changes, stellar intrinsic variations have to be ruled out by applying suitable diagnostics.
 (i) Checking for photometric stability (expected for velocity variations due to orbital motions).
 (ii) For Keplerian motion the phase of the variation stays constant in time, whereas for intrinsic variations due to a spot on the star surface, it is supposed to change over a few rotational periods of the star (the typical lifetime of spots is 10–100 days). Counter examples exist, however (see below).
 (iii) Checking for the invariability of the bisector of the spectral lines (or cross-correlation function), i.e. no variation in the shape of the lines is expected in the case of a planet. On the other hand, spots asymmetrise the lines, inducing correlated radial-velocity and line-shape variations.

A textbook case of *false detection* is provided by the star HD 166435 (Queloz *et al.* 2001), which has a period of a few days and turns out also to vary photometrically with

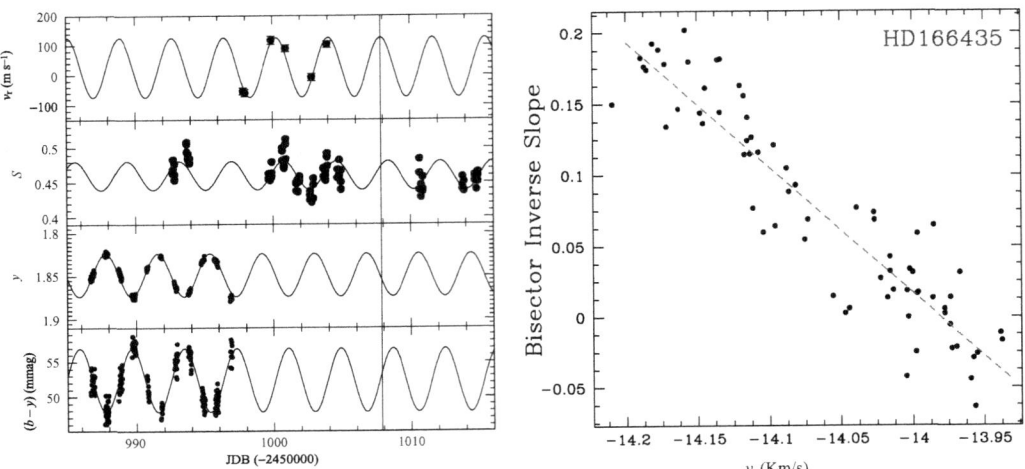

FIGURE 2.8. The star HD 166435 presents 'stable' radial-velocity variations that could be interpreted as due to a planetary companion. However, radial velocities vary in phase with the star luminosity (left) and the CCF bisector span, which furthermore correlates with the measured radial velocity (right).

the same period, along with the bisector of the cross-correlation function (Figure 2.8). The planetary explanation was thus rejected even if the variation period seems to stay stable over hundreds of cycles (>8 years). Such a posteriori verifications are fundamental, taking into account that the lower limit of the radial-velocity jitter related to stellar activity is still poorly known.

Acoustic modes

Stars (spheres of hot gas) propagate very well, in their interiors, acoustic waves generated by turbulent convection near the surface. Frequencies and amplitudes of the acoustic waves, also called oscillation modes or p-modes, depend on the physical conditions prevailing in the layers crossed by the waves and provide a powerful seismological tool for probing star structure. Helioseismology (monitoring of the oscillation modes of the Sun) has been used since the 1970s. It led to major revisions in the solar 'standard' model and allowed, for example, measurements of the Sun's inner rotation, the size of the convective zone, and the structure and composition of the external layers. Solar-like oscillation modes generate periodic motions of the stellar surface with periods in the range 3–30 min but with extremely small amplitudes. The corresponding amplitudes of the stellar surface velocity modulations are in the range 10–100 cms^{-1}.

During the commissioning of HARPS the star α Cen B (smaller than the Sun) was monitored for seven hours. The radial-velocity measurement sequence shows a dispersion of 51 cms^{-1} completely dominated by 4-minute stellar oscillations (Figure 2.9). The power spectrum of this sequence clearly exhibits a series of peaks around 4 mHz, corresponding to individual acoustic modes of the star, with amplitudes in the range 10–20 cms^{-1}. The positive interference of several oscillation modes may lead to amplitudes much larger than the amplitude of single modes. In this case, the global contribution of the modes amounts to $\sim$0.44 ms^{-1} leaving the remaining noise to be split into photon noise (0.17 ms^{-1} on single exposures), and all other possible error sources (centring and guiding errors, influence of the atmosphere, instrumental errors; $\sim$0.2 ms^{-1} in total).

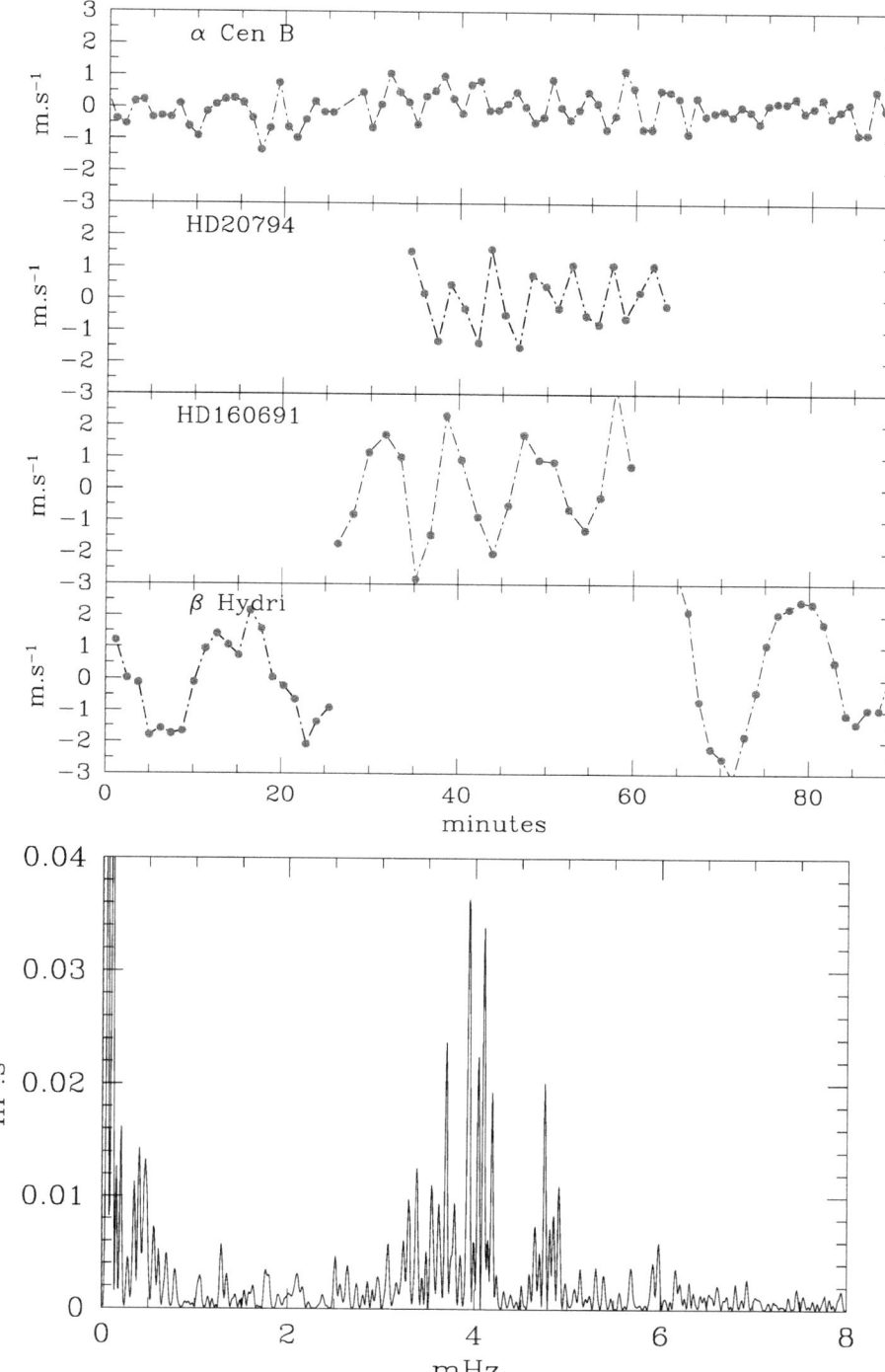

FIGURE 2.9. Top: short series of radial-velocity measurements for stars of different spectral types obtained with HARPS. Oscillations due to acoustic modes are clearly observed. Amplitudes and periods of these oscillations correlate with stellar properties (mass, radius), and also probably with the evolution stage of the star. Bottom: Fourier transform of the α Cen B series revealing the 4-minute modes.

After the amazing results obtained on α Cen B, a small set of solar-type stars was monitored with HARPS (Figure 2.9). On *each* of the sequences the stellar oscillations are clearly visible. These measurements indicate that the period and amplitude of the oscillation modes are directly related to the stellar properties. As the stellar size is the main parameter, the evolutionary stage of the star probably also plays an important role in setting the amplitude and frequency of the oscillations. This has to be investigated now in more detail. A very promising new approach of the study of stars is developing.

Only a couple of years ago the behaviour of the stars below 3 ms^{-1} was completely unknown. Asteroseismology observations carried out by HARPS have made it clear that the precision achieved is no longer set by instrumental characteristics but rather by the stars themselves (Mayor *et al.* 2003; Bouchy *et al.* 2005b). As a consequence, any exposure with a shorter integration time than the oscillation period of the star, or even than mode-interference variation timescales, might fall arbitrarily on a peak or on a valley of these mode interferences and thus introduce additional radial-velocity 'noise'. To minimize these effects as much as possible, stars for very high-precision radial-velocity measurements have to be chosen as slowly rotating, non-evolved, and low-activity stars. Moreover, in order to average out stellar oscillations, observations have to be designed to last at least 15 minutes on G and K-dwarfs. Subgiants and giants, which can show larger amplitudes and longer periods, should, however, be more affected.

2.3. Orbital properties of extrasolar planets

As for standard binary systems, radial-velocity measurements allow for the orbital element determination of planetary systems. The distributions of these orbital elements, as well as planetary mass distributions, are thought to retain traces from the physical processes active during the formation and evolutionary stages of the systems. Good quality statistical distributions should thus allow us to discriminate between the different scenarios proposed for planet formation (including our own Solar System) and disentangle the effects due to formation from those induced by subsequent evolution.

Resulting from the increase in the baseline of the 'historic' planet-search programmes and the launch of new large surveys (e.g. the HARPS planet search; Mayor *et al.* 2003) or metallicity-biased searches for hot jupiters (Da Silva *et al.* 2006; Fischer *et al.* 2005), the list of known extrasolar planets has grown from the few candidates found ten years ago to more than 200 planets, a number giving some confidence in the constraints drawn from the observed statistical distributions of planet properties. The most remarkable feature of the sample is undoubtedly the variety of the orbital characteristics, which challenges the conventional views of planetary formation. A global *visual* illustration of these properties is given in Figure 2.10, displaying orbital eccentricities as a function of planet–star separations. Several of the planet properties (close proximity to the star, large eccentricity, high mass) are clearly apparent in the figure. The goal now is to interpret the observed orbital distributions in terms of constraints for the planet-formation models.

The determination of the statistical properties of giant planets depends on surveys that are statistically well defined (e.g. volume limited) and that have well-understood detection thresholds in the different planet, primary star and orbital parameters. Several programmes aim at meeting these requirements, like the volume-limited CORALIE planet-search programme (Udry *et al.* 2000) or the magnitude-limited FGKM Keck survey (Marcy *et al.* 2005). In the diagrams below, we show most of the detected planet candidates, whatever sample they are coming from. However, the discussed properties have been checked to be observed in individual well-defined programmes as well. In the following, we emphasize the emerging properties of planet-host stars and characteristics

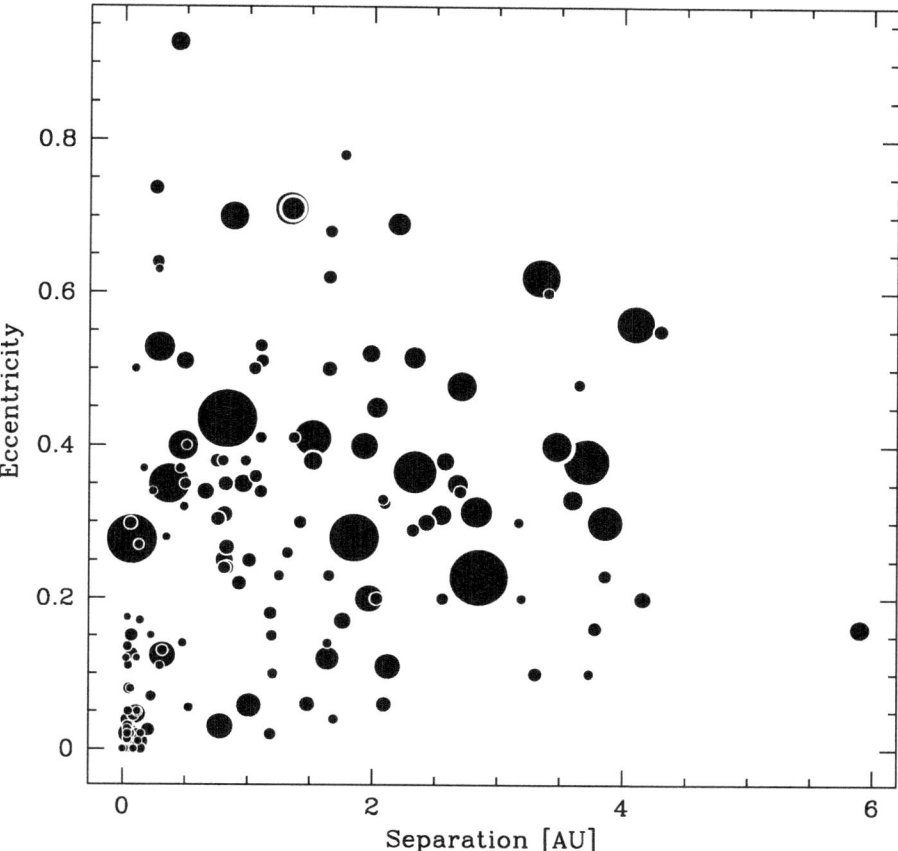

FIGURE 2.10. Separation–eccentricity diagram for the complete sample of presently known extrasolar planets. The size of the dots is proportional to the minimum mass of the planet candidates ($m_2 \sin i \leq 18\ M_{\text{Jup}}$).

of the different orbital-element distributions of exoplanetary systems and discuss their implications for our understanding of planet formation and evolution.

2.3.1. *Giant extrasolar planets in numbers*

The most direct property obtained from a planet-search programme is the rate of planet occurrence around the stars surveyed. Because of the typical precision and duration limits of the surveys, this rate can only be given for planets with masses larger than m_{lim} and periods smaller than P_{lim}. The minimum rate is obtained just by counting the fraction of stars hosting planets in the interval considered. For planets more massive than 0.5 M_{Jup}, Marcy et al. (2005) found in the Lick+Keck+AAT sample $16/1330 = 1.2\%$ of hot jupiters ($P \leq 10$ d) and 6.6% of planets within 5 AU. In the volume-limited CORALIE sample (including stellar binaries), we count $9/1650 = 0.5\%$ of hot jupiters and $63/1650 = 3.8\%$ of planets within 4 AU. As binaries were usually removed from planet-search programmes (along with rotating stars), restricting ourselves to stars *suitable* for planet search (not binary and with $v \sin i \leq 6$ kms^{-1}), we obtain a fraction $9/1120 = 0.8\%$ of planets within 0.1 AU and $63/1120 = 5.6\%$ for planets up to 4 AU.

This rate can be better approximated by estimating, through Monte Carlo simulations, the detection efficiency of the survey for a planet of given mass and period. This has not been done yet for the largest surveys. However, for the ELODIE programme, although dominated by small number statistics errors, Naef et al. (2005) estimate in this way a corrected fraction $0.7 \pm 0.5\%$ of hot jupiters with $P \leq 5$ days and $7.3 \pm 1.5\%$ of planets with periods smaller than 3900 days.

These results are all in good agreement within the uncertainties. They also agree with results obtained by Cumming et al. (1999) for the Lick survey and by Endl et al. (2002) for the planet-search programme with the ESO Coudé-echelle spectrometer. With the continuously increasing span of the surveys and the improvement in our ability to detect smaller-mass planets, we expect the estimated fraction of stars hosting planets to increase from these minimum values up to several tens of percent, taking into account that the number of planets detected is a rising function of (i) decreasing planet mass and (ii) increasing period (see Sections 2.3.2 and 2.3.3).

2.3.2. *Mass distribution of planetary companions*

Already after the detection of a handful of extrasolar planets it became clear that these objects could hardly be considered as the low-mass tail of stellar companions in binary systems (due to unfavourable orbital inclinations). The strongly bimodal aspect of the secondary-mass distribution to solar-type primaries (Figure 2.11) has rapidly been considered as the most conspicuous evidence of different formation mechanisms for stellar binaries and planetary systems. The interval between the two populations (the *brown-dwarf desert*) corresponding to masses between ~ 20 and ~ 60 $M_{\rm Jup}$ is almost empty, at least for the 'short' periods probed by radial-velocity measurements (Halbwachs et al. 2000). There is, however, a very probable overlap of the two distributions around ~ 10–20 $M_{\rm Jup}$ and, at this point, it is not easy to differentiate *low-mass brown dwarfs* from *massive planets* just from their mass estimates, without further information on the formation/evolution of these systems.[4] For example, two multi-planet systems (HD 168443 and HD 202206, see Table 2.2) host 'planetary' components with $m_2 \sin i = 17$ $M_{\rm Jup}$, whereas Chauvin et al. (2005) published the first image of planet-mass object, the 5 $M_{\rm Jup}$ 2M1207[5] orbiting a 25 $M_{\rm Jup}$ brown-dwarf primary.

Towards the low-mass side of the planetary mass distribution, a clear power-law type rise is first observed (Figure 2.11). Marcy et al. (2005) proposed a dependence of the form $dN/dM \propto M^{-1.05}$ for their FGKM sample. Such a distribution is practically unaffected by the unknown $\sin i$ distribution (Jorissen et al. 2001). The left part of the diagram, then, is strongly biased due to the observational limitation inherent in the radial-velocity technique, which is less sensitive to lower masses. We thus expect a large population of still undetected planets with masses below that of Saturn. The same trend is predicted by accretion-based planet-formation models. In particular, they foresee very low mass 'solid' planets in large numbers (Ida & Lin 2004a, 2005; Alibert et al. 2004, 2005).

2.3.3. *Period distribution of giant extrasolar planets*

Figure 2.12 displays the period distribution for the known exoplanets orbiting dwarf primaries. Among the most peculiar candidates are the numerous giant planets orbiting

---

[4] A dedicated working group of the IAU has proposed a *working definition* of a 'planet' based on the limit in mass at 13 $M_{\rm Jup}$ for the ignition of deuterium burning.

[5] The mass estimate of these young objects are very model-dependent and should be considered with care. It would however be surprising if the masses were wrong by a factor larger than 2.

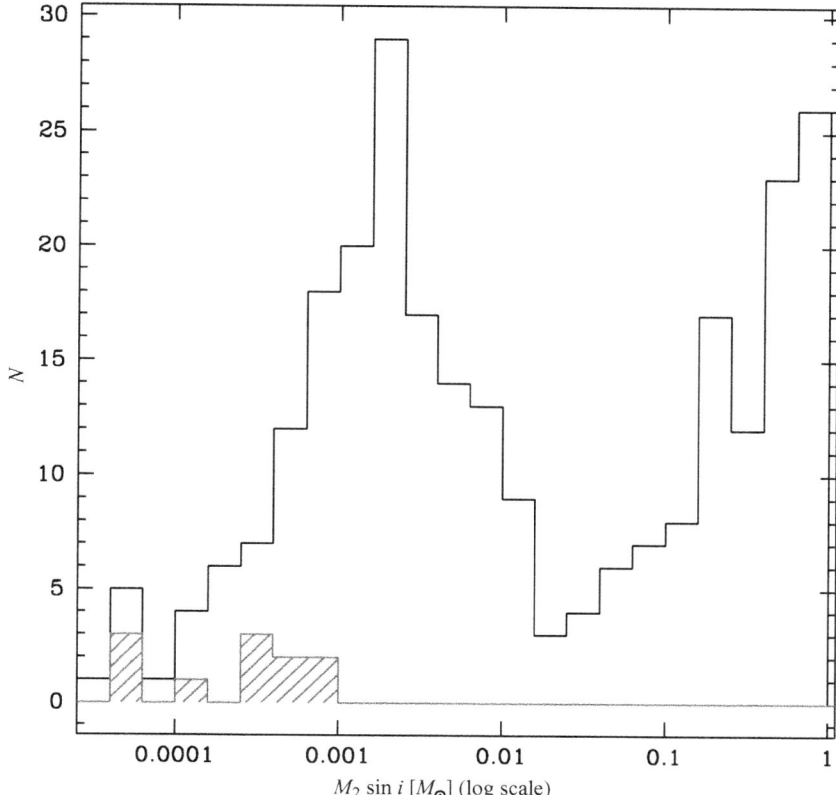

FIGURE 2.11. Secondary mass distribution of solar-type primaries. The stellar binaries are from Halbwachs *et al.* (2003). The hatched histogram represents HARPS planets (Section 2.4).

very close to their parent stars ($P < 10$ days). The accommodation of this observation to the prediction of the *standard model* (e.g. Pollack *et al.* 1996), in which planets form first from ice grains[6] in the outer region of the system where the temperature of the stellar nebula is not too high, requires that the planets undergo a subsequent migration process bringing them close to the central star (e.g. Lin *et al.* 1996; Ward 1997).[7] The observed piling up of planets with periods around 2–3 days is then seen as the result of the migration process and of a required stopping mechanism that prevents the planets from disappearing into the stars (e.g. Udry *et al.* 2003 and references therein for a more detailed discussion).

Another interesting feature of the period distribution is the rise in the number of planets with distance to the parent star. This is not an observational bias, as equivalent mass candidates are more easily detected at shorter periods, with the radial-velocity technique. The decrease in the distribution farther out is very probably due to the finite and

---

[6] Such grain growth provides the supposed requisite solid core around which gas could rapidly accrete (Safronov 1969) over the lifetime of the protoplanetary disc ($\sim 10^7$ y).

[7] Alternative points of view invoke *in situ* formation (Bodenheimer *et al.* 2000), possibly triggered through disc instabilities (Boss 2002; Mayer *et al.* 2005). Note, however, that even in such cases subsequent disc–planet interactions leading to migration are expected to take place as soon as the planet has formed.

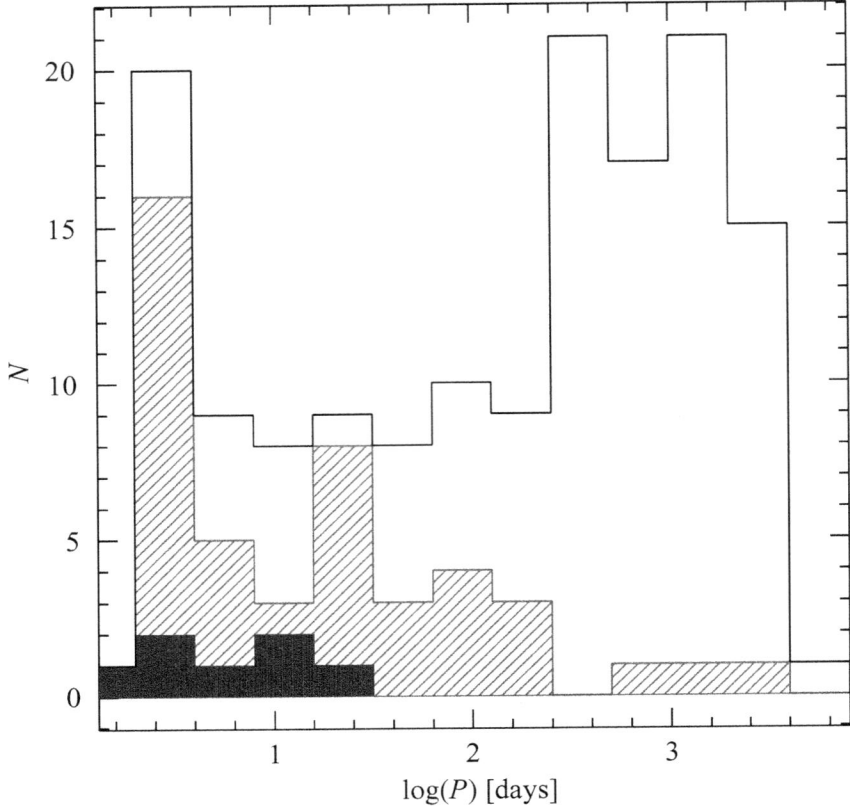

FIGURE 2.12. Period distribution of known extrasolar gaseous giant planets detected by radial-velocity measurements and orbiting dwarf primary stars (open histogram). The hatched part locates 'light' planets with $m_2 \sin i \leq 0.75\ M_{\rm Jup}$. For comparison, the period distribution of Neptune-mass planets with $p < 30\ d$ (Section 2.4) is given by the filled histogram.

still limited duration of most of the radial-velocity surveys. The overall distribution can then be understood as consisting of two parts: a main distribution rising with increasing period (as for binary stars; Duquennoy & Mayor 1991), the maximum of which being still undetermined and, on top of it, a peak of planets having migrated inwards. The visible shortage of planets with periods between 10 and 100 days is then just the interval between these two main features.

A *minimum* flat extrapolation of the distribution to larger distances would about double the rate of planets (Marcy et al. 2005). This suggests that a large population of as yet undetected Jupiter-mass planets may exist between 5 and 20 AU. This is of prime importance for the direct-imaging projects under development on large telescopes, such as SPHERE (the VLT Planet Finder) or the Gemini Planet Imager (Beuzit et al. 2007).

Benefits from hot jupiters

Interestingly, it is possible to take advantage of the proximity of the hot jupiters to their parent stars to gain more information on the system, especially if the planet transits the stellar disc. Amongst the possible complementary studies, we can mention: detection of the reflected light of the star on the planet (Collier Cameron et al. 2000, Leigh et al.

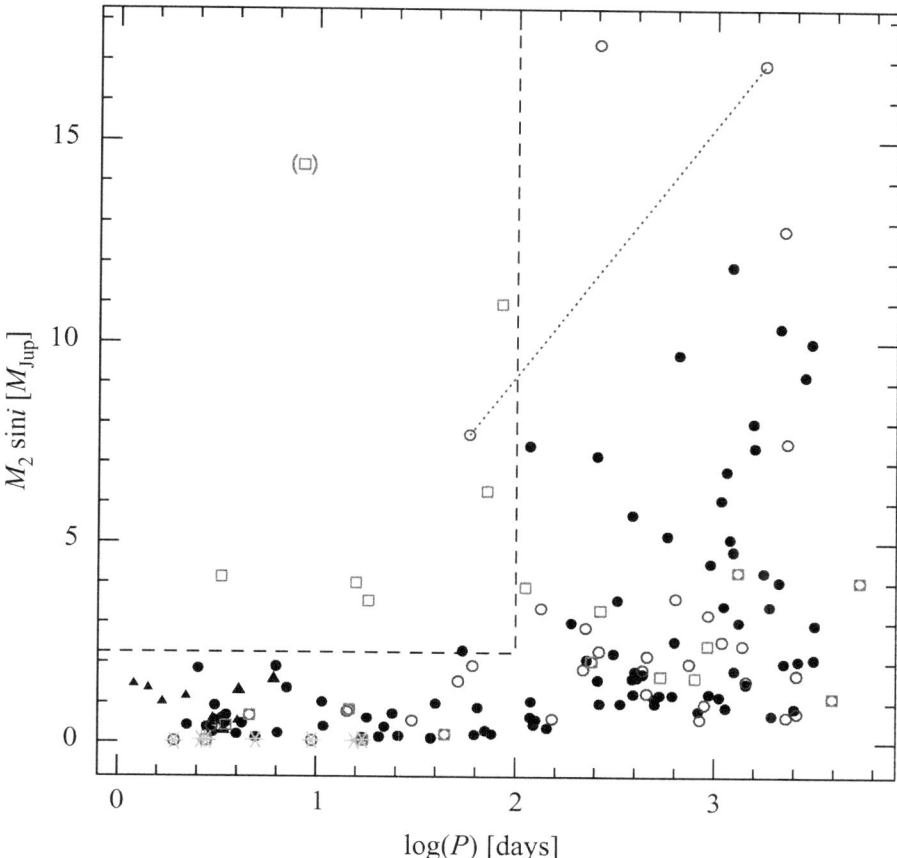

FIGURE 2.13. Period–mass distribution of known extrasolar planets orbiting dwarf stars. Open squares represent planets orbiting one of the components of a binary system, whereas circles are for 'single' stars. Open circles represent planets in multi-planet systems. Starred symbols are for Neptune-mass planets. Dashed lines are limits at 2.25 $M_{\rm Jup}$ and 100 days. The dotted line connects the two 'massive' orbiters of HD 168443.

2003), detection of planetary spectral signatures in the stellar light crossing the planet atmosphere (Chapter 3), modulation of the visibility curve in astrometric measurements (Coudé du Foresto 2000) and, in the case of transits, the measure of the planet radius from photometric transit observations (Chapter 3; Section 2.7) or the determination of constraints for the geometry of the system from spectroscopic transit measurements (e.g. the results of Queloz et al. 2000 and Winn et al. 2005 for HD 209458).

2.3.4. Period–mass distribution

Study of the orbital-period distribution has shown the importance of considering migration processes to explain the observed configuration of planetary systems. When coupling period and mass, further striking features appear in the distributions. This is illustrated in Figure 2.13, showing the mass–period diagram for the known exoplanets in orbit around dwarf primaries.

The first noticeable characteristic in the diagram is the paucity of massive planets in short-period orbits (Zucker & Mazeh 2002; Udry *et al.* 2002; Pätzold & Rauer 2002). This is not an observational bias, as these candidates are the easiest ones to detect. Even more striking, when we neglect the multiple-star systems (see Section 2.3.5), a complete void of candidates[8] is observed in the diagram for masses larger than ~ 2 $M_{\rm Jup}$ and periods smaller than ~ 100 days.

Several processes have been proposed to explain the lack of massive planets on short-period orbits. In the context of the migration scenario, they mainly follow two different approaches: (i) type II migration (after a gap opens in the disc) is shown to be less effective for massive planets, i.e. massive planets stay farther out than lighter ones, or (ii) when the planet reaches the central regions, some process related to planet–star interactions provokes mass transfer from the planet to the star, decreasing the mass of the former (e.g. Trilling *et al.* 1998), or leads massive planets to fall into the central star (Pätzold & Rauer 2002).

Another interesting feature of the period distribution is the rise in the largest mass of the detected planets with distance to the parent star (Figure 2.14; Udry *et al.* 2003). This is not an observational bias, as those candidates are more easily detected at shorter periods. This can be understood in the context of the migration scenario as well. More massive planets are expected to form farther out in the protoplanetary disc, where more material is available in the longer path around the central star. The bigger the planet, then, the more difficult it is to initiate migration, as a larger portion of the disc has to be disturbed to overcome the inertia of the planet. This view is also supported by noting that the peak of hot jupiters is mostly composed of 'light' planets ($m_2 \sin i \leq 0.75$ $M_{\rm Jup}$) that migrate more easily (Figure 2.12).

The possibility that multi-planet chaotic interactions send the lighter candidates in the inner regions (or out) of the system, whereas the massive ones stay in the outer part, may also be invoked (Rasio & Ford 1996; Weidenschilling & Marzari 1996; Marzari & Weidenschilling 2002). The frequency of planets ending very close to the central star seems then, however, to be small and the observed distributions of periods and eccentricities difficult to reproduce (Ford *et al.* 2001, 2003).

The above discussion suggests that the migration rate of planets is decreasing with increasing mass. This agrees with simulations of migrating planets in viscous discs (Trilling *et al.* 1998, 2002; Nelson *et al.* 2000). We thus expect a large number of massive planets to be in long-period orbits and so to be still undetected because of the limit duration of the present surveys. Lower-mass planets probably exist in long-period orbits as well; they are, however, more difficult to detect. The latter represent primary targets for higher-precision surveys (see Section 2.4), whereas the youngest among the former are interesting targets for direct imaging of planetary-type objects (Beuzit *et al.* 2007).

2.3.5. *Giant planets in multiple stellar systems*

Among the 200 or so extrasolar planets discovered to date, more than 30 are orbiting a component of a double or multiple star system (Patience *et al.* 2002; Eggenberger *et al.* 2004; Mugrauer *et al.* 2004, 2005). Although the sample is not large, some differences between the orbital parameters and the masses of these planets and those of planets orbiting single stars are emerging in the mass–period (Figure 2.13) and eccentricity–period

---

[8] The only remaining point is HD 168443 b, member of a possible multi-brown-dwarf system (Marcy *et al.* 2001; Udry *et al.* 2002).

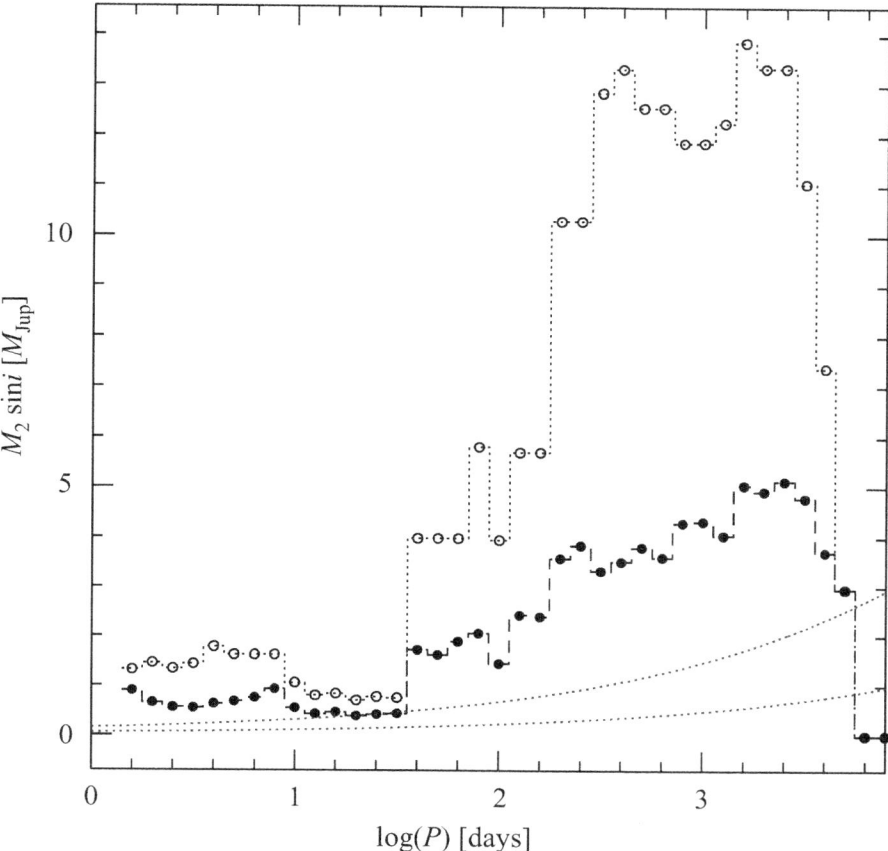

FIGURE 2.14. Mean (filled circle) or highest (average of the three highest values; open circles) mass of planets in period smoothing windows of width log P[days] $= 0.2$. A clear increase of planet largest masses with period is observed, even if massive planets at a given period are easier to detect at smaller periods. Detection limits at 10 and 30 m s^{-1} ($M_1 = 1M_\odot, e = 0$) are represented by the dotted lines.

(Figure 2.15) diagrams. As pointed out by Zucker & Mazeh (2002), the most massive short-period planets are all found in multiple star systems. The planets in multiple star systems also tend to have a very low eccentricity when their period is shorter than about 40 days (Eggenberger et al. 2004).[9] These observations suggest that some kind of migration responsible for the observed low-eccentricity values has been at work in the history of these systems. The properties of the five short-period planets in multiple star systems seem, however, difficult to explain with the current models of planet formation and evolution, at least if we want to invoke a single mechanism to account for all the characteristics of these planets (Eggenberger et al. 2004).

Even if the orbital parameters of the binaries hosting planets are not exactly known, we have some information, such as the projected separations of the systems or stellar

---

[9] The only exception is the 'massive' companion (14.4 $M_{\rm Jup}$) of HD 162020, which very possibly is a low-mass brown dwarf (Udry et al. 2002).

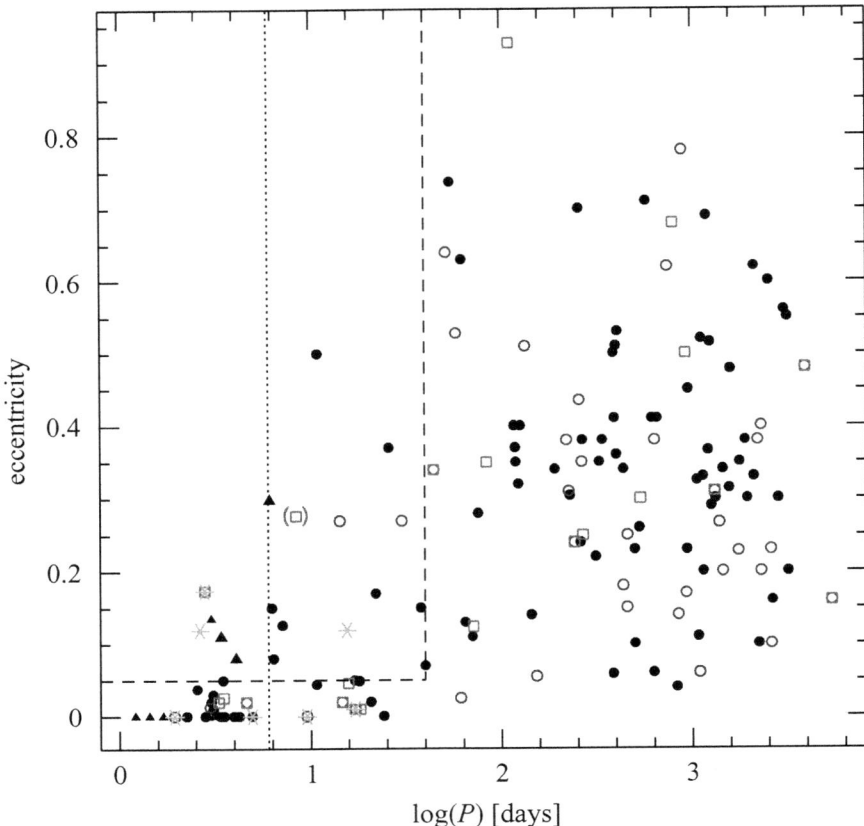

FIGURE 2.15. Period–eccentricity diagram of the known extrasolar planets. Open squares represent planets orbiting one of the components of a binary system, whereas circles are for 'single' stars. Open circles represent planets in multi-planet systems. The dotted line is indicative of a tidal circularization period around 6 days (Halbwachs et al. 2005) and the dashed lines limit the $e > 0.05$ and $P < 40$ days domain. Planets detected in metallicity-biased or photometric-transit surveys are indicated by filled triangles. Starred symbols are for Neptune-mass planets. The ($\square$) symbol indicates HD 162020 with $m_2 \sin i = 14.4\ M_{\rm Jup}$.

properties. However, no obvious correlation between the properties of these planets and the known orbital characteristics of the binaries, or of the star masses, have been found yet.

Searches for extrasolar planets using the radial velocity technique have shown that giant planets exist in certain types of multiple star systems. The number of such planets is, however, still low, in part because close binaries are difficult targets for radial-velocity surveys and were consequently often rejected from the samples. Even if the detection and characterization of planets in binaries are more difficult to carry out than the study of planets around single stars, it is worth doing it because of the new constraints and information it can provide about planet formation and evolution. In this context, we are following different approaches: (i) aiming at detecting short-period planets in long-period spectroscopic binaries, and (ii) looking for statistical differences between samples of stars with and without planets through adaptive-optics searches for faint companions (Eggenberger et al. 2004, 2005; Udry et al. 2004a, 2004b).

Due to the limitations of available observational techniques, most detected objects are giant (Jupiter-like) planets. The existence of smaller mass planets in multiple star systems is thus still an open question.

2.3.6. Giant planet eccentricities

Extrasolar planets with periods larger than ∼6 days have eccentricities significantly larger than those of giant planets in the Solar System, much more like typical eccentricities of binary stars. They almost span the full available range between 0 and 1 (Figure 2.15). This median eccentricity is $\langle e \rangle = 0.29$. Planets with $P \leq 6$ days are probably tidally circularized (see below).

The origin of the eccentricity of extrasolar giant planets has been sought in several directions: gravitational interaction between multiple giant planets (Rasio & Ford 1996; Weidenschilling & Marzari 1996; Lin & Ida 1997); between planets and planetesimals in the early stages of the system formation (Levison et al. 1998); or in the secular influence of an additional, passing (Zakamska & Tremaine 2004) or bounded companion in the system (see Tremaine & Zakamska 2004, for a comprehensive review of the question).

The latter effect is of particular interest in some cases. Among the giant planet candidates, several eccentric orbits show a drift in their mean velocity, indicating the presence of a long-period companion whose gravitational perturbation may be suspected of being responsible for the observed (high) planetary eccentricity, such as for the planet orbiting 16 Cyg B (Mazeh et al. 1997). From a statistical point of view, however, Takeda & Rasio (2005) have shown that such a process produces an excess of both very high ($e \geq 0.6$) and very low ($e \leq 0.1$) eccentricities, calling for at least one additional mechanism to reproduce the observed distribution. In fact, none of the proposed mechanisms to explain planet eccentricities is able by itself to reproduce the observed distribution.

For small periastron distance, giant planets are supposed to undergo tidal circularization. For short periods, nearly all gaseous giant planets are in quasi-circular orbits[10] ($e \leq 0.05$, Figure 2.15; Halbwachs et al. 2005). High-eccentricity orbits also bring the planet close to the star, where tidal circularization is efficient, and thus will also evolve towards more circular systems. In a diagram showing eccentricity as a function of the period of the circular orbit at the end of the process (assuming angular momentum conservation), a clear circularization period of around six days is observationally determined (Halbwachs et al. 2005).

In the data, some (non-significant) trends are observed between eccentricity and period, and between eccentricity and mass. The more massive planets (more massive than 5 $M_{\rm Jup}$) exhibit systematically higher eccentricities than do the planets of lower mass (Marcy et al. 2005). This cannot be a selection effect (larger induced radial-velocity variation). If planets form initially in circular orbits, the high eccentricities of the most massive planets are puzzling. Such massive planets have the largest inertial resistance to perturbations that are necessary to drive them out of their initial circular orbits. Note that the more massive planets are also found further out (Section 2.3.3) and thus that the two trends mentioned are coupled. The long-period planets have usually been observed for one period only and are rarely well covered in phase. This often leads to an overestimate of the derived eccentricity in the Keplerian fit (Butler et al. 2000). This effect could also be partly responsible for the observed trend.

---

[10] The only exceptions, with eccentricities of just over 0.1, are HD 88133, HD 149143 (Fischer et al. 2005; Da Silva et al. 2006) and TrES-1 (Alonso et al. 2004), all detected in surveys biased for short-period orbits (metallicity-biased or photometric-transit searches). At this stage it is not possible to reject an additional companion in the system that may be responsible for the observed slight eccentricity values.

Finally, we can point out from Figure 2.15 that a few long-period, low-eccentricity candidates are emerging from the surveys. They form a small subsample of so-called *solar-system analogues*.

2.4. Below the mass of Neptune

After a decade of enthusiastic discoveries in the field of extrasolar giant planets, coming mainly from large high-precision radial-velocity surveys of solar-type stars, the *quest for other worlds* has now passed a new barrier. Most of the planets detected are gaseous giants similar to our own Jupiter, with typical masses of a few hundred Earth masses. However, in about one year, seven very light candidates with masses in the Uranus–Neptune range (15–20 Earth masses) have been detected (Table 2.1). Because of their small masses and locations in the system, close to their parent stars, they may well be composed mainly of a large rocky/icy core, being formed without, or having lost, the extended gaseous atmosphere expected to grow during the planet migration toward the centre of the system.

These planetary companions, together with recently detected sub-Saturn mass planets in intermediate-period orbits, start to populate the lower end of the secondary-mass distribution, a region strongly affected by detection incompleteness (Figure 2.11). The discovery of these very low-mass planets close to the detection threshold of radial-velocity surveys suggests that this kind of objects may be rather frequent. But already the simple existence of such planets may provide headaches for theoreticians. Indeed, statistical considerations predict that planets with masses between 1 and 0.1 $M_{\rm Sat}$ and semi-major axes of 0.1 to 1 AU must be rare (the so-called *planet desert*; Ida & Lin 2004a). For the moment, recent discoveries contradict these predictions (although very little is known about the actual populating of the planet desert). In any case, the continuous detection of planets with even lower masses will set new constraints on possible planetary system formation and evolution models.

This new step forward has been made possible primarily thanks to the development of a new generation of instruments capable of radial-velocity measurements of unprecedented quality. The leading horse among them is undoubtedly the HARPS spectrograph (Section 2.2.2) with a radial-velocity accuracy at the level of 1 ms^{-1} over months/years (Mayor *et al.* 2003; Lovis *et al.* 2005), and even better on a short term basis (Bouchy *et al.* 2005b). New instruments, such as the HRS spectrograph on the HET in Texas or the improved HIRES spectrograph on the Keck telescope, aim at reaching the same level of precision as well.

Another fundamental change that allowed this progress in planet detection towards the very low masses is the application of a careful observing strategy to reduce as far as possible the perturbing effect of stellar oscillations hiding the tiny radial-velocity signal induced in solar-type stars by Neptune-mass planets (Section 2.2.3).

Such a strategy is now applied to stars in the 'high-precision' part of the HARPS and Keck planet-search programmes. An illustration of the results obtained is given by the histogram of the radial-velocity dispersion of the HARPS high-precision survey (Figure 2.16). The distribution mode is just below 2 ms^{-1}, and the peak decreases rapidly towards higher values. More than 80% of the stars show a dispersion smaller than 5 ms^{-1}, and more than 35% have dispersions below 2 ms^{-1}. It must be noted that the computed dispersion includes photon-noise error, wavelength-calibration error, stellar oscillations and jitter, and, in particular, it is 'polluted' by known extrasolar planets (hatched area in Figure 2.16) and still undetected planetary companions. The recently announced 14 $M_\oplus$

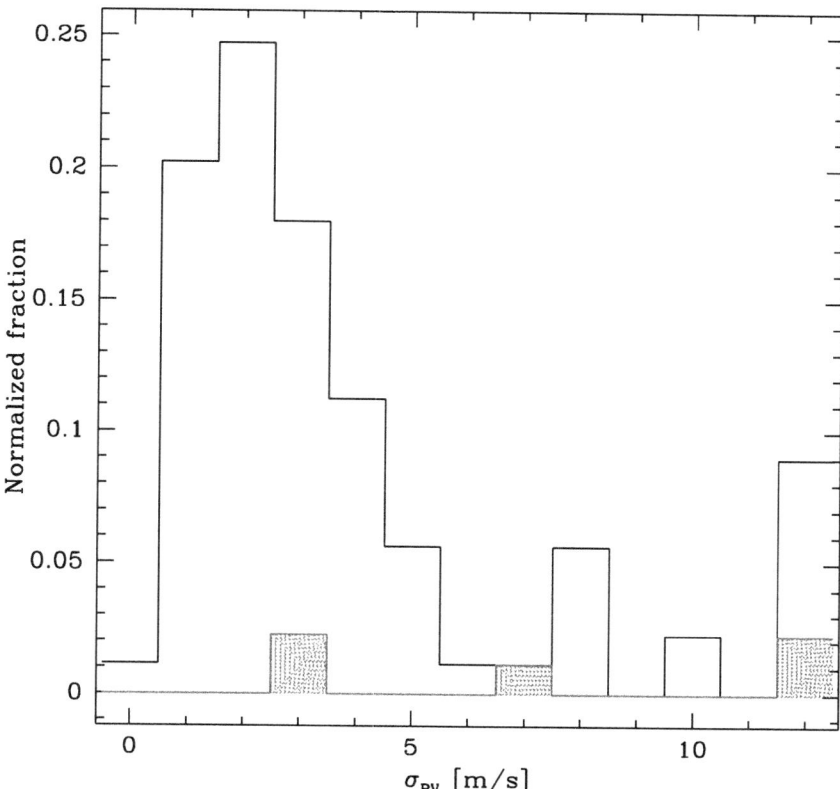

FIGURE 2.16. Histogram of the observed radial-velocity dispersion ($\sigma_{\rm RV}$) of the stars in the HARPS 'high-precision' subprogram. The position of the planets detected with HARPS is indicated by the hatched area.

planets orbiting μ Ara, HD 4308 and Gl 581 (Table 2.1, Figure 2.17) are part of this HARPS 'high-precision' subsample.

2.4.1. *Gaseous- vs. solid-planet properties at short periods*

Although the number of known Neptune-mass planets is small, it is interesting to see how their orbital parameters compare with the properties of giant extrasolar planets. Because of the tiny radial-velocity amplitude that small-mass planets induce in primary stars, limiting possible detections to short periods, a meaningful comparison can only be done with giant planets of $P \leq {\sim}20$ days.

The distribution of short-period giant planets peaks strongly at periods of around three days (Figure 2.13). In contrast, despite the above-mentioned detectability bias, the period distribution of Neptune-mass planets is rather flat up to 15 days. We also observe that orbits of Neptune-mass planets have low eccentricities (Figure 2.15). In particular, for periods between 9 and 15 days (three of the seven candidates), the mean eccentricity value is much smaller than that of giant planets. On the contrary, at periods smaller than six days, orbits are supposed to be tidally circularized, especially if these planets are 'solid'. However, among Neptune-mass planets, the highest observed eccentricities are for 55 Cnc e ($P = 2.8$ d and $e = 0.17$) and Gl 436 ($P = 2.6$ d and $e = 0.12$). The former is a member of a multi-planet system, which might explain the non-zero eccentricity of

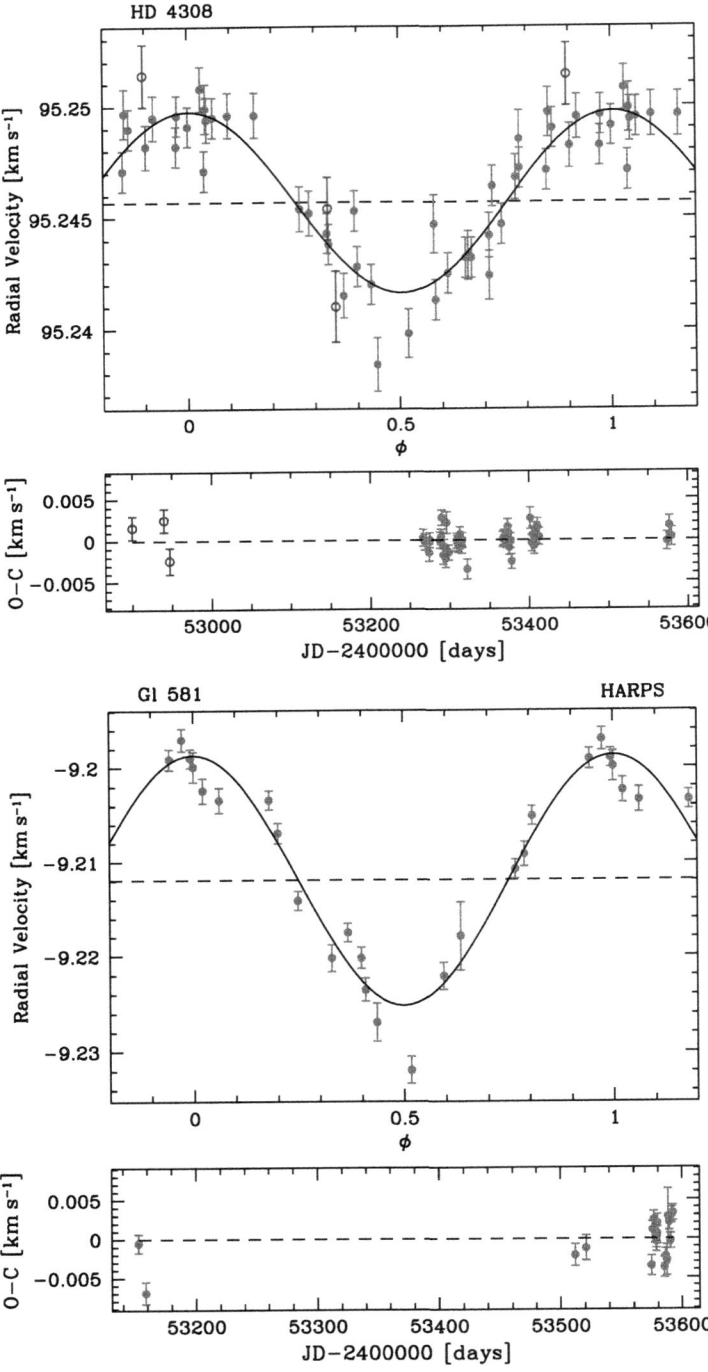

FIGURE 2.17. Phase-folded HARPS radial velocities of the Neptune-mass planet hosts HD 4308 and Gl 581, superimposed on the best Keplerian solutions.

TABLE 2.1 Summary table of the parameters of the recently discovered Neptune-mass planets. The parameter $q = m_2 \sin i/m_1$ is the planet to primary mass ratio. The lowest $m_2 \sin i$ of 6 $M_\oplus$ is obtained for Gl 876 d while the lowest q of $4.2 \cdot 10^{-5}$ is achieved for μ Ara c. References: [1] Santos et al. (2004a); [2] McArthur et al. (2004); [3] Udry et al. (2006); [4] Vogt et al. (2005); [5] Rivera et al. (2005); [6] Butler et al. (2004); [7] Bonfils et al. (2005a)

Planet	P [days]	e	$m_2 \sin i\ [M_\oplus]$	$q\ [10^{-5}]$	$(o\text{-}c)$ [ms^{-1}]	Reference
μ Ara c	9.55	0.0	14	4.2	0.9	[1]
55 Cnc e	2.81	0.17	14	4.7	5.4	[2]
HD 4308 b	15.6	0.12	14	5.4	1.3	[3]
HD 190360 c	17.1	0.01	18	6.0	3.5	[4]
Gl 876 d	1.94	0.0	6	6.0	4.6	[5]
Gl 436 b	2.64	0.12	21	16.0	5.3	[6]
Gl 581 b	4.96	0.0	17	17.1	2.5	[7]

the inner small-mass planet (Section 2.3.6). The problem is more difficult for the latter case. Another difference between the two types of planets can be found in the parent-star metallicity distribution (see Section 2.6.2).

Although not significant, these small differences may suggest that giant gaseous and 'solid' planets form two distinct populations with different properties. More detections are needed, however, to consider this question in a more convincing way.

2.5. Multi-planet systems

In mid-2006, 18 of the 160 planet host stars harbour multi-planet systems rather than single planets. Another system, HD 217107, shows an additional curved drift in the residuals of the one-planet Keplerian solution that is compatible with a second planetary companion. The most prolific of the multi-planet systems is 55 Cnc, with four detected planets. Upsilon Andromedae, HD 37124, and GJ 876 each have three known planets. μ Arae (HD 160691) also presents an additional long-period radial-velocity signal very probably due to a third planet in the system, despite the fact that the corresponding period is not covered yet (Gozdziewski et al. 2005). Finally, there are a total of eleven known double-planet systems. The orbital characteristics of these systems are summarized in Table 2.2. They can be roughly divided into two categories: *hierarchical* systems with well separated planets and *resonant* systems with rational period ratios. Illustrative examples of such systems are given in Figure 2.18.

Among planet-bearing stars, $\sim$12% are known multiple planet systems. Thus, the probability of finding a second planet is enhanced by a factor of almost two over the $\sim$7% probability of finding the first planet. The fraction of known multi-planet systems is certainly a lower limit. One difficulty is that low amplitude trends from more distant, longer-period planets are easily absorbed into single-planet Keplerian models. Detection of additional planets is also easier in systems where the more distant planet produces larger velocity amplitudes. However, the mass histogram (Figure 2.11) shows that high-mass planets are not very common. A second challenge for systems with small orbital period ratios like Gl 876 ($P_2/P_1 = 2/1$) is that dynamical interactions between planets (Section 2.5.2) can make Keplerian fitting of the observations more difficult and delay the whole system characterization. As a result, while one orbital period is sufficient for a single-planet system with velocity amplitudes larger than 10 m s^{-1}, longer phase

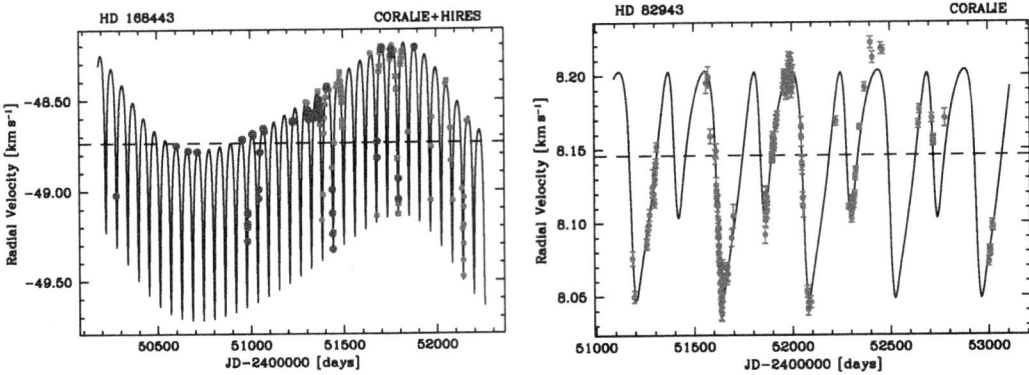

FIGURE 2.18. Radial velocities superimposed on the best two-Keplerian solutions for the *hierarchical* HD 168443 (left) and *resonant* HD 82943 systems (right; $P_2/P_1 = 2/1$).

coverage is generally required to disentangle additional components. The longest-running high precision survey is the 15-year planet search at Lick Observatory. This sample of 100 stars includes four multi-planet systems (55 Cnc, v And, Gl 876, 47 UMa) corresponding to half of the planet-hosting stars from the sample. For the somewhat younger ELODIE planet-search programme in Haute-Provence, started in 1994 and enlarged in 1996, 25% of the stars with detected planets host more than one planet.

In the light of the challenges that preclude the detection of multi-planet systems and given the high fraction of multi-planet systems in the older long-running search programmes, it seems likely that most stars form *systems of planets* rather than isolated, single planets. New techniques, complementary to radial velocities, to discover exoplanets with imaging, interferometry or astrometry will very probably exploit the sizeable fraction of multiple planet systems when designing their programmes.

2.5.1. Mean motion resonance systems

Among the known (mid-2006) multi-planet systems, at least eight (nearly half) are in mean motion resonances (MMRs) and four of these are in the low order 2:1 resonance. Except for HD 37124, which has an uncertain Keplerian model, orbital ratios less than or equal to 4:1 are all very close to integral period ratios (2:1, 3:1, or 4:1). Beyond the 4:1 MMR, the orbital period ratios quickly stray from integral values.[11] This suggests that if planets are close enough, it is likely that resonance capture will occur. Conversely, resonance capture seems less effective if the orbital period ratio is larger (i.e. the planets do not make a close approach), although longer orbital periods are not as precisely determined.

For period ratios less than 20, we do not see any correlation between the mass ratio of the planets and the orbital period ratios. For period ratios larger than 20, only large mass ratios for the planets are observed. However, this is likely to be a selection effect since longer period planets need to be more massive in order to have the same detectability as shorter-period, lower-mass planets.

---

[11] The outer two planets orbiting v And are in a 16:3 MMR and HD 12661 may be in a 13:2 MMR. No mean motion resonances are observed close to the exact ratio of 5:1 or 6:1. However, uncertainties in the orbital solution for HD 12661 allow for the possibility of a 6:1 MMR and the stability study of HD 202206 (Correia *et al.* 2005) suggests that the system is trapped in the 5:1 resonance. In this latter case the 5:1 resonance could indicate that the planet formed in a circumbinary disc as the inner 'planet' has a minimum mass of 17 $M_{\rm Jup}$.

TABLE 2.2 Orbital parameters of multi-planet systems. Resonances are indicated in the 'Rem' column

Star ID	P [days]	e	$m_2 \sin i$ [$M_{\rm Jup}$]	a [AU]	Rem
HD 75732 b	14.67	0.02	0.784	0.115	55 Cnc
HD 75732 c	43.9	0.44	0.22	0.24	3:1 (c:b)
HD 75732 d	4517	0.33	3.92	5.26	
HD 75732 e	2.81	0.17	0.045	0.038	
HD 9826 b	4.617	0.012	0.69	0.06	υ And
HD 9826 c	241.5	0.28	1.89	0.83	
HD 9826 d	1284	0.27	3.75	2.53	16:3 (d:c)
HD 37124 b	154.5	0.06	0.61	0.53	
HD 37124 c	843.6	0.14	0.60	1.64	
HD 37124 d	2295[a]	0.20	0.66	3.19	8:3 (d:c)
GJ 876 b	60.94	0.025	1.93	0.208	2:1 ± 0.02 (b:c)
GJ 876 c	30.10	0.27	0.56	0.13	
GJ 876 d	1.938	0.0	0.023	0.021	
HD 160691 b	629.6	0.26	1.67	1.5	μ Ara
HD 160691 c	9.55	0.0	0.044	0.09	
HD 160691 d	2530[b]	0.43	1.22	4.17	4:1 ± 0.25 (c:b)
HD 12661 b	262.5	0.35	2.37	0.83	
HD 12661 c	1684	0.02	1.86	2.6	6.43:1 ± 0.8 (c/b)
HD 217107 b	7.12	0.13	1.35	0.10	
HD 217107 c	>10000[b]	–	>10	>20	
HD 168443 b	58.12	0.53	7.64	0.29	
HD 168443 c	1740	0.22	17.0	2.85	
HD 169830 b	225.6	0.34	2.86	0.81	
HD 169830 c	1769	0.0	3.05	3.20	
HD 190360 b	2891	0.36	1.5	3.92	
HD 190360 c	17.1	0.01	0.057	0.128	
HD 202206 b	256.2	0.43	17.5	0.83	
HD 202206 c	1297	0.28	2.41	2.44	5.1 ± 0.07 (c:b)
HD 38529 b	14.3	0.25	0.837	0.13	
HD 38529 c	2182	0.35	13.2	3.68	
HD 73526 b	187.5	0.39	2.07	0.66	
HD 73526 c	376.9	0.40	2.30	1.05	2:1 ± 0.01 (c:b)
HD 74156 b	51.6	0.64	1.86	0.29	
HD 74156 c	2025	0.583	6.19	3.4	
HD 82943 b	219.5	0.39	1.82	1.03	
HD 82943 c	439.2	0.02	1.75	1.62	2:1 ± 0.01 (c:b)
HD 95128 b	1089	0.061	2.54	2.09	47 UMa
HD 95128 c	2594	0.1	0.76	3.73	
HD 108874 b	395.4	0.07	1.36	1.05	
HD 108874 c	1606	0.25	1.02	2.68	4:1 ± 0.1 (c:b)
HD 128311 b	458.6	0.25	2.18	1.10	
HD 128311 c	928	0.17	3.20	1.77	2:1 ± 0.03 (c:b)

[a] See Vogt et al. (2005) for an alternative orbital solution.
[b] Incomplete orbit.

Kley et al. (2004) have modelled the resonant capture of planets and find that for the 2:1 MMR, their models predict a larger mass for the outer planet, and higher eccentricity for the inner planet. We find that the orbital eccentricity is higher for the inner planet in three of the four 2:1 resonance systems. In the fourth system, HD 73526, the eccentricities for both components are comparable. We find that the outer planet is more massive (assuming coplanar orbits) in Gl 876 and HD 128311. The outer planet is only slightly more massive in HD 73526 and is slightly less massive in the Keplerian model for HD 82943 (Mayor *et al.* 2004).

The orbital parameters of multi-planet systems seem indistinguishable from those of single-planet systems. For example, this is illustrated in Figures 2.15 and 2.13, which compare the mass–period and eccentricity distributions of multiple and single planet systems.

2.5.2. *Dynamics: planet–planet interaction and stability*

The presence of two or more interacting planets in a system dramatically increases our potential ability to constrain and understand the processes of planetary formation and evolution. Planet–planet interactions can reasonably be divided into three categories: interactions during the planet formation, ongoing secular or resonant interactions, which can be observed on timescales of decades or less, and long-term dynamical interactions that shape the system on a timescale comparable to the star's lifetime. Short-term dynamical interactions are of particular interest because of the directly observable consequences.

Among these interactions, the observed $P_i/P_j = 2/1$ resonant systems are very important because, when the planet orbital separations are not too large, planet–planet gravitational interactions become non-negligible during planet 'close' encounters, and will noticeably influence the system evolution on a timescale of the order of a few times the long period. The radial-velocity variations of the central star will then differ substantially from velocity variations derived assuming the planets are executing independent Keplerian motions (Figure 2.19). We observe a temporal variation of the *instantaneous* orbital elements. In the most favourable cases, the orbital-plane inclinations, not otherwise known from the radial-velocity technique, can be determined since the amplitude of the planet–planet interaction directly scales with their true masses.

In the case of multiple planets, only approximate analytic solutions of the gravitational equations of motion exist, and one must resort to numerical integrations to model the data. Several studies have been conducted in this direction for the Gl 876 system (Laughlin *et al.* 2005; Rivera *et al.* 2005) hosting two 2/1 resonant planets at fairly small separations. The results of the *Newtonian* modelling of the Gl 876 system have validated the method, notably improving the determination of the planetary orbital elements and also unveiling the small-mass planet embedded in the innermost region of the system. The time coverage of the measurements is, however, still too small for the method to provide strong constraints on the plane inclinations. The valley of the acceptable solutions is shallow. In particular, the derived planet inclination (Rivera *et al.* 2005; planets supposed coplanar) does not correspond to the astrometric result obtained with the *HST* (Benedict *et al.* 2002). Further radial-velocity measurements will undoubtedly improve the situation.

Another useful application of the dynamical analysis of a multi-planet system is the determination of the system 'structure' in terms of orbit content, or in other words the determination of the location of the resonances in the system. For example, in the HD 202206 system (Correia *et al.* 2005), the large mass of the inner planet provokes high perturbations in the orbit of the outer one. The system is thus in a very chaotic region of parameter space, but the existence of a 5/1 mean motion resonance close to the best solution fitted to the observations points out the more realistic set of parameters that 'stabilizes' the orbits of the two 'planets' (Figure 2.19).

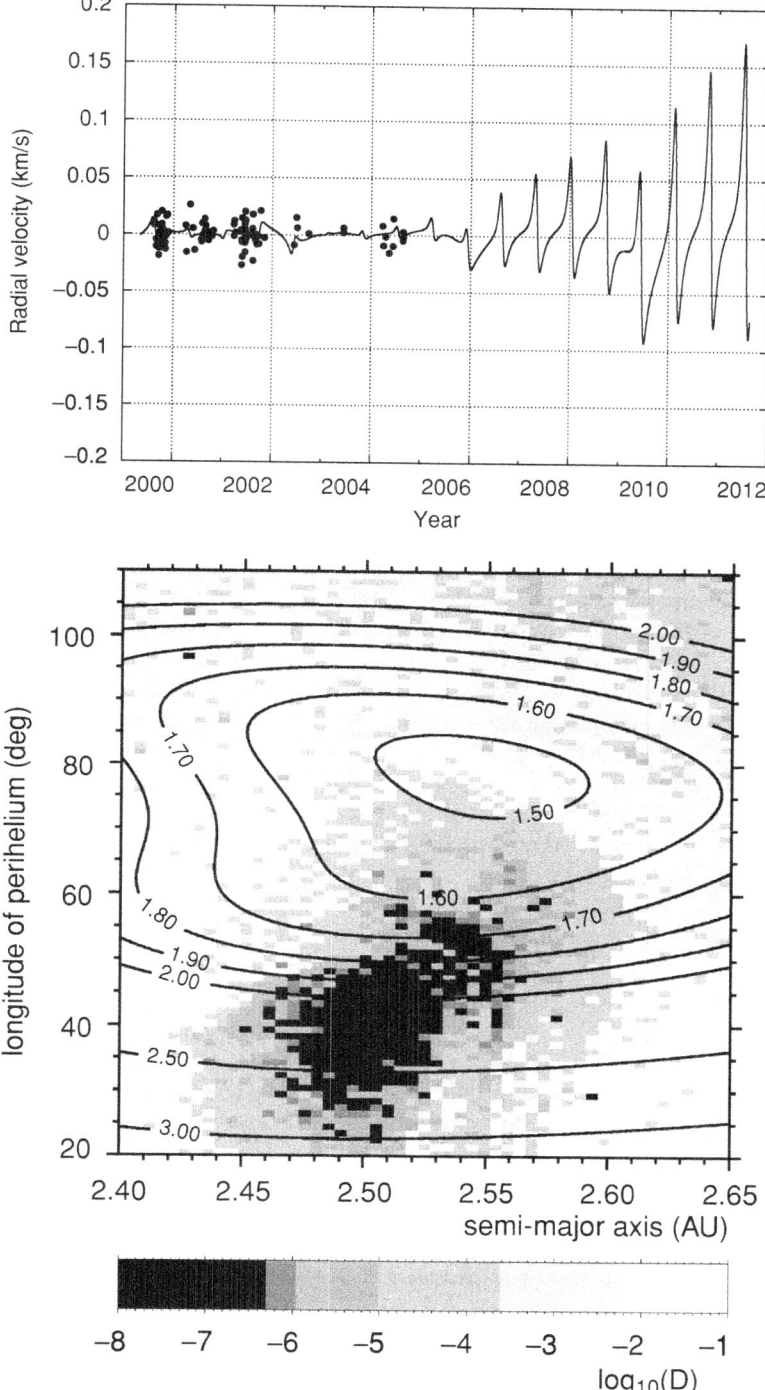

FIGURE 2.19. Top: temporal differences between the radial velocities predicted by the two-Keplerian models and the numerical integration of the system HD 202206 (Correia et al. 2005). Residuals of the CORALIE measurements around the Keplerian solution are displayed as well. Bottom: stability map of the system in a two-parameter plane of the second planet. The minimum χ^2 of the best-fit solution (lines are iso-χ^2 contours) is in a chaotic region, close to a stability island corresponding to a 5/1 resonance.

2.6. Primary star properties

To find solutions for the many problems raised by extrasolar planet properties, we need observational constraints. As seen above, these can come from analysis of the orbital parameters of the known-planet sample. Further information also comes from the study of the planet hosts themselves. In particular, the mass and metallicity of the parent stars seem to be of prime importance for planet-formation models (Ida & Lin 2004b, 2005; Benz et al. 2006).

2.6.1. Metallicity of stars hosting giant planets

Very soon after the discovery of the first extrasolar giant planets, stellar spectroscopists noticed that the planet hosts were systematically metal-rich (Gonzalez 1997, 1998; Gonzalez et al. 1999; Gonzalez & Laws 2000; Fuhrmann et al. 1997, 1998; Santos et al. 2000, 2003). The following detections indeed confirmed this trend, and the possible origin of this metallicity excess was a matter of great debate. One explanation argues that high metallicity enhances planet formation because of increased availability of small particle condensates (silicates, dust), the building blocks of planetesimals. Another argument suggests that enhanced stellar metallicity comes from a pollution of the stellar convective zone resulting from late-stage accretion of gas-depleted material. A third explanation invoking the possibility that planet migration is somewhat controlled by the dust content of the disc – and thus leads to an observed bias in favour of close-in planets around metal-rich stars – seems to be reasonably ruled out by current models (Livio & Pringle 2003). The two main proposed mechanisms result in different stellar structures; in the first case, the star is metal-rich throughout, while in the latter case, the convective zone has significantly higher metallicity than the stellar interior.

The early observation of the planet-metallicity correlation was made with only a handful of planet-hosting stars. Moreover, the comparison metallicity distributions came from volume-limited studies carried out by different researchers at a time when systematic offsets of 0.1 dex in metallicity results were common. Eventually, systematic homogeneous studies of all the stars in planet-search surveys were completed (Santos et al. 2001b), by automatic or 'statistical' (e.g. CCF surface calibration, Section 2.2.2) metallicity estimates, with the further requirement that the stars have enough observations to detect a Jupiter-like planet with an orbital period out to several years (Fischer et al. 2003; Santos et al. 2004b, 2005; Fischer & Valenti 2005). The known metallicity distributions of large planet-search surveys then allow us to estimate the fraction of stars with known planets per metallicity bin, i.e. the probability for a star of a given metallicity to host a planet. The probability obtained of finding a planet is a steeply rising function of the metallicity of the star (Figure 2.20). More metal-rich stars have a higher probability of harbouring a planet than lower metallicity objects. Current numbers seem to suggest that at least 25% of the stars with twice the metal content of the Sun ([Fe/H] ≥ 0.3) are orbited by a planet, while this number decreases to below 5% for solar-metallicity objects.

Figure 2.20 shows the percentage of stars with planets from 1040 stars on the Lick, Keck and AAT planet surveys (solid line, Fischer & Valenti 2005) and the percentage of stars with planets from ~1000 stars on the CORALIE survey (non-binary and with more than five observations; dashed line, Santos et al. 2004b). The occurrence of planets as a function of metallicity was fitted by Fischer & Valenti (2005) with a power law:

$$\mathcal{P}(\text{planet}) = 0.03 \times \left(\frac{(N_{\text{Fe}}/N_{\text{H}})}{(N_{\text{Fe}}/N_{\text{H}})_\odot}\right)^2.$$

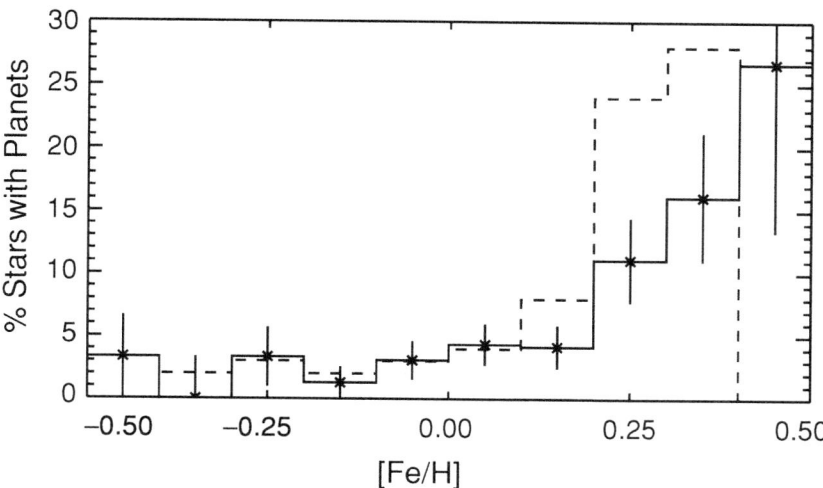

FIGURE 2.20. The percentage of stars with exoplanets is shown as a function of stellar metallicity. The dashed line shows the results of Santos et al. (2004b) for ∼1000 CORALIE non-binary stars and the solid line shows the analysis of 1040 Lick, Keck and AAT stars (Fischer & Valenti 2005).

Thus, the probability of forming a gas giant planet is roughly proportional to the square of the number of metal atoms.

Metallicity seems to play a crucial role in the formation and/or evolution of giant planets, at least for the kind of planets radial-velocity searches have revealed up to now. However, these trends do not imply that giant planets cannot be formed around more metal-poor objects, but rather that the probability of formation among such systems is lower. Indeed, there is some hint that for lower [Fe/H] values, the frequency of planets may remain relatively constant as a function of metallicity (Santos et al. 2004b). Whether this reflects the presence of two different regimes, or simply a low metallicity tail, is currently under debate, while more data will be needed to tackle this problem.

The self-consistent analysis of high-resolution spectra for many stars on planet-search surveys also allows us to distinguish between the two enrichment hypotheses. Metallicity was not observed to increase with decreasing convective zone depth for main-sequence stars, suggesting that accretion was not responsible for self-pollution of planet-bearing stars.[12] Even more importantly, the analysis of subgiants in the sample shows that subgiants with planets have high metallicity, while subgiants without detected planets have a metallicity distribution similar to main-sequence stars without detected planets. Since significant mixing of the convective zone takes place along the subgiant branch, subgiants would have diluted accreted metals in the convective zone. The fact that high metallicity persists in subgiants with planets demonstrates that these stars were metal-rich throughout. The existence of a planet metallicity correlation supports core accretion over gravitational instability as the formation mechanism for gas giant planets with orbital periods as long as four years.

---

[12] This argument is, however, questioned by Vauclair (2004) invoking *thermohaline convection* (metallic fingers) that might dilute the accreted matter inside the star and thus reconcile the overabundances expected in the case of accretion of planetary material with the observations of stars of different masses.

Up to now, no clear correlation between metallicity and orbital parameters has been observed (Santos *et al.* 2003).

No metallicity bias

Particular concern has been expressed by the community regarding the fact that a higher metallicity will imply that the spectral lines are better defined. This could mean that the final precision in radial-velocity would be better for the more 'metallic' objects, leading thus to an increasing detection rate as a function of increasing [Fe/H]. When examining the mean photon-noise error for stars with different [Fe/H] having V magnitudes between 6 and 7 in the CORALIE sample, no particular trend is seen in the data. The very slight tendency (metal-rich stars have, in average, measurements with only about 1–2 ms^{-1} better precision than metal-poor stars) is definitely not able to induce the strong effect seen in the [Fe/H] distribution of planet host stars, especially when we compare this difference with the usual velocity amplitude induced by the known planetary companions (a few tens of ms^{-1}). In fact, in the CORALIE survey we always set the exposure times in order to have approximately the same photon-noise error. This also seems to be the case for the Lick/Keck planet-search programmes (Fischer, private communication).

Metallicity-biased planet-search programmes

The observed relation between stellar metal content and planet occurrence (Section 2.6.1) has favoured the launch of metallicity-biased planet-search programmes targeting short-period planets to look for hot jupiters, perfect candidates for a photometric transit-search follow-up. These surveys are successful, with five candidates detected in slightly more than one year (Fischer *et al.* 2005; Da Silva *et al.* 2006; Sato *et al.* 2005; Bouchy *et al.* 2005a; Section 2.7). Two of them, HD 149026 and HD 189733, are transiting their parent stars, the latter being the best candidate for a direct planet detection because of its very favourable planet-to-star flux ratio in the IR, and the former allowing for the determination of very unusual planet radius and mean density (the planet is found to have an unexpectedly large core). This result clearly illustrates the importance of such programmes for our understanding of planet interiors. However, when examining possible statistical trends between orbital and stellar parameters to derive constraints for planet formation models, we have to keep in mind the built-in bias of this subsample of exoplanets. In particular these planets must be removed when considering correlations with the star metallicity.

2.6.2. *Metallicity of stars hosting Neptune-mass planets*

It is well-established that the detected giant planets preferentially orbit metal-rich stars. What is the situation for the newly found Neptune-mass planets? If, as proposed by several authors, the new *hot neptune* planets are the remains of evaporated ancient giant planets (e.g. Lecavelier *et al.* 2004; Baraffe *et al.* 2004, 2005, and references therein), they should also follow the metallicity trend of their giant progenitors. This does not seem to be the case, considering that the seven known planets with $m_2 \sin i \leq 21\ M_\oplus$ (Table 2.1) have metallicities of 0.33, 0.35, 0.02, 0.14, −0.03, −0.25, and −0.31, respectively.[13] Although the statistics are still poor, the spread of these values over the nearly full range of planet-host metallicities (Figure 2.21) suggests a different relation between metal content and planet frequency for the icy/rocky planets with respect to the giant ones.

---

[13] The metallicity of the three M-dwarfs comes from the photometric calibration derived by Bonfils *et al.* (2005b).

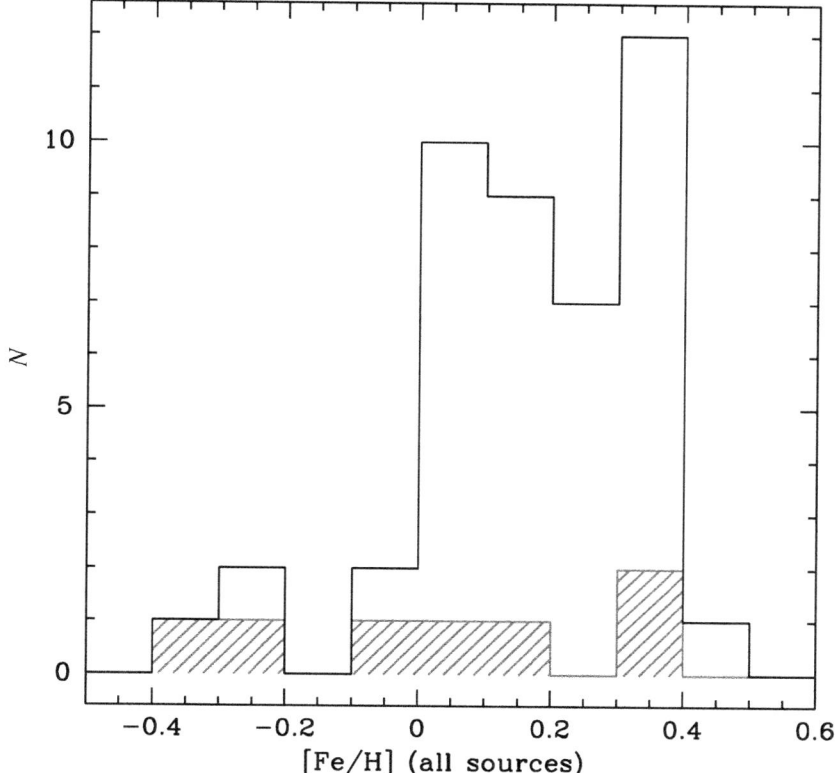

FIGURE 2.21. Comparison of the metallicity distributions between the sample of extrasolar giant planet hosts for planets with periods shorter than 20 days (open histogram) and stars with Neptune-mass planets (shaded histogram).

However, we have to note here that three of the candidates orbit M-dwarf primaries. Recent Monte Carlo simulations by Ida & Lin (2005) show that planet formation around small-mass primaries tends to form planets with lower masses in the Uranus/Neptune domain. A similar result that favours lower-mass planets is also observed for solar-type stars in the case of the low metallicity of the protostellar nebula (Ida & Lin 2004b; Benz et al. 2006). Future improvements in the planet-formation models and new detections of very-low-mass planets will help to better understand these two converging effects.

2.6.3. *Primary-mass effect*

The mass of the primary star is also an important parameter for planet formation processes. On the side of small-mass stars, results from ongoing surveys indicate that giant gaseous planets are rare around M-dwarfs in comparison to FGK primaries. The only known system with two giant planets is Gl 876 (Table 2.2). In particular, no hot jupiter has been detected close to an M-dwarf. This result, however, still suffers from small number statistics and is not statistically very robust yet. On the other hand, as seen above, three of the five planets found to orbit an M-dwarf have masses below 21 $M_\oplus$. The probability of finding a planet around an M-dwarf seems, then, directly dependent on the domain of planet masses considered.

For primaries more massive than the Sun, new surveys targeting earlier, rotating A–F-dwarfs (Galland et al. 2005a,b) and programmes surveying G–K giant stars (Setiawan et al. 2005; Sato et al. 2004; Hatzes et al. 2005) are starting to provide interesting candidates. The detected planets are generally massive, but it is still too early to conclude on a 'primary-mass' effect, as those programmes are still strongly observationally biased (larger-mass primaries and short history of the surveys).

2.7. Follow-up of transiting planets

In recent years ground-based photometric-transit searches have produced an increased number of planetary transiting candidates. The most successful of these searches has been the OGLE survey with close to 180 possible transiting planets (Udalski et al. 2002a,b). These new detections stimulated intensive follow-up observations to detect the radial-velocity signature induced by the orbiting body. Surprisingly, these studies revealed that most of the systems were rather eclipsing binaries of small stars (M-dwarfs) in front of F–G-dwarfs, eclipsing binaries in blended multiple stellar systems (triple, quadruple), or grazing stellar eclipses, all mimicking photometric planetary transits (Bouchy et al. 2005c; Pont et al. 2005). These spectroscopic follow-ups demonstrated, however, the difficulty of the interpretation of shallow transit light curves without complementary radial-velocity measurements. The magnitude of the OGLE candidates ranges from $V \sim 16$ to 17.5. It is close to the faint capability of an accurate fibre-fed spectrograph with thorium calibration like FLAMES on the VLT, and beyond the capability of slit spectroscopy with iodine self-calibration. This implies that deeper photometric transit surveys would run into the difficulty of confirming by Doppler follow-up the planetary nature of the transiting object.

Today (mid-2006), we know of six planets detected from transit surveys and confirmed by radial velocities. Five have been found by the OGLE project (Udalski et al. 2002a,b) and one by the TrES network (Alonso et al. 2004). Three of the OGLE planets have periods smaller than 2 days (very hot jupiters). Such short periods, although easy to detect, are not found in the radial-velocity surveys, suggesting that those objects are about ten times less numerous than hot jupiters ($2.5 \leq P \leq 10$ days; Gaudi et al. 2005). In addition to the photometrically-detected candidates, three planets identified by radial-velocity measurements have been found transiting in front of their parent stars.

Transit photometry combined with high-precision radial-velocity measurements provide accurate mass and radius for the planets (Table 2.3), and subsequently the planet mean density (Figure 2.22), important values to constrain planetary interior models, as well as planet evolution history.[14] The derived density of transiting extrasolar planets covers a fairly wide range of values from 0.3 to 1.3 g cm^{-3} (Figure 2.22). HD 209458 presents an anomalously large radius and low density. These characteristics are clearly not shared by all close transiting planets since objects with similar mass are found to have different densities. This demonstrates a surprising diversity and reflects our lack of detailed understanding of the physics of irradiated giant planets.

The distribution of planets in a *period vs. mass* diagram shows an intriguing correlation (Figure 2.22). Transiting planets seem to lie on a well-defined line of mass decreasing with

---

[14] It is interesting to note here that the majority of planets for which we know both mass and radius have been found by transit survey despite the fact that ~ 200 planets have been identified by radial-velocity searches. This is a consequence of the low probability of finding a transiting configuration among the planets found by radial-velocity surveys, while most of the transiting candidates can be followed-up by radial-velocity measurements. On the other hand, the three planets transiting the brightest stars have been found first by radial velocities as transit surveys are mainly targeting crowded fields with fainter stars.

TABLE 2.3 List of planets with both radius (from transit) and mass estimate (from accurate radial velocities). Data from: Alonso et al. (2004); Moutou et al. (2004); Pont et al. (2004, 2005); Bouchy et al. (2005a,c); Sato et al. (2005); Winn et al. (2005)

Object	Period [days]	Mass [$M_{\rm Jup}$]	Radius [$R_{\rm Jup}$]
OGLE-TR-10	3.10	0.57±0.12	1.24±0.09
OGLE-TR-56	1.21	1.45±0.23	1.23±0.16
OGLE-TR-111	4.02	0.53±0.11	1.00±0.10
OGLE-TR-113	1.43	1.35±0.22	1.08±0.06
OGLE-TR-132	1.69	1.19±0.13	1.13±0.08
TrES-1	3.03	0.73±0.04	1.08±0.05
HD 209458	3.52	0.66±0.01	1.35±0.005
HD 189733	2.22	1.15±0.04	1.26±0.03
HD 149026	2.88	0.33±0.02	0.73±0.06

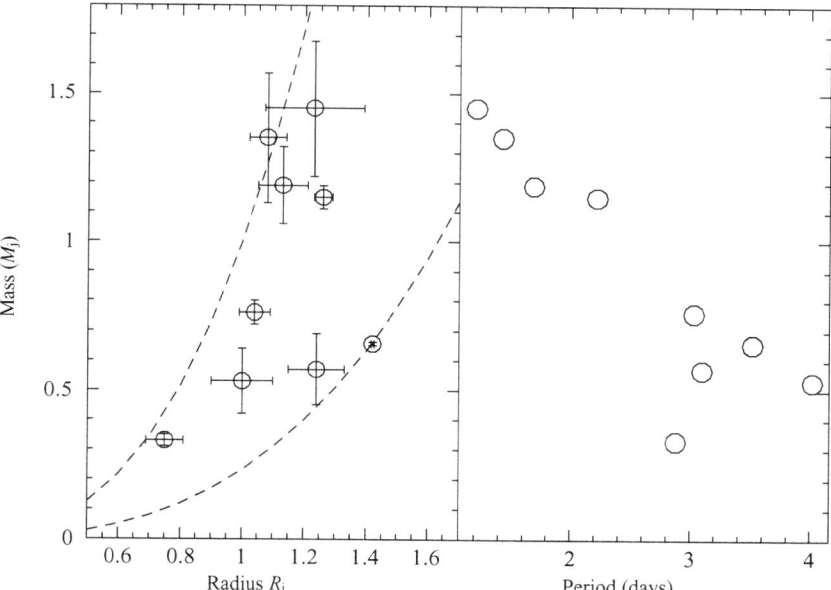

FIGURE 2.22. Mass–radius and mass–period diagrams of transiting planets with radius and accurate mass estimates. In the left panel, the dashed lines indicate isodensity contours of 0.3 and 1.3 g cm^{-3}.

increasing orbital period. This puzzling observation, pointed out by Mazeh et al. (2005), could be the consequence of mechanisms such as thermal evaporation (Lecavelier et al. 2004; Baraffe et al. 2004, 2005) or Roche-limit mass transfer (Ford & Rasio 2005). It is worth noting the location of HD 149026, below the relation, which could be an effect of its different structure consisting of a large core (Sato et al. 2005; Charbonneau et al. 2006). Even more surprising in the diagram is the complete lack of candidates above the relation. Why are we missing more massive transiting planets at $P \simeq 3$–4 days? No convincing explanation has yet been proposed for this puzzling observation.

2.8. Future of radial velocities

An important lesson learnt from the past few years is that the radial-velocity technique has yet not reached its 'limits' in the domain of exoplanets. In fact, the future of radial-velocities is still bright.

(1) Recent discoveries indicate that a population of Neptune- and Saturn-mass planets remains to be discovered below 1 AU. The increasing precision of the radial-velocity surveys will help answer this question in the near future, thereby providing us with useful new constraints on planet formation theories. With the precision level now achieved for radial-velocity measurements, a new field in the search for extrasolar planets is open, allowing the detection of companions of a few Earth masses around solar-type stars. Very low-mass planets (<10 $M_\oplus$) might be more frequent than the previously found giant worlds.

(2) As described above, radial-velocity follow-up measurements are mandatory in order to have access to the masses of transiting companions and then to their mean densities. They thus ascertain the planetary nature of the companions and provide important parameters to constrain planetary atmosphere and interior models. This is important in view of the results expected of the space missions *COROT* and *Kepler* that should provide hundreds of transit candidates of various sizes and masses in the coming years. If one considers a transit signal with known orbital period, measuring its mass is less demanding on both the number and the accuracy of the required radial velocity measurements. For example, a 2 $M_\oplus$-planet on a 4-day orbit induces a radial-velocity amplitude of about 80 cm s^{-1} that will be possible to detect with only 'a few' high-precision radial-velocity measurements, provided that the period of the system is known in advance. In this context, the most exciting aspect is the opportunity to explore the *mass–radius* relation down to the Earth-mass domain.

(3) *Towards the detection of Earthlike planets.* The threshold of the lowest mass planet detectable by the Doppler technique continues to decrease. Nobody has yet explored in detail the domain below the 1 m s^{-1} level. Results obtained with the HARPS spectrograph show that, even if stars are intrinsically variable in radial velocity (at different levels) due to acoustic modes, it is nevertheless possible in the short term to reach precisions well below 1 m s^{-1} (10 cm s^{-1}) by applying a suitable observational strategy. One open issue remains, however, unsolved: the behaviour of the stars on longer timescales, where stellar jitter and spots may impact the final achievable accuracy. In this case, an accurate pre-selection of the stars may help in focusing on good candidates and optimizing the observation time. In addition, bisector analysis and follow-up of activity indicators such as $\log(R_{\rm HK})$, as well as photometric measurements, would allow identification of potential error sources.

Nevertheless, the discovery of an extrasolar planet by means of the Doppler technique requires that the radial-velocity signal induced by the planet is significantly higher than the dispersion, or alternatively requires the recording of a large number of data points. This is particularly important to rule out artefacts, given the relatively high number of free parameters in the orbital solution, in particular for multi-planet systems. A large number of measurements will overcome this problem but will demand an enormous investment in observing time. Thus, as long as we accept paying the price in terms of telescope time, radial-velocity measurements on specially designed spectrographs (high-level temperature and pressure control) should in principle be able to detect Earthlike planets in the habitable zone around well-selected 'quiet' stars (Pepe *et al.* 2005).

REFERENCES

Alibert, Y., Mordasini, C. & Benz, W., 2004, *A&A* **417**, L25.
Alibert, Y., Mordasini, C., Benz, W. & Winisdoerffer, C., 2005, *A&A* **434**, 343.
Alonso, R., Brown, T. M., Torres, G., et al., 2004, *ApJ* **613**, L153.
Baliunas, S. L., Donahue, R. A., Soon, W. H., et al., 1995, *ApJ* **438**, 269.
Baraffe, I., Selsis, F., Chabrier, G., et al., 2004, *A&A* **419**, L13.
Baraffe, I., Chabrier, G., Barman, T., et al., 2005, *A&A* **436**, L47.
Baranne, A., Queloz, D., Mayor, M., et al., 1996, *A&AS* **119**, 373.
Beckwith, S., 2000, in *Planetary Systems in the Universe*, eds. A. Penny et al., *IAU Symposium* **202**, 22.
Beckwith, S. & Sargent, A., 1993, in *Protostars and Planets III*, eds. E. Levy & J. Lunine, Tucson, AZ: University of Arizona Press, 521.
Beckwith, S. & Sargent, A., 1996, *Nature* **383**, 139.
Benedict, G. F., McArthur, B. E., Forveille, T., et al., 2002, *ApJ* **581**, L115.
Benz, W., Mordasini, C., Alibert, Y. & Naef, D., 2006, in *Tenth Anniversary of 51 Peg b: Status and Prospects for Hot Jupiters Studies*, eds. L. Arnold, F. Bouchy & C. Moutou, Paris: Frontier Group, p.13.
Beuzit, J.-L., Mouillet, D., Oppenheimer, B. & Monnier, J., 2007, in *Protostars and Planets V*, eds. B. Reipurth, D. Jewitt & K. Keil, Tucson, AZ: University of Arizona Press.
Bodenheimer, P., Hubickyj, O. & Lissauer, J., 2000, *Icarus* **143**, 2.
Bonfils, X., Forveille, T., Delfosse, X., et al., 2005a, *A&A* **443**, L15.
Bonfils, X., Delfosse, X., Udry, S., et al., 2005b, *A&A* **442**, 635.
Boss, A., 2002, *ApJ* **576**, 462.
Bouchy, F., Udry, S., Mayor, M., et al., 2005a, *A&A* **444**, L15.
Bouchy, F., Bazot, M., Santos, N. C., Vauclair, S. & Sosnowska, D., 2005b, *A&A* **440**, 609.
Bouchy, F., Pont, F., Melo, C., et al., 2005c, *A&A* **431**, 1105.
Butler, P., Marcy, G., Williams, E., et al., 1996, *PASP* **108**, 50.
Butler, P., Marcy, G., Vogt, S. & Fischer, D., 2000, in *Planetary Systems in the Universe*, eds. A. Penny et al., *IAU Symposium* **202**, 1.
Butler, P., Vogt, S., Marcy, G., et al., 2004, *ApJ* **617**, 580.
Charbonneau, D., Winn, J., Latham, D., et al., 2006, *ApJ* **636**, 445.
Chauvin, G., Lagrange, A.-M., Dumas, C., et al., 2005, *A&A* **438**, L25.
Collier Cameron, A. C., Horne, K., Penny, A. & James, D., 2000, in *Planetary Systems in the Universe*, eds. A. Penny et al., *IAU Symposium* **202**, 1.
Correia, A., Udry, S., Mayor, M., et al., 2005, *A&A* **440**, 751.
Coudé du Foresto, V., 2000, in *From Extrasolar Planets to Cosmology*, eds. J. Bergeron & A. Renzini, *ESO Astrophysics Symposia*, 560.
Cumming, A., Marcy, G. & Butler, P., 1999, *ApJ* **526**, 890.
Da Silva, R., Udry, S., Bouchy, F., et al., 2006, *A&A* **446**, 717.
Duquennoy, A. & Mayor, M., 1991, *A&A* **248**, 485.
Eggenberger, A., Udry, S. & Mayor, M., 2004, *A&A* **417**, 353.
Eggenberger, A., Udry, S., Mayor, M., et al., 2005, in *Multiple Stars Across the H-R Diagram*, eds. S. Hubrig, M. Petr-Gotzens & A. Tokovinin, *ESO Astrophysics Symposium*.
Endl, M., Kürster, M., Els, S., et al., 2002, *A&A* **392**, 671.
Fischer, D. & Valenti, J., 2005, *ApJ* **622**, 1102.
Fischer, D., Valenti, J. & Marcy, G., 2003, *ApJ* **622**, 1102.

FISCHER, D., LAUGHLIN, G., BUTLER, P., et al., 2005, *ApJ* **620**, 481.
FORD, E. & RASIO, F. A., 2005, in *Protostars and Planets V*, Proceedings of the Conference held October 24–28, 2005 in Hilton Waikoloa Village, Hawai'i. LPI Contribution 1286, p. 8360.
FORD, E., HAVLICKOVA, M. & RASIO, F. A., 2001, *Icarus* **150**, 2.
FORD, E., RASIO, F. A. & YU, K., 2003, in *Scientific Frontiers in Research on Extrasolar Planets*, eds. D. Deming & S. Seager, ASP Conference Series, **294**, 181.
FUHRMANN, K., PFEIFFER, M. & BERNKOPF, J., 1997, *A&A* **326**, 1081.
FUHRMANN, K., PFEIFFER, M. & BERNKOPF, J., 1998, *A&A* **336**, 942.
GALLAND, F., LAGRANGE, A.-M., UDRY, S., et al., 2005a, *A&A* **443**, 337.
GALLAND, F., LAGRANGE, A.-M., UDRY, S., et al., 2005b, *A&A* **444**, L21.
GAUDI, B. S., SEAGER, S. & MALLEN-ORNELAS, G., 2005, *ApJ* **623**, 472.
GONZALEZ, G., 1997, *MNRAS* **285**, 403.
GONZALEZ, G., 1998, *A&A* **334**, 221.
GONZALEZ, G. & LAWS, C., 2000, *AJ* **119**, 390.
GONZALEZ, G., WALLERSTEIN, G. & SAAR, S., 1999, *ApJ* **511**, 111.
GOZDZIEWSKI, K., KONACKI, M. & MACIEJEWSKI, A., 2005, *AJ* **622**, 1136.
HALBWACHS, J.-L., ARENOU, F., MAYOR, M., UDRY, S. & QUELOZ, D., 2000, *A&A* **355**, 581.
HALBWACHS, J.-L., MAYOR, M., UDRY, S. & ARENOU, F., 2003, *A&A* **397**, 159.
HALBWACHS, J.-L., MAYOR, M. & UDRY, S., 2005, *A&A* **431**, 1129.
HATZES, A. P., GUENTHER, E., ENDL, M., et al., 2005, *A&A* **437**, 743.
IDA, S. & LIN, D., 2004a, *ApJ* **604**, 388.
IDA, S. & LIN, D., 2004b, *ApJ* **616**, 567.
IDA, S. & LIN, D., 2005, *ApJ* **626**, 1045.
JORISSEN, A., MAYOR, M. & UDRY, S., 2001, *A&A* **379**, 992.
KLEY, W., PEITZ, J. & BRYDEN, G., 2004, *A&A* **414**, 735.
LAGRANGE, A.-M., BACKMAN, D. & ARTYMOWICZ, P., 2000, in *Protostars and Planets IV*, eds. V. Mannings, A. Boss & S. Russel, Tucson, AZ: University of Arizona Press, 639.
LAPLACE, P., 1796, *Exposition du Système du Monde*. Paris: Imprimerie du Circle-Soual.
LAUGHLIN, G., BUTLER, P., FISCHER, D., et al., 2005, *ApJ* **622**, 1182.
LECAVELIER DES ETANGS, A., VIDAL-MADJAR, A., MCCONNELL, J. C. & HEBRARD, G., 2004, *A&A* **418**, L1.
LEIGH, C., COLLIER CAMERON, A., UDRY, S., et al., 2003, *MNRAS* **346**, L16.
LEVISON, H. F., LISSAUER, J. & DUNCAN, M. J., 1998, *AJ* **116**, 1998.
LIN, D. & IDA, S., 1997, *ApJ* **477**, 781.
LIN, D., BODENHEIMER, P. & RICHARDSON, D. C., 1996, *Nature* **380**, 606.
LIVIO, M. & PRINGLE, J. E., 2003, *MNRAS* **346**, 42.
LOVIS, C., MAYOR, M., BOUCHY, F., et al., 2005, *A&A* **437**, 1121.
MARCY, G., BUTLER, P., VOGT, S., et al., 2001, *ApJ* **555**, 418.
MARCY, G., BUTLER, P., FISCHER, D., et al., 2005, *Progr. Theor. Phys. Suppl.* **158**, 24.
MARZARI, F. & WEIDENSCHILLING, S. J., 2002, *Icarus* **156**, 570.
MAYER, L., WADSLEY, J., QUINN, T. & STADEL, J., 2005, *MNRAS* **363**, 641.
MAYOR, M. & QUELOZ, D., 1995, *Nature* **378**, 355.
MAYOR, M., PEPE, F., QUELOZ, D., et al., 2003, *The ESO Messenger* **114**, 20.
MAYOR, M., UDRY, S., NAEF, D., et al., 2004, *A&A* **415**, 391.
MAZEH, T., KRYMOLOWSKI, Y. & ROSENFELD, G., 1997, *ApJ* **477**, L103.
MAZEH, T., ZUCKER, S. & PONT, F., 2005, *MNRAS* **356**, 955.

McArthur, B., Endl, M., Cochran, W., et al., 2004, ApJ **614**, L81.
McCaughrean, M., Stapelfeld, K. & Close, L., 2000, in *Protostars and Planets IV*, eds. V. Mannings, A. Boss & S. Russel, Tucson, AZ: University of Arizona Press, 485.
Moutou, C., Pont, F., Bouchy, F. & Mayor, M., 2004, *A&A* **424**, L31.
Mugrauer, M., Neuhäuser, R., Mazeh, T., Alves, J. & Guenther, E., 2004, *A&A* **425**, 249.
Mugrauer, M., Neuhäuser, R., Seifahrt, A., Mazeh, T. & Guenther, E., 2005, *A&A* **440**, 1051.
Naef, D., Mayor, D., Beuzit, J.-L., et al., 2005, in *13th Cool Stars, Stellar Systems and the Sun*, ESA SP-560, 833.
Nelson, R., Papaloizou, J., Masset, F. & Kley, W., 2000, *MNRAS* **318**, 18.
Noyes, R., Hartmann, L. W., Baliunas, S., Duncan, D. & Vaughan, A., 1984, *ApJ* **279**, 763.
Patience, J., White, R. J., Ghez, A., et al., 2002, *ApJ* **581**, 654.
Pätzold, M. & Rauer, H., 2002, *ApJ* **568**, L117.
Pepe, F., Mayor, M., Ruprecht, G., et al., 2002, *The ESO Messenger* **110**, 9.
Pepe, F., Mayor, M., Queloz, D., et al., 2005, *The ESO Messenger* **120**, 22.
Pollack, J. B., Hubickyj, O., Bodenheimer, P., et al., 1996, *Icarus* **124**, 62.
Pont, F., Bouchy, F., Queloz, D., et al., 2004, *A&A* **426**, L15.
Pont, F., Bouchy, F., Melo, C., et al., 2005, *A&A* **438**, 1123.
Queloz, D., Eggenberger, A., Mayor, M., et al., 2000, *A&A* **359**, L13.
Queloz, D., Henry, G. W., Sivan, J. P., et al., 2001, *A&A* **379**, 279.
Rasio, F. A. & Ford, E., 1996, *Science* **274**, 954.
Rivera, E. J., Lissauer, J., Butler, P., et al., 2005, *ApJ* **634**, 625.
Saar, S. H., Butler, P. & Marcy, G., 1998, *ApJ* **498**, L153.
Safronov, V. S., 1969, *Evoliutsiia Doplanetnogo Oblaka*, Moscow: Izdatel'stvo Nauka.
Santos, N. C., Israelian, G. & Mayor, M., 2000, *A&A* **363**, 228.
Santos, N. C., Mayor, M., Naef, D., et al., 2001a, in *11th Cambridge Workshop on Cool Stars, Stellar Systems and the Sun*, eds. R. J. Garcia Lopez, R. Rebolo & M. Zapaterio Osorio, ASP Conference Series **223**, 1562.
Santos, N. C., Israelian, G. & Mayor, M., 2001b, *A&A* **373**, 1019.
Santos, N. C., Mayor, M., Naef, D., et al., 2002, *A&A* **392**, 215.
Santos, N. C., Israelian, G., Mayor, M., Rebolo, R. & Udry, S., 2003, *A&A* **398**, 363.
Santos, N. C., Bouchy, F., Mayor, M., et al., 2004a, *A&A* **426**, L19.
Santos, N. C., Isrealian, G. & Mayor, M., 2004b, *A&A* **415**, 1153.
Santos, N. C., Israelian, G., Mayor, M., et al., 2005, *A&A* **437**, 1127.
Sato, B., Ando, H., Kambe, E., et al., 2004, *ApJ* **597**, L157.
Sato, B., Fischer, D., Henry, G., et al., 2005, *ApJ* **633**, 465.
Setiawan, J., Rodmann, J., da Silva, L., et al., 2005, *A&A* **437**, L31.
Takeda, G. & Rasio, A. F., 2005, *ApJ* **627**, 1001.
Tremaine, S. & Zakamska, N., 2004, in *The Search for Other Worlds*, AIP Conference, **713**, 243.
Trilling, D. E., Benz, W., Guillot, T., et al., 1998, *ApJ* **500**, 428.
Trilling, D. E., Lunine, J. & Benz, W., 2002, *A&A* **394**, 241.
Udalski, A., Paczynski, B., Zebrun, K., et al., 2002a, *Acta Astron.* **52**, 1.
Udalski, A., Zebrun, K., Szymanski, M., et al., 2002b, *Acta Astron.* **52**, 115.
Udry, S., Mayor, M., Naef, D., et al., 2000, *A&A* **356**, 590.
Udry, S., Mayor, M., Naef, D., et al., 2002, *A&A* **390**, 267.

UDRY, S., MAYOR, M. & SANTOS, N. C., 2003, *A&A* **407**, 369.
UDRY, S., EGGENBERGER, A., MAYOR, M., et al., 2004a, *RMxAC* **21**, 207.
UDRY, S., MAYOR, M., BEUZIT, J.-L., et al., 2004b, *RMxAC* **21**, 215.
UDRY, S., MAYOR, M., BENZ, W., et al., 2006, *A&A* **447**, 361.
VAUCLAIR, S., 2004, *ApJ* **605**, 874.
VOGT, S., BUTLER, P., MARCY, G., et al., 2005, *ApJ* **632**, 638.
WARD, W., 1997, *Icarus* **126**, 261.
WEIDENSCHILLING, S. J. & MARZARI, F., 1996, *Nature* **384**, 619.
WINN, J., NOYES, R. W., HOLMAN, M., et al., 2005, *ApJ* **631**, 1215.
WRIGHT, J., 2004, *AJ* **128**, 1273.
ZAKAMSKA, N. & TREMAINE, S., 2004, *AJ* **128**, 869.
ZUCKER, S. & MAZEH, T., 2002, *ApJ* **568**, L113.

3. Characterizing extrasolar planets

TIMOTHY M. BROWN

Transiting extrasolar planets provide the best current opportunities for characterizing the physical properties of extrasolar planets. In this chapter, I first describe the geometry of planetary transits, and methods for detecting and refining the observations of such transits. I derive the methods by which transit light curves and radial velocity data can be analyzed to yield estimates of the planetary radius, mass and orbital parameters. I also show how visible-light and infrared spectroscopy can be valuable tools for understanding the composition, temperature and dynamics of the atmospheres of transiting planets. Finally, I relate the outcome of a participatory lecture-hall exercise relating to one term in the Drake equation, namely the lifetime of technical civilizations.

3.1. Introduction

Finding extrasolar planets is good; learning something about their intrinsic properties is much better. Planets that are known only from their radial velocity signatures can be studied only in a limited sense: we can put a fairly reliable lower limit on their masses, and we can know the size and shape of their orbits. Transiting planets offer opportunities for more complete characterization: we can measure their radii with some precision, and in principle we can learn something of their temperature structure and of their atmospheres. For this reason, this review deals almost entirely with transiting planets. The plan of the paper is as follows: Section 3.2 introduces the basic geometrical and astrophysical ideas relating to transiting planets, and establishes the relationships among them. Section 3.3 describes the various approaches to detecting transiting planets, and outlines the advantages, disadvantages and biases of each. After a brief summary of ways of discriminating against astrophysical false alarms, it discusses the ways in which precise photometry can be used to estimate the radii and orbital properties of transiting planets. Section 3.4 returns to the problem of false alarms, examining in some detail the expected detection rates for various stellar systems that may look like transiting planets, and investigating how such systems will affect future transit detection efforts, both ground- and space-based. Section 3.5 covers transit spectroscopy as a way of probing the atmospheres of extrasolar planets. It first describes the relevant physical and geometrical ideas, and then applies these notions to observations of HD 209458b. Section 3.6 describes 'Fun Day', a break from quantitative astrophysics in which faculty and students brain-stormed about the nature of technical progress in the twentieth century, to provide some context for projections about the lifetime of technical civilizations.

3.2. Basic ideas concerning transiting planets

Transits by extrasolar planets occur when such a planet, as a result of its orbital motion, passes across the face of its parent star as seen from the viewpoint of a distant observer. What a given observer sees depends to a large degree on his or her (or its) orientation relative to the planet's orbital plane. Indeed, among a large set of observers

Extrasolar Planets, eds. Hans Deeg, Juan Antonio Belmonte and Antonio Aparicio.
Published by Cambridge University Press.
© Cambridge University Press 2007.

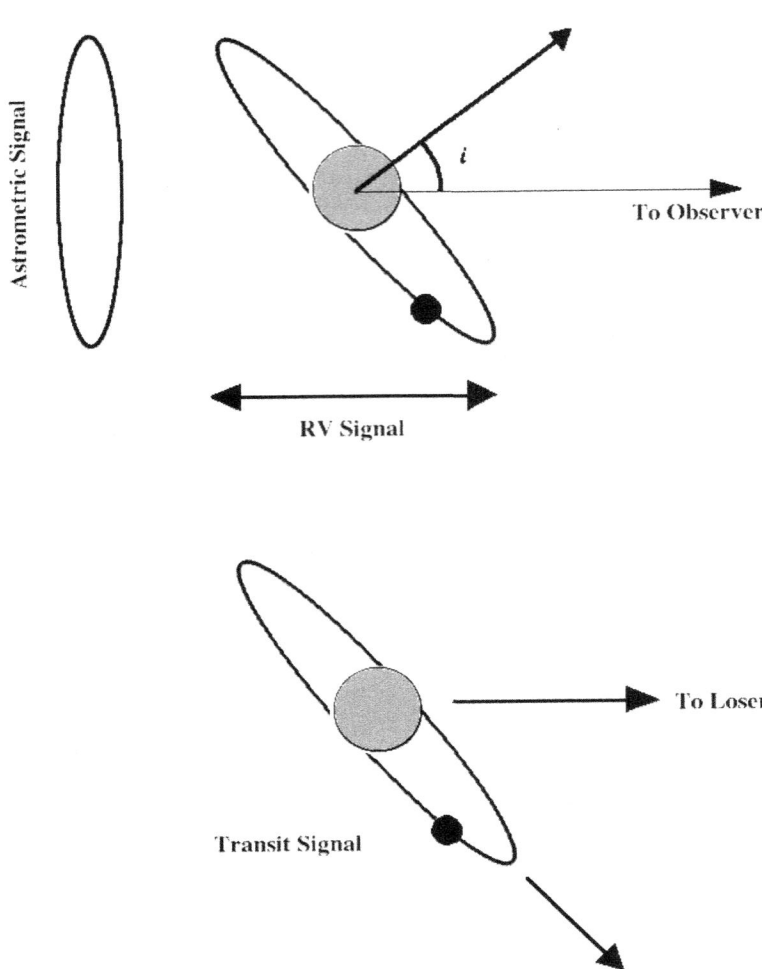

FIGURE 3.1. Top: geometry of astrometric and radial velocity (RV) signals from the reflex velocity of a planet's parent star. Bottom: geometry of planetary transits, with transits seen (winners) in the plane of the orbits, but no transits seen by observers (losers) who are noticeably out of the orbital plane.

randomly positioned in space, for a given planet most are 'losers' who see no transits at all; a small fraction are 'winners' who see transits, and what they see depends delicately on where they are, and on the parameters of the planetary system. This section describes the most important factors contributing to these dependencies.

3.2.1. *Astrometric, radial velocity and transit signals*

The three most commonly discussed indirect methods for detecting extrasolar planets are the astrometric, radial velocity and transit techniques. The astrometric and radial velocity techniques depend upon the reflex motion of the planet's parent star relative to the system's centre of mass. The transit method depends upon the motion of the planet relative to the parent star. These are illustrated in Figure 3.1.

The astrometric signal consists of the ellipse traversed by the star, projected on to the plane of the sky. The worst that can happen in this case is that one can observe the orbit edge-on, in which case the elliptical motion projects into a back-and-forth straight-line motion on the sky. There is no vantage point from which the motion vanishes, so *all* observers are winners if they have the (difficult!) technical wherewithall to detect tiny motions of the parent star.

The radial velocity (RV) signal consists of the projection of the star's reflex velocity along the line of sight connecting the star and the observer. If i is the inclination between the line of sight and the orbital angular momentum vector, then the observed radial velocity signal is proportional to $\sin i$. Thus, the RV signal *can* vanish if $\sin i = 0$, i.e. if the orbit is seen face-on. This is rather unlikely, however: assuming randomly chosen orbital orientations, the probability of having $\sin i$ as small as (say) 0.1 is only about 0.5%.

The transit signal is another story entirely. For transits to be observed, it is necessary for the projected discs of the star and planet to overlap, which (for a circular orbit with semi-major axis a) implies that

$$a \cos i \leq (R_* + R_p), \tag{3.1}$$

where R_* and R_p are the radii of the star and planet, respectively. Since typically we have $a \gg R_* \gg R_p$, this means that most observers see no transit at all. For example, in the case of the Earth orbiting the Sun, the probability that a randomly positioned distant observer *will* see a transit is only about 0.5%. For Jupiter orbiting the Sun, it is only about 0.1%.

3.2.2. *Motivations for transit searches*

So if the odds for seeing transits by distant planets are so poor, why should one bother? There are several answers to this question.

First, one can see a lot of stars; low occurrence probabilities are not necessarily fatal if the number of stars examined can be made large.

Second, as already mentioned, transiting planets are unusually valuable because of the many things that can be learned about them that cannot be known for non-transiting planets.

Finally, knowing (as we do from RV surveys) that there is a population of Jupiter-sized planets in small, short-period orbits makes a huge difference in the prospects for transit searches. For planets such as 51 Peg b, the orbital period is only a few days, and a is typically 0.05 AU. The small semi-major axis means that transits are much more likely to be visible, and the short orbital period means that one need not observe for very long in order to see a transit, if one is going to occur. Thus, all ground-based transit searches target the short-period hot jupiters, and are optimized to detect this kind of object.

3.2.3. *Signal characteristics*

It is useful to have some simple expressions and order-of-magnitude estimates for the most important characteristics of planetary transits. Figure 3.2 shows the geometry of a planetary transit and a schematic of the corresponding light curve. In the simplest approximation, the dip in the transit light curve is described by three parameters: its depth, its duration and its repetition time.

Ignoring details such as limb darkening, the transit depth, $\delta I/I \equiv d$, is simply given by the ratio of the planetary to stellar areas:

$$\frac{\delta I}{I} = \frac{R_p^2}{R_*^2}. \tag{3.2}$$

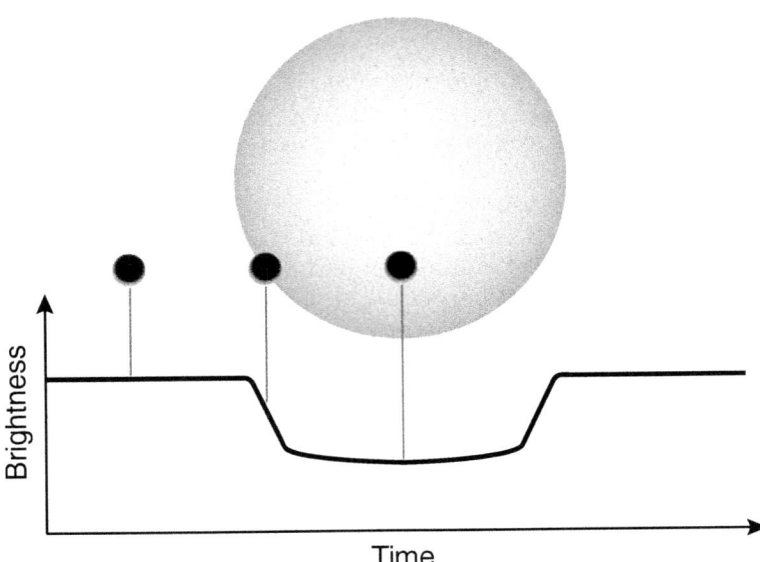

FIGURE 3.2. Planet transiting a star as seen (with immense magnification) by a distant observer. Small black circles show the planet's location against the stellar disc at successive times, with the solid curve below showing the observed light curve.

Numbers for various interesting cases are easy to derive. For instance, Jupiter crossing the Sun, or Neptune crossing an M5-dwarf, both give $d \simeq 0.01$. Neptune crossing the Sun gives $d \simeq 10^{-3}$, and the Earth crossing the Sun gives $d \simeq 10^{-4}$.

The repetition time $\tau_{\rm rep}$ for planetary transits is simply the orbital period:

$$\tau_{\rm rep} = P_{\rm orbit}. \tag{3.3}$$

An implication is that, if one has observed a star for a total time equal to $T_{\rm obs}$, with observation intervals randomly distributed in time, then the probability that one has seen all or part of a transit is roughly $\Pr = T_{\rm obs}/P_{\rm orbit}$. In practical observing situations, observations are usually distributed far from randomly in time, with the result that the detection probability depends upon $P_{\rm orbit}$ in a complicated way. This will be discussed further below.

The maximum duration, $l_{\max}$, of a transit is the time required for the planet's orbital speed to carry it across the width of the stellar disc. A useful expression for this is

$$l_{\max} = 13 \frac{R_*}{R_\odot} P_{\rm orbit}^{1/3} M_*^{-1/3}, \tag{3.4}$$

where $l_{\max}$ is the transit duration in hours, M_* is the stellar mass in units of the solar mass and $P_{\rm orbit}$ is in years. For hot jupiters with periods of about three days, this implies transit durations of about three hours. Equation (3.4) applies for central transits. Non-central ones are shorter by a factor $\sqrt{1 - h^2/R_*^2}$, where h is the impact parameter, equal to the smallest distance between the stellar and planetary centres, as projected on the plane of the sky. About 87% of all transits have durations greater than half of the maximum possible, and the average duration is $(\pi/4)l_{\max}$.

As indicated above, the fraction of stars showing transits (even if all have planets) will be rather small. How many stars must one examine in order to have a good chance of detecting a transiting planet?

From Eq. (3.1), one may infer that the probability of seeing transits from a planet orbiting at distance a from a star of radius R_* is $\text{Pr}_{\text{observable}} = R_*/a$. As a crude approximation, let us suppose that the transiting planets found will be predominantly the hot jupiters, with $a \leq 0.1$ AU. Assuming parent stars similar to the Sun (with $R_* \simeq 0.005$ AU, and with a uniformly distributed between 0 and 0.1 AU, one finds that $\text{Pr}_{\text{observable}} \simeq 0.1$. From RV observations (Marcy et al. 2005), we may guess that the fraction of single Sun-like stars hosting hot jupiters is about 0.8%. In wide-field searches for transiting planets, one ordinarily points the telescope at some place in the sky that is thought to be promising and monitors whatever stars are there to be found. Among field stars, one might expect that 50% will be giants, and among the dwarfs, 50% will be close double or multiple stars, which probably cannot harbour planets in small, stable orbits (Duquennoy and Mayor 1991). Thus, the probability that a given star will show planetary transits is the product of the above factors, or $0.008 \times 0.1 \times 0.5 \times 0.5 = 2 \times 10^{-4}$. This means that one must observe on the order of 5000 stars for every transiting planet found. (And experience suggests that even this number is too small by factors of a few, for reasons that will be discussed below.) In any case, it is clear that a transit detection experiment must be sized to deal with tens of thousands of stars, all with photometric precision better than about 1%.

Among randomly chosen field stars, transiting Jupiter-sized planets are therefore fairly rare. How common are objects that are not transiting planets, but that produce similar photometric signals? The most important source of such astrophysical false alarms turns out to be eclipsing binary star systems. About 2% of field stars are eclipsing binaries, and about half of these show no evidence of secondary eclipses. (An interesting subset of these are systems in which the components have nearly equal effective temperatures, so that the primary and secondary eclipses are indistinguishable.) The typical central eclipse depth is roughly 40%, which of course depends upon the relative radii of the stellar components and (more importantly, for the current purpose) on the orbital inclination. Roughly 10% of eclipsing systems will therefore show fairly extreme grazing eclipses in which 10% or less of the maximum possible eclipse depth is attained. Such eclipses will have observed depths of 4% or less and may reasonably be confused with planetary transits. This means that roughly 10^{-3} of observed field stars are likely to be shallow eclipsing binaries, outnumbering the transiting planets by a factor of ten or more. Some systematic method for testing and identifying targets as eclipsing stars (instead of planets) is therefore essential for a successful transit-search program.

3.2.4. Transit search system choices and requirements

The estimates in the previous section suggest that finding transits by extrasolar planets requires:

(1) The sample of stars observed (i.e. stars for which all of the following requirements are met) should include at least a few times 10^4 objects.
(2) Each star in the sample should be measured with a precision (repeatability, not necessarily absolute accuracy) of 0.5% or better, i.e. 5 mmag.
(3) Photometric samples should be obtained often enough that a dozen or so samples occur during a transit, so that something can be said about its shape. This implies at least one sample every 15 minutes.
(4) The total time on the sky for one sample of stars should be at least two full orbital periods, for the systems judged most likely to be seen.

This implies 200 hours or more of observing time, ideally all within one season. Experience with transit searches suggests that these are bare minimum requirements, and it is advisable to exceed *all* of these by a factor of two or more, if at all possible.

3.2.5. Strategic choices in transit-detection systems

In designing a system to search for transiting extrasolar planets, one is faced at the outset with several large-scale decisions. Unlike the engineering-level design choices that come later, the answers that one chooses to these strategic questions depend mostly upon one's goals; different choices make sense for different kinds of searches. The principal issues that arise are these: (1) Should one first search for stars that display photometric transits, followed by spectroscopic follow-up as necessary, or vice versa? (2) Should photometric searches be conducted with small-aperture, wide-field telescopes, or with larger telescopes with more restricted fields of view? (3) Should one search for transits by distant planets that are the size of (say) Jupiter, or Neptune, or the Earth?

If detecting transiting planets is the goal, then one may imagine either searching for stars showing photometric transit signatures and rejecting the large proportion of resulting false alarms via observations with larger telescopes, or searching for stars that show a radial velocity signature indicating a planet, and then performing time-series photometry to see if transits occur. The choice between these strategies rests largely on the availability of time on intermediate- and large-sized telescopes.

Suppose that one searches for transits first with a tiny (10 cm aperture) wide-field telescope. In this case, the estimates in the previous section suggest that 1000 hours of observing time (give or take a factor of a few) will be required for each transiting planet found. Along the way, one can expect to find a hundred or so false alarms, requiring on the order of 100 hours of 1 m telescope time to identify them. About ten of these objects will each require an hour of 10 m telescope time, to make the final separation between planets and binary stars.

If, however, one looks first for radial-velocity planets, a plausible approach is to observe a large number of target stars, each at only a few epochs. In this way, one may locate stars with planets in small orbits, which are relatively likely to have orbital orientations that allow transits to occur. With such directed observations, it may take about 20 hours of 10 m observing time to locate each planet, and perhaps 400 hours for each transiting planet. To separate the transiting planets from those that do not, each of the stars with planets must be observed photometrically, with smaller telescopes, requiring about 20 hours of 1 m telescope time. Observations by tiny telescopes (which are cheap in any case) are not required.

The telescope-aperture/field-of-view choice relates both to the brightness of the desired target stars and to their stellar types. Small-aperture systems target transits among fairly bright stars (magnitude 10–12), so to observe a large number of stars they must have large fields of view. Searches such as OGLE (Udalski et al. 2002a,b, 2003) using larger (1 m) telescopes typically target stars that are several magnitudes fainter (14–17) than for small-aperture searches. Because of the rapidly increasing star density at fainter magnitudes (especially in some directions, such as towards the galactic bulge), an adequate number of target stars can be followed even with the relatively small fields implied by the larger telescopes. At the large-aperture end of this tradeoff, the EXPLORE experiment (Mallén-Ornelas et al. 2003; Yee et al. 2003) used 4 m-class telescopes with large mosaic detectors. If the target field were to lie fairly far from the galactic plane (which the EXPLORE fields do not), the result would be that the majority of the 16–18 magnitude stars that provide good photometry would be nearby red dwarfs. In this faint magnitude range, stars that are significantly brighter than M-dwarfs would have to lie well outside the disc of the galaxy; contamination by giants would be fairly small. This ability to select the stellar population being observed is one of the most interesting aspects of large-aperture surveys.

Finally, all of the above implicitly assumes that the target objects are Jupiter-sized planets. Are searches for smaller planets sensible? A jupiter transiting a Sun-sized star makes a photometric signal of about 1%. A neptune transiting a medium-sized M-star would make a transit of similar depth. Such transits would of course be detectable; the difficulty with designing an experiment to locate them is simply getting enough M-dwarfs in the field of view to make a detection likely. Similarly, an earth transiting a bottom-of-the-main-sequence M-star would be detectable; the problem is to monitor a large number of such intrinsically faint stars. An earth transiting a Sun-like star would produce a signal of only 0.01%. Such small transits are not feasible to measure from the ground, but they fall within the planned capabilities of space missions such as *COROT* (Baglin *et al.* 2000; Baglin 2003) and *Kepler* (Borucki *et al.* 2003).

3.3. Detection and measurement techniques

Assuming that one has settled on a strategy to use in searching for transiting planets, how does one go about meeting the requirements spelled out in the previous section?

3.3.1. *Noise sources*

The first problem is obtaining photometric time series with adequate quality. Several noise sources require consideration.

The irreducible minimum noise is set by photon-counting statistics applied to the signal from the target star. For a broad-band (150 nm bandpass) observation, a rough rule of thumb says that the rms photon noise in magnitudes is about $0.06 t^{-1/2} D^{-1}$ for a star of magnitude 10, where t is the observing time in seconds and D is the telescope aperture in cm. For example, for a 10 cm telescope and a 100 s exposure, photon noise on a 12th magnitude star is about 1.5 mmag. This value is small compared to the 10 mmag or so expected from the transit of a Jupiter-sized planet across a Sun-like star. Thus, deep searches could be conducted with very small telescopes if target photon noise were the only important noise source. Alas, it is not.

In many cases, counting statistics on photon noise from the sky background becomes important. The magnitude of this noise source depends on many factors, including wavelength, site conditions, lunar phase and telescope parameters. For transit searches, it is usually desirable to put as many target stars as possible on to a detector with some predetermined number of pixels. This in turn implies that each pixel should ideally cover a sizeable piece of sky, typically a modest fraction of the typical distance between target stars. Since the total light from the 'dark' sky comes more from the airglow than it does from all stars fainter than sixth magnitude (Cox 2000), this means that in untargeted surveys, the sky contribution will always be important for all but the brightest stars. For example, in a moonless sky from a good site, sky noise is comparable to target photon noise for a telescope with $D = 10$ cm at f/3, and $R \simeq 11.5$.

Noise from atmospheric scintillation is independent of stellar brightness, so it sets a limit on the attainable S/N for bright objects. At a typical site at 2000 m altitude, a rough approximation for its amplitude in magnitudes is $0.07 D^{-2/3} t^{-1/2} X^2$, where X is the airmass. (Strictly speaking, the exponent applied to X depends on whether one's line of sight is orientated along or across the wind vector at the altitude where the scintillation occurs (Dravins *et al.* 1998; Young 1969, 1974). But for order-of-magnitude estimates, the above formula is sufficient.) For most transit surveys, scintillation noise is dominated by the various kinds of photon-counting noise.

Uncorrected atmospheric extinction is highly variable from site to site and from night to night; depending on circumstances, it can be negligible or dominant. It can usually be substantially reduced by referencing the magnitude of a target star to the average of an ensemble of nearby stars. As the telescope's field of view grows, however, this procedure becomes less effective.

Finally, the effects of image motion acting on uncorrected gain variations is usually negligible, but near bad CCD columns or other detector artefacts it can become important. Crowded fields also add noise to photometric reductions, with the amount of excess noise depending on the details of the photometric algorithm used.

3.3.2. *Aperture vs stellar magnitude*

The aperture of a transit-detection system is an important system parameter, because of its influence on the number and stellar type of the stars to which the system will be most sensitive. Ignoring interstellar extinction, the distance d to which a star's brightness can be measured at constant S/N scales with the telescope aperture D. The solid angle Ω on the sky that the system can observe at one time can be written in terms of the telescope aperture, *f*-number f, and detector edge length w as approximately $\Omega = w^2/(D^2 f^2)$. The volume of space sampled is therefore $V = (d^3 w^2)/(D^2 f^2) = (Dw^2)/f^2$. If space were uniformly filled with stars, the number of observable targets would be proportional to V. Thus, ignoring other considerations, the best transit-finding system has a huge detector, a tremendous aperture, and a tiny *f*-number, with detector size and *f*-number playing the biggest roles.

Of course, the finite size of the galaxy also is important. For a system with $D = 10$ cm, most targets will be roughly magnitude 12, which for Sun-like stars implies a distance modulus of about 7.5 magnitudes, and a distance of about 300 pc. This is roughly the thickness of the thin disc, so that one will see many fewer targets than might be expected if the line of sight lies at high galactic latitude. For sight lines that lie close to the galactic plane, the number of useful target stars will be larger, but the sample will be increasingly contaminated with distant giants as the galactic plane is approached.

Large-telescope (1 m and above) surveys look at stars that lie much farther away (as in the case of the OGLE survey (Udalski *et al.* 2002a), which views stars about half way to the galactic bulge), or else at stars that are intrinsically very faint. Examining stars towards the bulge has the advantages of very dense star fields, which allow the study of vast numbers of stars if the problems associated with crowding can be handled.

3.3.3. *Observing time and duty cycle*

For planets that lie less than 0.1 AU from their stars, the orbital periods are on the order of 100 hours, and the duration of a transit is typically a few hours. If one assumes that observations are in short blocks that are randomly spaced over an interval spanning many orbits, then the probability Pr that one of these blocks contains at least part of a transit is approximately $\Pr \simeq 1 - \exp(-T_{\text{obs}}/P_{\text{orbit}})$, where T_{obs} is the summed duration of the observed time blocks. Therefore, one expects that in order to have a fairly good chance of seeing N transits, one must stare at the sky for something like N orbital periods, or (to see two or three transits) typically several hundred hours.

In practice the situation is a bit more complicated, because the diurnal cycle limits one's ability to sample all orbital phases, especially for periods that are near-integral multiples of a solar day. This is illustrated in Figure 3.3, which shows, as a function of orbital period, the probability of seeing at least parts of two transits using a data set actually acquired by the TrES network in the fall of 2003. In its entirety, this data set included 765 hours of observations.

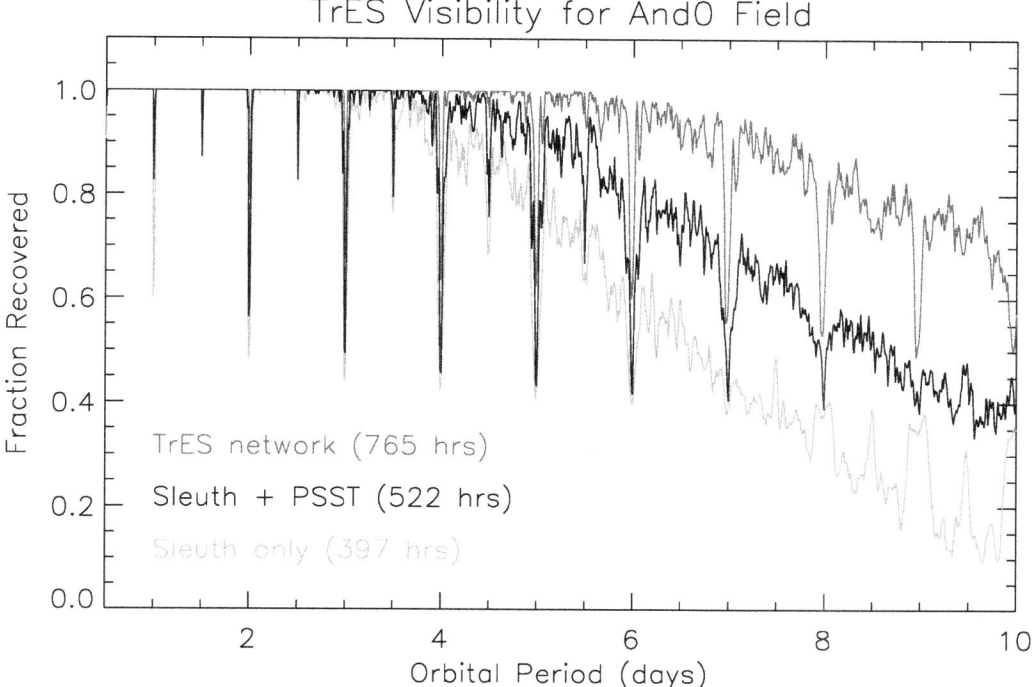

FIGURE 3.3. The fraction of planets with all possible orbital phases that could in principle be recovered from a data set consisting of various combinations of the sites in the TrES network, shown as functions of the planet's orbital period in days. Figure courtesy of Francis T. O'Donovan and D. Charbonneau.

The figure illustrates several important points. First, even for periods as short as four days, the extensive data set from Sleuth alone (almost 400 hours of observations) begins to show significant incompleteness. The probability of detecting parts of three transits would, of course, be even worse. Second, for periods longer than eight days, the likelihood of seeing even two transits becomes small for a single site, and begins to drop significantly even for the full network. Last, for periods that are nearly an integral number of days, the probability of seeing any number of transits from a single site tends to converge to roughly 1/3, independent of the period. This last behaviour can be understood as follows. If the period is exactly an integral number of solar days, then transits occur at the same solar time every day. If this time happens to be during darkness at one's site, then transits are seen every night, and many are observed. If it happens during daylight, then no transits are ever observed.

Because of the progression of the seasons, a given field of view in the sky is visible for long intervals each night only for typically two months or so each year. Thus, in order to accumulate the requisite hours of observing time in the available seasonal window, it is highly desirable to observe from more than one site, ideally with sites widely distributed in longitude. The best arrangement, of course, would be to have a world-wide network with enough stations (one needs about six) so that bad weather would seldom take down all sites in a given sector of the globe. In that case, one could obtain nearly continuous observations, and see (with a high probability) as many as $S_{\rm tot}/P_{\rm orbit}$ transits, where $S_{\rm tot}$ is the total span of the data set. The TrES network is one step towards this ideal.

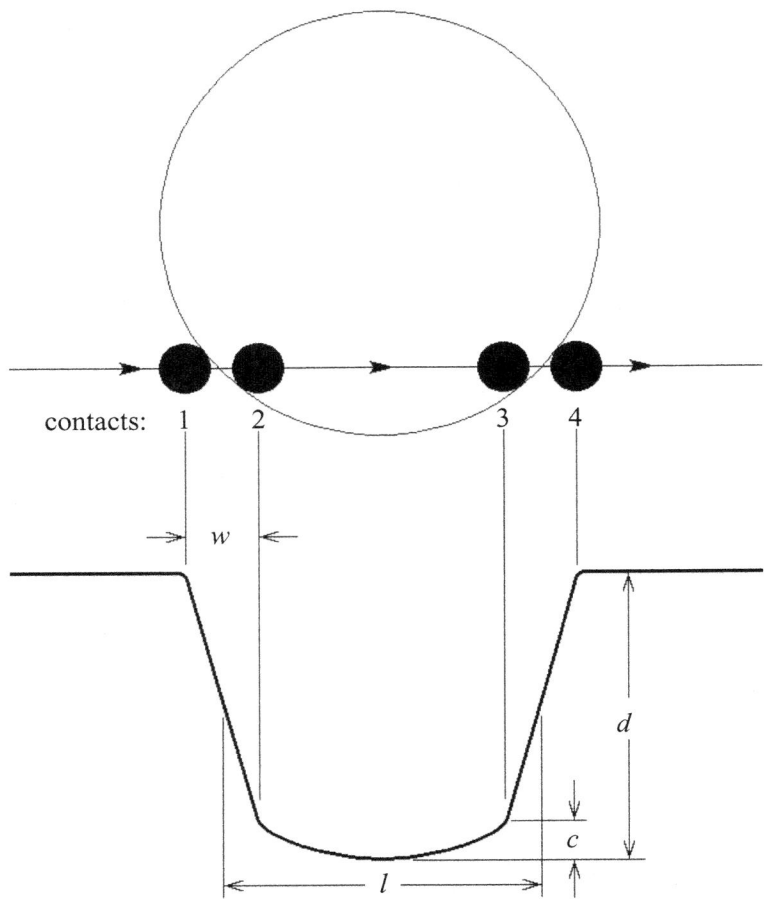

FIGURE 3.4. Observable properties of a transit light curve: (1) total duration l, (2) transit depth d, (3) ingress/egress time w, (4) central curvature c.

3.3.4. *Measuring gross characteristics of transiting planets*

Once a transiting planet has been identified, what can be learned from its light curve? As illustrated in Figure 3.4, there are four conceptually distinct parameters that can be measured in a transit light curve; two of these are relatively easy to measure, while the other two are harder, because they correspond to more subtle features in the curve. These features are (1) the transit duration l, (2) the transit depth d, (3) the ingress/egress duration w, and (4) the central curvature c.

The various observable parameters can be computed if one knows four physical parameters of the star/planet system, namely (1) the stellar radius R_s, (2) the planetary radius R_p, (3) the orbital inclination i, and (4) the stellar limb darkening coefficient b. Unfortunately, the dependence of the observable quantities on the physical ones is not simple. Thus, l depends upon the planet's orbital speed, which can be inferred from the orbital period and the stellar mass, but also on R_s, R_p and i. The depth d depends mostly on R_p^2/R_s^2, but also slightly on the inclination i and the limb darkening b. The ingress/egress time w depends upon the orbital speed, and on R_p and i. Finally, the curvature c depends upon the limb darkening b and slightly on the inclination i.

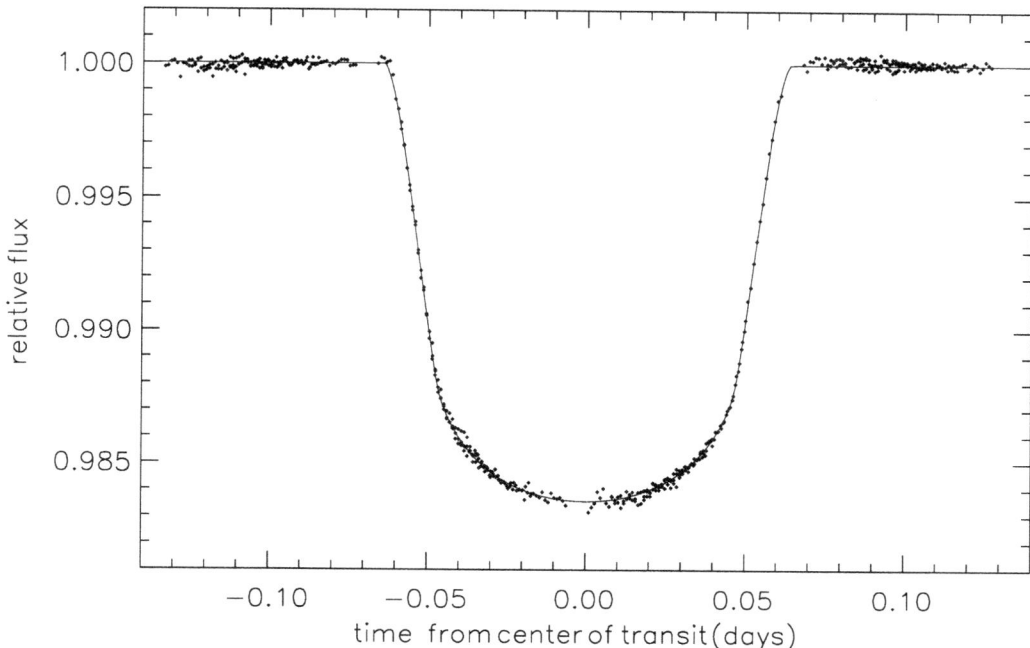

FIGURE 3.5. Transit light curve of HD 209458b obtained using the STIS spectrometer on *HST*, averaged over wavelengths between 582 nm and 638 nm. The data shown represent the phased superposition of observations of three different transits. (Brown *et al.* 2001)

When presented with photometry showing transits with only modest S/N (as, for instance, typical discovery light curves), it can be difficult to estimate more observable parameters than l and d. In that case, knowing the orbital period and having a reasonable guess about the stellar mass, one is able to make pretty good estimates of R_p/R_s and of the inclination i, and to place some limits on the mean density of the parent star. Formulas for deriving these quantities can be found in Seager & Mallén-Ornelas (2003).

If, however, one has photometry with much lower noise (such as the *HST* light curve displayed in Figure 3.5), then it is productive to fit for all of the physical parameters R_p, R_s, i, b independently. Combined with RV data giving the ratio of planetary to stellar masses, meaningful estimates of the planet's density (and hence composition) can then be made. With data of *HST* quality, one typically finds that the largest uncertainty in estimating the physical system parameters arises from uncertainty in the orbital speed, which in turn depends upon the accuracy of one's guess concerning the stellar mass.

3.4. False alarms

Astrophysical false alarms are the bane of photometric transiting planet searches. By definition, these arise from objects (usually binary or multiple stars) that do not contain transiting planets, but that produce photometric variations that closely resemble planetary transits. The problem with false alarms is that they are very numerous compared to transiting planets, at least among the stellar populations usually searched by transit methods. To get some idea of the problem, consider the statistics for the field, six degrees square, containing the transiting planet TrES-1 (Alonso *et al.* 2004; Alonso 2005).

For this field, the TrES network generated light curves for 12 000 stars. Among these, 16 were deemed to be 'transit candidates'. That is, they had dips in their light curves with depths and durations consistent with transiting planets and they showed none of the obvious hallmarks of stellar-mass binary components (i.e. secondary eclipses or obvious out-of-transit light variations with the transit period, or small harmonics thereof). Of the sixteen candidates, six proved to be grazing eclipsing binaries (EBs), seven were EBs that were blended, either with the images of brighter nearby foreground stars, or as hierarchical triples in which the least luminous of the three stars generated the eclipses. There remained two candidates that could not be characterized in detail, and one star with a transiting planet. And recall that this field was anomalous, in that it actually did contain a transiting planet. Thus, depending on how fussy one is in declaring stars to be 'candidates', one finds that false alarms outnumber real transits by a factor of something like 20 to 100. How can they be identified and discarded?

Most schemes for identifying false alarms rely on a multi-stage approach, aiming to minimize the total observational expense of the identification. The procedure used to winnow TrES candidates involves five steps: (1) careful examination of the light curve folded with the orbital period, (2) low-precision radial velocity measurements, (3) multi-colour, high spatial resolution photometry, (4) analysis of line asymmetry variations in high-resolution spectra, and (5) high-precision radial velocity measurements.

3.4.1. *Eclipse shape and out-of-transit variations*

Most planetary transits produce eclipse light curves that are fairly flat in their central portions, whereas grazing eclipses of two stars (i.e. objects with similar radii) make eclipse shapes that are nearly triangular. Also, planets (even in close orbits) are not massive enough to produce substantial tidal distortion or gravity darkening in their host stars; in short-period binary stars, these effects are common. Finally, planets are too faint in visible light to produce measurable secondary eclipses, but binary star systems often do so. These differences in light curve morphology form the basis for cheap and surprisingly effective methods for identifying false alarms.

The simplest morphological test one can do is the search for a secondary eclipse: one simply folds the observed light curve with the photometric period and looks for evidence of a dip located about half a period after the primary eclipse. If evidence of a dip is found, then the object is almost certainly a binary star and not a star/planet pair. One caveat is that pretty often one sees binary stars that are closely matched in surface brightness. In this case, the primary and secondary eclipses may be indistinguishable, so that one guesses a period that is too short by a factor of exactly two, and no secondary eclipse is seen.

An effective test based on transit shapes was devised by Seager and Mallén-Ornelas (2003), who presented an analytic relation connecting the primary star's mean density with a planet's observable transit shape parameters (depth, total duration and duration of the flat-bottomed part of the transit). The stellar density thus derived can be compared with that of a main-sequence star having the same effective temperature as the target star; if the comparison is poor, this is evidence that the transits arise from a stellar companion, rather than a planetary one (Alonso 2005).

A variety of physical effects (tidal distortion, gravity darkening and radiative heating by a companion) cause binary stars to show light variations at times other than during eclipses. The brightness fluctuations caused by these mechanisms tend to be smooth and sinusoidal, with periods that are equal to or integer submultiples of the orbital period (Drake 2003; Sirko and Paczyski 2003; Tingley 2004). Given the long photometric time series that are typically used to identify transit candidates, it is usually possible to measure amplitudes of such sinusoidal variations with precision of a mmag or better. This allows identification of stellar binaries in a large fraction of the cases, at least for short-period orbits. In fact, Alonso (2005) found that among the sixteen transit candidates found in the TrES network's Lyr0 field, only one of them passed both the stellar density and the out-of-transit variation tests, and this was the true planet-bearing star TrES-1.

3.4.2. Low-precision radial velocity

The most obvious distinguishing characteristic of grazing eclipsing binaries (the most numerous source of false alarms) is the large reflex orbital velocity the faint companion imposes on the brighter star. For example, a very low-mass (100 $M_{\rm Jup}$) M-star in a 4-day orbit around a 1.4-$M_\odot$ F-star produces a reflex velocity of 10 km/s. Thus, measurements with fairly coarse (1 km/s) radial velocity precision suffice to identify stellar-mass companions even with periods of months, well down into the brown dwarf domain.

A few systems for measuring radial velocities are optimized for obtaining low-precision velocities cheaply and quickly; the CfA 'digital speedometers' are good examples (Latham 1992). These systems are fed by 1.5 m telescopes and use 1970s imaging technology, but their efficient data pipelines and operating procedures make them extremely cost-effective sources of the necessary radial velocity data.

3.4.3. Multi-colour and high spatial resolution photometry

A substantial number of false alarms arise from groups of three or more stars, two of which form an eclipsing binary. The remaining component may be physically linked to the other two or not; in either case the problem is that light from the third star dilutes the eclipses from the other two, possibly resulting in eclipses that have the same duration, depth and shape as planetary transits.

If the third star is not gravitationally bound to the other two, then most often one finds that it is separated from them by a modest distance (several arcsec) in the sky. Because wide-field transit surveys necessarily have poor spatial resolution (about 20 arcsec for TrES), even not particularly close stellar pairings appear as single stars. Groupings that appear to have planet-like transits usually consist of a faint, large-amplitude eclipsing binary lying near a much brighter foreground star. Such groups can easily be identified using CCD photometry taken with higher spatial resolution, in which the putative bright target star and the fainter eclipsing one can be seen as separate, and the true depth of the eclipses can be distinguished. Digitized sky survey images (e.g. at archive.eso.org/dss/dss and archive.stsci.edu/dss/) are extremely useful in helping to determine whether a given target star has faint, moderately close companions.

For gravitationally bound triples, the angular separation between the components is usually too small to be resolved without great effort. (A typical physical separation of 10 AU at 300 pc distance translates into an angular separation of 0.033 arcsec.) In this case, one may still learn something from photometry, because the colour of the bright star in the system may be different from that of the star that is being eclipsed. This implies that the eclipse depth must be different when observed in different colours, unlike the light curve of a planetary transit. The depth difference is proportional to the colour difference between the two brightest stars in the triple system; if all are main-sequence objects and the brightness ratio between them is two or more stellar magnitudes, then this colour difference will be large enough for the depth effect to be measured fairly readily. It is fairly common, however, for the brightest star in the triple to be a slightly evolved F-star. If so, then (because of the shape of the evolutionary tracks on the colour–magnitude diagram) it can happen that the eclipsed star is two or more magnitudes fainter than the brightest star, yet has almost exactly the same colour (Mandushev *et al.*

2005). In this case, the transit depth is independent of colour, and one must resort to other means to identify the system as a triple.

Until recently, most observations of the sort described in this section have been obtained using general-purpose telescopes of 1 m class, on an as-needed basis. However, as the demand for this kind of follow-up photometry has increased, several workers have found it useful to assemble special-purpose systems with smaller aperture (30 cm), so that rapid follow-up observations of promising candidates can be obtained (Kotredes et al. 2004).

3.4.4. Spectrum line asymmetries

Mandushev et al. (2005) describe a sobering cautionary example of a hierarchical triple in which both the photometric light curve and the radial velocity signal of an eclipsing binary were compromised by the presence of the third, bright member of the triple system. In this case, low-precision radial velocity measurements showed a Keplerian signal with the photometric period, and with an amplitude of 3 km/s, indicating a 30-$M_{\rm Jup}$ brown-dwarf companion. The multi-colour light curve for this system was completely consistent with this brown-dwarf interpretation.

On further examination, however, it was remarked that the brightest (non-eclipsing) star in the system rotates at 30 km/s, and the tidally-locked rotation speed for the eclipsed star would probably be about the same. If the unseen eclipsing component were a low-mass M-star, then the reflex velocity of the eclipsed component would be about the same as these rotational speeds. Since the eclipsed star is about ten times fainter than the brightest component (and has a very similar spectrum), its Doppler-shifted line profiles would simply move around within the rotationally-broadened profiles of the brightest component, causing an apparent Doppler shift of only 3 km/s (see Figure 3.6). In addition to the Doppler shift, however, one would also see small changes in the symmetry of the spectrum lines as the contribution from the eclipsed component moved from one side of the broadened profile to the other.

Once this possibility was identified, earlier spectroscopic data on the star were searched for time-varying asymmetric line profiles, changing in phase with the orbital period. Such asymmetry changes do indeed occur, allowing the system to be correctly identified as a triple (two F-stars, one a subgiant and the other on the main sequence), with an M-star companion eclipsing the fainter F-star. The moral is that many kinds of star systems occur in the galaxy, and transit searches are remarkably adept at turning up odd ones. Great caution in performing follow-up observations on alleged planets is therefore a good idea.

3.4.5. High-precision radial velocity

Measuring radial velocities with precision of tens of m/s or better (nowadays approaching 1 m/s in the best cases) requires specially designed spectrographs and, for most stars of interest, large telescopes. Obtaining such data is therefore relatively difficult and expensive, so that most teams engaged in transit searches refrain from doing so except for stars that have passed all of the other tests intended to reject false alarms. There is, however, no substitute for measuring the reflex velocity of a supposed planet's parent star to show that the transiting object has planetary and not stellar mass. For geometrical reasons explained in Section 3.2, transit searches tend to find close-in planets. Thus, the expected radial velocities typically are relatively large. This means that with telescopes of 10 m class, it is possible to measure useful radial velocities even for stars as faint as $R = 17$ and so confirm the planetary status of the OGLE transiting objects (which circle distant, roughly Sun-like stars located in the direction of the galactic bulge) (Konacki

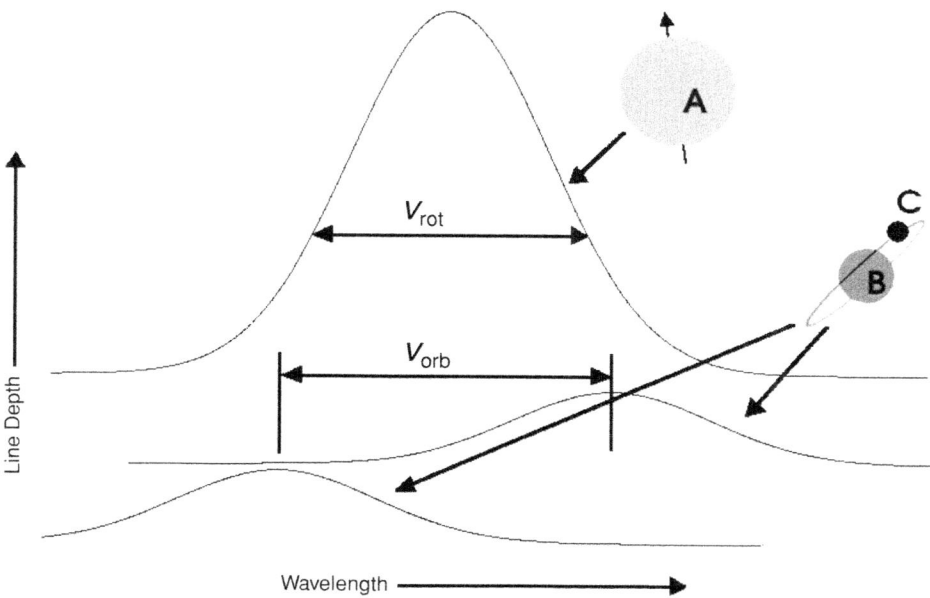

FIGURE 3.6. The line profiles (shown as positive-going Gaussians) resulting from a triple star system in which the brightest component (A) is rapidly rotating, and the eclipsed star (B) has an orbital motion that is comparable in magnitude to the rotation speed of component A. Displacement of the weak line profiles from B causes a shift in the centroid of the line profile from A, but also changes its symmetry in phase with the orbital motion.

et al. 2003, 2004; Bouchy et al. 2004; Pont et al. 2004). For brighter stars, such as the $10 \leq R \leq 12$ targets that emerge from the wide-field surveys, accuracies of order 10 m/s can be obtained with reasonable observing effort, yielding planetary masses with accuracy that is limited by the accuracy with which one can estimate the mass of the parent star.

3.5. Transit spectroscopy

Transiting planets are more useful than non-transiting ones in an essential way: the optics of a transit permit measurements of some physical properties of the planet in ways that are not possible for non-transiting planets. Indeed, it seems that in principle the kinds of information that can be obtained from transits are limited primarily by the investigator's imagination (in practice, they are limited by the available signal-to-noise ratio). One form of observation that has already proved its worth is that of transit spectroscopy, in which a wavelength-dependent absorption signature is imposed by the outer parts of the planetary atmosphere on the starlight that passes through on its way to the Earth.

3.5.1. Fundamentals

The basic idea is illustrated in Figure 3.7. A ray of light connecting the star and the Earth, passing near the planet's limb, may be wholly, partly, or negligibly absorbed, depending upon its wavelength and upon the ray's minimum height in the planetary atmosphere. At wavelengths where the atmospheric gas is relatively transparent, the ray may travel fairly deep without being absorbed. At wavelengths where the gas is more opaque (near strong

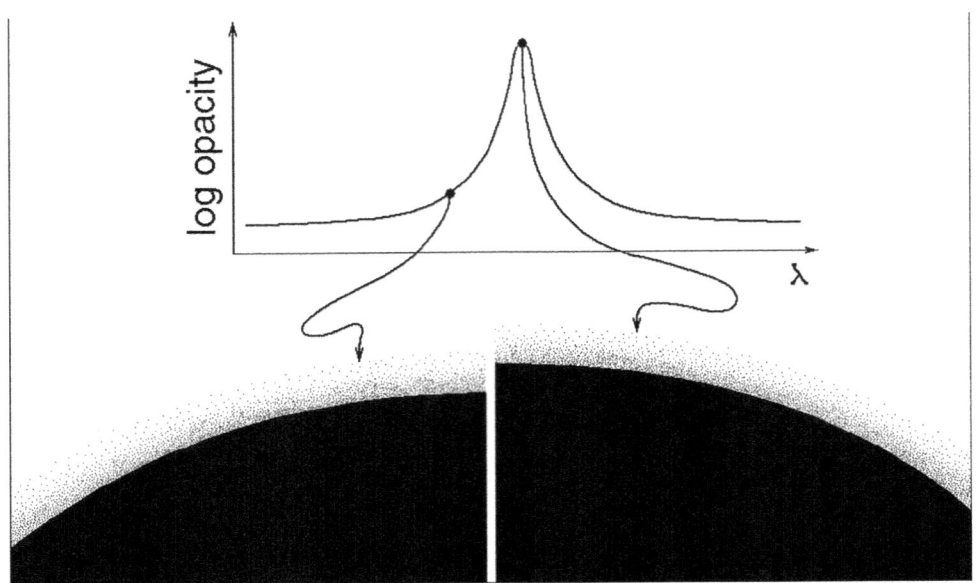

FIGURE 3.7. The apparent radius of a planet depends upon its atmospheric opacity at the wavelength chosen for observation. At wavelengths where the opacity is large, so is the radius. At wavelengths with lower opacity, rays can pass through the atmosphere at heights where the density is higher, making the apparent radius smaller (Brown, 2001).

molecular transitions, for instance), the ray is completely absorbed unless its maximum depth is much smaller, where the gas density is lower. The effect is that the observer measures a larger planetary radius at wavelengths where the atmospheric opacity is high. This causes transits to have greater depth at wavelengths with high opacity; one may then infer the opacity's wavelength variation by measuring the corresponding change in transit depth.

Two special cases of the above ideas are both useful and simple to treat. In the case of clouds made of large (compared to light wavelengths) particles, one may suppose that the atmosphere is opaque to tangential rays all the way to the top of the cloud deck. In this case the planet has a minimum possible radius, namely that of the cloud tops, and no opacity effects that rely on rays passing below this minimum radius can be observed. Another simple case is that in which some chemical species has a wavelength-dependent opacity, and it is well mixed into the planetary atmosphere with a relative abundance that changes negligibly with height. Then suppose that in some wavelength range, the minimum opacity occurs at $\lambda_{\min}$ and is equal to $\kappa_{\min}$, and that the apparent radius of the planet seen at $\lambda_{\min}$ equals R_0. Then the radius R_1 measured at a wavelength where the species produces a higher opacity, say κ_1, is given to a good approximation by $R_1 = R_0 + H \ln(\kappa_1/\kappa_{\min})$, where H is the density scale height of the atmosphere. This simple relation follows because, in order to compensate for, say, a factor e increase in the opacity due to a wavelength change, one must decrease the gas density by the same factor, and this is done by increasing the radius by one scale height.

3.5.2. *Expected spectra*

One can use the results from the last section to estimate the magnitude of radius change (or, equivalently, depth of transit) that one might expect from hot jupiters and from distant terrestrial planets. In both cases, one may guess that the maximum opacity ratio between opaque and transparent wavelengths might be a factor of 10^4. This implies radius changes of $\ln 10^4 \simeq 10$ scale heights.

Recall that the density scale height is given by $H = kT/mg$, where k is Boltzmann's constant, T is the temperature, m is the mean molecular mass and g is the gravitational acceleration. For the Earth (with $T = 300$ K, $g \simeq 10^3$ cm s^{-2}, and atmospheric mean molecular weight about 28 AMU, one finds $H \simeq 8$ km. This is about 10^{-3} of the Earth's radius, so the increase in cross-sectional area resulting from a 10-scale-height radius change is about 2×10^{-2}. The total depth of a transit by an Earth-sized planet across a Sun-sized star is about 8×10^{-5} of the stellar light, so the variations due to the atmosphere will be at most about 1.6×10^{-6}. A small signal!

For hot jupiters, T is larger by a factor of about 3, the mean molecular weight is smaller by a factor of about 14, and g is similar to that at the Earth's surface. The resulting H is much larger, in some cases exceeding 400 km. Since the radius of these planets is roughly 8×10^4 km, a 10-scale-height radius change translates into a cross-sectional area change of 10%. Since the transit depths are much larger to begin with (typically 0.015 of the stellar intensity), the transit depth changes relative to the total stellar light may in principle exceed 10^{-3}. Thus, the transit spectra of hot-jupiter atmospheres are (or at least may be) possible to observe, using existing ground- and space-based instruments. The results of attempts to do so (some successful, some not) will be described below.

Theoretical transit spectra of hot jupiters have now been computed by several groups; Figure 3.8 shows one such from Brown (2001), displaying not only the general nature of the spectrum, but also the effect on the spectrum of changing cloud height. In these spectra, the continuum opacity at cloud-free heights arises from Rayleigh scattering, from collision-induced absorption by H_2 molecules, and from the broad pressure-broadened wings of resonance lines of the alkali metals. Aside from the lines of Na and K, the strongest features in the spectrum come from common molecules. At the high temperatures and low pressures found in hot-jupiter atmospheres, the dominant constituents are expected to be H_2, He, N_2, CO and H_2O. A small amount of carbon may also be found in CH_4, with more in cooler atmospheres and at higher pressures. These molecules produce millions of lines, some quite strong, organized in band structures that dominate the appearance of the near infrared (NIR) transit spectrum. In the wavelength range between 700 nm and 2 µm, the most prominent bands are those of H_2O, but CO and CH_4 can also be seen. Although the water bands in particular are strong, observing them from the ground is quite difficult, because of interference from water vapour in the terrestrial atmosphere.

3.5.3. Observations of transit spectra

To date, clear-cut transit-spectrum detections on extrasolar planets have been obtained only for the sodium D lines and the Ly α resonance line, both on HD 209458b. Detections of other species are so far of marginal significance, or provide only upper limits on the transition equivalent widths.

The *HST* observations that yielded the light curve (Figure 3.5) also provided moderate-resolution spectroscopy. Charbonneau *et al.* (2002) used these data to measure excess sodium absorption during the transits; while the central depth of the transit in nearby continuum wavelengths was about 1.6%, the depth in a band 1.2 nm wide, centred on the Na D doublet, was deeper by 0.023% ± 0.0057%. This degree of excess absorption is large enough to be statistically significant, but it is only about half that predicted for a fiducial model atmosphere, with cloud tops at 0.03 bar. It therefore seems likely

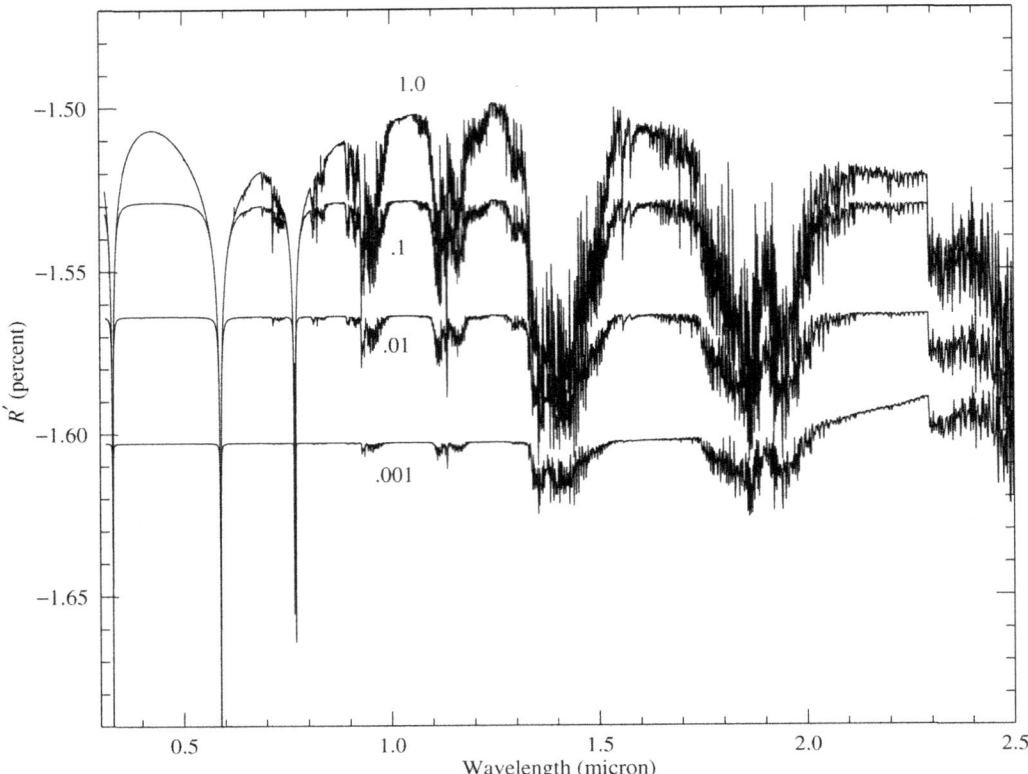

FIGURE 3.8. Theoretical visible and near-infrared transit spectra of HD 209458b, computed assuming a range of cloud heights. Deep bands of H_2O and CO can be seen, as well as the pressure-broadened resonance lines of Na and K. The contrast of all these features depends upon the cloud height, decreasing as the height of the cloud tops increases (Brown, 2001).

either that (1) some additional source of continuous opacity (clouds, say) is cutting off transmission of tangential light rays at heights of 1 mb or so, or that (2) some mechanism (photoionization, or non-LTE effects, small primordial metallicity, or unknown chemistry) has reduced the abundance of Na atoms in the planetary atmosphere by a factor of perhaps 100 relative to the solar abundance.

Vidal-Madjar et al. (2003) used the HST to observe transits of HD 209458b in the ultraviolet, measuring the transit depth in the Ly α resonance line of atomic hydrogen. They found that the depth of the transit seen in this line is larger than that in visible light by a factor of almost 10. This enhancement is so large that it is best explained in terms of a massive hydrogen exosphere larger than the planet's Hill sphere, that must consequently be constantly blown away by the stellar wind and constantly replenished by ablation of the planet. The implied mass loss rate from the planet is at least 10^{10} g/s; at this loss rate, the survival time of the planet would be of order 1000 Hubble times, so the exosphere could be much more massive than the lower limit and still not destroy the planet in a Solar System lifetime.

There have been several ground-based searches for excess absorption in the CO bandhead near 2.3 μm wavelength. The best and most recent of these is that by Deming et al. (2005), who used NIRSPEC at the Keck-II telescope to observe three transits of HD 209458b. They found no convincing evidence for excess absorption by CO, but were able to set an upper limit on the absorption that excludes the 'fiducial' model considered by Charbonneau et al. (2002). These NIR observations also are consistent with a cloud layer reaching up to pressures of a few mbar. An alternative explanation is that the atmosphere of HD 209458b is so cool that a large fraction of its carbon resides in CH_4, but, as we shall see below, this possibility is ruled out by spaceborne observations in the thermal infrared.

3.5.4. Observations of secondary transits

In recent months, two groups of workers have announced exciting observations of secondary transits using the *Spitzer* infrared space telescope. Primary transits occur when a planet passes across the face of its star as seen from the Earth; the secondary transit occurs one-half orbit later, when the planet passes behind its star. At the time of the secondary transit, the light emitted by (or reflected from) the planet is blocked, leading to another dip in the brightness of the whole system. Since the surface brightness of the planet is smaller than that of the star, the secondary transit is less deep than the primary. For hot jupiters observed in the thermal infrared, however, this difference is relatively small, only a factor of a few (the ratio of the stellar to planetary effective temperatures).

Charbonneau et al. (2005) observed a transit of the planet TrES-1 at wavelengths near 4 and 8 μm, using the *Spitzer* IRAC imaging photometer. Transits of significant depth were detected at both wavelengths; these may be interpreted directly as flux densities coming from the planet's day side, just before the transit ingress and after the egress. Announced at the same time by Deming et al. (2005) was a detected transit of HD 209458b observed at 24 μm, using *Spitzer*'s MIPS instrument. Taken together, these three measurements mark not only the first direct detection of photons from extrasolar planets, but also represent the first crude emission spectra of these objects.

The theoretical interpretation of these measurements has been swift and comprehensive (Burrows et al. 2005; Fortney et al. 2005; Seager et al. 2005). There is general agreement that the observed fluxes are roughly in agreement with expectations, provided that the planets' Bond albedos are not very close to unity. (If the albedos were very large, the planets would absorb too little radiation to account for their observed IR fluxes.) Moreover, it seems likely that the 4 μm flux from TrES-1 is depressed below that of a black body at the planet's effective temperature because of strong absorption by the fundamental vibration–rotation band of CO. There is disagreement, however, as to whether the fluxes are somewhat larger than the models predict (Burrows et al. 2005), or somewhat smaller (Fortney et al. 2005), or about the same (Seager et al. 2005). The first of these authors find agreement with observations is best if one assumes little transport of heat from the planet's day side to its night side; the second ones find that the heat transport must be efficient, and also that the planet's metallicity must be several times the solar value. This kind of disagreement is probably the result of comparing models that have different deficiencies, interpreting observations that have not yet been validated by repetition. At this stage, one should therefore be cautious about accepting conclusions about the detailed conditions obtaining in the planetary atmospheres. Even so, the advent of meaningful wavelength-resolved measurements of IR fluxes emitted by extrasolar planets is a clear milestone in the effort to understand these objects, and it is cause for excitement about the future.

3.6. The future of technical civilization

Discussions of intelligent life elsewhere in the Universe are often framed in terms of the Drake equation, which provides a means of thinking about the issues involved:

$$N = R_* f_p n_e f_l f_i f_c L, \qquad (3.5)$$

where N is the number of civilizations we might expect to be able to communicate witt at any given time, R_* is the rate of star formation in our Galaxy, f_p is the fraction of those stars that have planets, n_e is the average number of planets that can potentially support life per star that has planets, f_l is the fraction of the above that actually go on to develop life, f_i is the fraction of the above that actually go on to develop intelligent life, f_c is the fraction of the above that are willing and able to communicate, and L is the expected lifetime of such a civilization for the period that it can communicate across interstellar space (see Shklovsky & Sagan 1966). Studying extrasolar planets provides information about the factors f_p, n_e, and f_l; for one session at the Winter School we engaged in a participatory project to encourage thinking about the last factor L, which is the lifetime of a technological civilization.

Asking what number to put in for L seemed unlikely to lead to any useful result, so we concentrated on a more restricted question. This was based on the notion that by most measures (number of scientists, number of publications, total budget) science has been growing exponentially for the last 400 years. It is not so clear, however, whether the *social effects* of science have grown at the same rate. Has life (in the last century, say) become exponentially better (or even exponentially different) as a result of technical progress? There are at least two distinct reasons to suspect that it might not be so. One, the 'End of Science' argument, is based on the John Horgan book of the same title (Horgan 1996). It holds that science itself is slowing down, because we are running out of genuinely fundamental concepts to discover. According to this thinking, most of the big concepts (fundamental forces, the big bang, the nature of life, germs, etc.) are now exposed, so that modern science consists more and more of filling in the details. The other argument is that, even if science is continuing to expand, the scales with which it deals (scales of time, or distance, or complexity, or whatever) are becoming so small or so large that they are less relevant to human concerns.

To put even this broad question into a more tractable form, we attempted to compare the socially effective technical innovations between the first and last halves of the twentieth century. To do this, we split into two subgroups, and each group was given 20 minutes to provide an ordered list of the ten technical innovations that were reduced to practice within their respective half-centuries, and that had the greatest impact on the way that ordinary citizens of technically advanced societies led their lives. According to the rules, the innovations could be effective either for good or ill, and we assigned dates according to when their feasibility had been demonstrated at least once, and their further application was merely a matter of money (possibly quite a lot of money). I acted as referee in disputes about dates (on the principle that it is unimportant that a referee be right; only that he be obeyed), and I was ably assisted by volunteers Suzanne Aigrain, Saskia Hekker, Avi Mandell and Omer Tamuz, who acted as discussion leaders and recorders for the two groups. Once the lists had been composed, we displayed them on the screen and spent the remainder of the hour discussing them. The lists that emerged (slightly edited) appear in Table 3.1.

TABLE 3.1 Socially-important technical innovations

1900–1950	1950–2000
Aviation	Internet
Refrigeration	DNA & genetic engineering
Radio/TV	Spaceflight
Vaccinations	Environmental protection
Antibiotics	Mobile communication
Nuclear weapons	Microchips
Mechanized agriculture	Personal computers
Universal suffrage	Home electronics
United Nations	Birth control
Plastics	Medical imaging

The discussion was, alas, not recorded. But based on what I can remember, and subsequent consideration of the lists, and viewing this all through the lens of my own preconceptions, I am emboldened to make a few summarizing remarks.

We noted that the restriction to technically-advanced (i.e. wealthy) societies was in some ways a misleading one, since a large fraction of people on the Earth do not have access even to clean water, let alone most of the items on the lists.

The sense of the discussion was pretty clearly that science has not stopped, and that its influence on society remains robust and growing. I think it is fair to say that this view was common both among the students and the teachers, two groups with rather different average ages and degrees of experience.

Another useful point was that people have a hierarchy of needs and desires, so that they can be expected to turn first to the most pressing ones (adequate food, a dry place to sleep) before moving on to the less urgent (high definition TV, scratch-n-sniff advertising). Also, it is worth distinguishing the idea of 'progress' from that of 'change'. Thus, there are human activities (the popular music and fashion industries are good examples) in which the *raison d'être* is novelty for its own sake, unrelated to any particular vision of progress.

During my own perusal of the lists that we produced, I had several thoughts which I have set down in the following paragraphs. These were guided in part by the discussion, but I cannot blame them on the group; they are simply my own opinions.

It seems to me that the 'Universal suffrage' and 'United Nations' items on the first-half list may be inappropriate, since they refer to social (not technical) innovations.

The 'Radio/TV' entry in the first-half list may reasonably be paired with a cluster of items on the second-half list: 'Microchips', 'Home electronics', 'Mobile communications', 'Personal computers', and 'Internet'. It is perhaps unclear what exactly is meant by 'Home Electronics', but presumably it refers to video recorders, Tivo systems, digital cameras, DVD players, and, yes (for Douglas Adams afficionados), digital watches. These, I think, can be viewed as evolutionary applications of the first-half 'Radio/TV' idea, but I would rank them lower on the primal importance scale than their progenitor. The same argument applies to 'Mobile communications' in comparison with the earlier 'Telegraph/telephone' entry (which did not make it into the first-half list because it happened in the nineteenth century).

The 'Personal computers' and 'Internet' entries are of somewhat different character. Neither of these technologies (as applied to daily life, not research) permits one to do much that was previously inconceivable, but they make some previously laborious tasks vastly easier, and they impose their own slant on familiar processes. Perhaps the best example of the latter idea is the way that the internet puts mass communication in the hands of the masses, by making it very cheap and fast to get the word (*any* word) around. The resulting trade-off between accessibility and veracity is unique to the medium; it is hard to say whether the result is better or worse than previous communication schemes, but it is certainly different.

The 'Aviation' and 'Spaceflight' entries provide another interesting comparison. Both were initiated early in their respective half-centuries (reasonable claims for pre-1950 spaceflight can be made, but for the sake of this argument it is unhelpful to quibble about dates). It seems clear, however, that the practical effects of aviation were vastly greater by 1950 than those of spaceflight by 2000. Any argument promoting spaceflight as the more successful technology would have to rest in large measure on its impact on the public imagination, and even that (it appears to me) is problematic. One of the most surprising lessons of the *Apollo* Moon landings was how quickly the public got bored with people landing on the Moon.

Both halves feature important medical technologies, with the pre-1950 'Vaccinations' and 'Antibiotics' vying with the post-1950 'Birth control', 'Medical imaging', and (loosely) 'DNA & bio-engineering'. Here, I imagine that 'Birth control' actually means 'The Pill', and the other roughly contemporaneous methods of contraception that are cheap, effective, and that (most significantly) put choices about conception into the hands of women. Thus interpreted, it seems to me fair to argue that 'Vaccinations', 'Antibiotics', and 'Birth control' are three of the most important medical technologies of the last century, and in terms of impact there is little point in trying to choose among them. 'Medical imaging' in the post-1950s sense may be a different matter. Clearly, there have been great advances in digitally processed three-dimensional imaging, and these have allowed diagnoses and treatments that were not previously feasible. But it is unclear (to me) how many patients these methods help, and how much they help, compared to what one could do with the older film-based X-ray technology. Perhaps the new imaging technologies are essential and pervasive, or perhaps they are marginal and incidental; it would be interesting to hear an expert's opinion about this. Finally, 'DNA & bio-engineering' has also had some notable successes, but in the broad health care picture, my impression is that it is still mostly an unrealized promise. In principle, it could utterly change human life, for better or worse. In practice, such big changes seem to be still some distance away.

General concerns of public welfare are reflected in the first half's 'Refrigeration' and 'Mechanized agriculture' entries, and in 'Environmental protection' from the second half. It is interesting to note that environmental protection has become necessary in large part because of the rapid population and consumption growth caused by the successes of refrigeration, mechanized agriculture, antibiotics and other such first-half advances.

In the domain of warfare, the first half of the century has 'Nuclear weapons', which is hard to top. The closest thing offered by the second half is again 'DNA & bio-engineering', which may in the future produce biological agents that are just as deadly and even harder to control than nuclear bombs. But, as with the positive results of genetic manipulation, this possibility is (thankfully) as yet unrealized.

Putting all of this together, I tend toward mild disagreement with the group opinion about the growing importance of science in daily life. It seems to me that the innovations prior to 1950 led to changes that are both deeper and broader than those afterwards, with fundamental changes taking place in many domains. One can hardly argue that technically driven change has stopped, but it takes more effort to remake the world than it once did. One can guess at huge future upheavals resulting from bio-engineering, or really successful artificial intelligence, or any of several other science fiction tropes. But, at the moment, these appear to be rather remote in time. For whatever it may be worth, my prediction is that the twenty-first century will be dominated by our attempts to organize and preserve intelligent life on a small planet, using more-or-less present-day technology. Time will tell.

REFERENCES

ALONSO, R., 2005, Ph.D. thesis, University of La Laguna.

ALONSO, R., BROWN, T., TORRES, G., et al., 2004, *ApJ* **613**, L153.

BAGLIN, A., 2003, *Adv. Space Res.* **31**, 345.

BAGLIN, A., VAUCLAIR, G. & The COROT Team, 2000, *J. Astrophys. Astron.* **21**, 319.

BORUCKI, W. J., KOCH, D. B., LISSAUER, J. J., et al., 2003, *Proc. SPIE* **4854**, 129.

BOUCHY, F., PONT, F., SANTOS, N. C., MELO, C., MAYOR, M., QUELOZ, D. & UDRY, S., 2004, *A&A* **421**, L13.

BROWN, T. M., 2001, *ApJ* **553**, 1006.

BROWN, T. M., CHARBONNEAU, D., GILLILAND, R. L., NOYES, R. W. & BURROWS, A., 2001, *ApJ* **552**, 699.

BURROWS, A., HUBENY, I. & SUDARSKY, D., 2005, *ApJ* **625**, L135.

CHARBONNEAU, D., BROWN, T. M., NOYES, R. W. & GILLILAND, R. L., 2002, *ApJ* **568**, 377.

CHARBONNEAU, D., ALLEN, L. E., MEGEATH, S. T., et al., 2005, *ApJ* **626**, 523.

COX, A. N., 2000, *Allen's Astrophysical Quantities*, fourth edn. Ed. Arthur N. Cox. New York: AIP Press; Springer.

DEMING, D., BROWN, T. M., CHARBONNEAU, D., HARRINGTON, J. & RICHARDSON, L. J., 2005, *ApJ* **622**, 1149.

DRAKE, A. J., 2003, *ApJ* **589**, 1020.

DRAVINS, D., LINDEGREN, L., MEZEY, E. & YOUNG, A. T., 1998, *PASP* **110**, 610.

DUQUENNOY, A. & MAYOR, M., 1991, *A&A* **248**, 485.

FORTNEY, J. J., MARLEY, M. S., LODDERS, K., SAUMON, D. & FREEDMAN, R., 2005, *ApJ* **627**, L69.

HORGAN, J. 1996, *The End of Science: Facing the Limits of Science in the Twilight of the Scientific Age*. Reading, MA: Addison-Wesley.

KONACKI, M., TORRES, G., JHA, S. & SASSELOV, D. D., 2003, *Nature* **421**, 507.

KONACKI, M., TORRES, G., SESSELOV, D. D., et al., 2004, *ApJ* **609**, L37.

KOTREDES, L., CHARBONNEAU, D., LOOPER, D. L. & O'DONOVAN, F. T., 2004, *The Search for Other Worlds*, AIP Conf. Proc. **713**, 173.

LATHAM, D. W., 1992, IAU Colloq. 135, *Complementary Approaches to Double and Multiple Star Research*, ASP Conf. Ser. **32**, 110.

MALLÉN-ORNELAS, G., SEAGER, S., YEE, H. K. C., MINNITI, D., GLADDERS, M. D., MALLÉN-FULLERTON, G. M. & BROWN, T. M., 2003, *ApJ* **582**, 1123.

MANDUSHEV, G., TORRES, G., LATHAM, D. W., et al., 2005, *ApJ* **621**, 1061.

MARCY, G. W., BUTLER, R. P., VOGT, S. S., et al., 2005, *ApJ* **619**, 570.

PONT, F., BOUCHY, F., QUELOZ, D., SANTOS, N. C., MELO, C., MAYOR, M. & UDRY, S., 2004, *A&A* **426**, L15.

SEAGER, S. & MALLÉN-ORNELAS, G., 2003, *ApJ* **585**, 1038.

SEAGER, S., RICHARDSON, L. J., HANSON, B. M. S., MENOU, K., CHO, J. Y.-K. & DEMING, D., 2005, *ApJ* **632**, 1122.

SHKLOVSKY, J. S. & SAGAN, C., 1966, *Intelligent Life in the Universe*, San Francisco: Holden-Day.

Sirko, E. & Paczyński, B., 2003, *ApJ* **592**, 1217.

Tingley, B., 2004, *A&A* **425**, 1125.

Udalski, A., Paczynski, B., Zebrun, K., et al., 2002a, *Acta Astronomica* **52**, 1.

Udalski, A., Zebrun, K., Szymanski, M., Kubiak, M., Soszynski, I., Szewczyk, O., Wyrzykowski, L. & Pietrzynski, G., 2002b, *Acta Astronomica* **52**, 115.

Udalski, A., Pietrzynski, G., Szymanski, M., et al., 2003, *Acta Astronomica* **53**, 133.

Vidal-Madjar, A., Lecavelier des Etangs, A., Désert, J.-M., Ballester, G. E., Ferlet, R., Hébrard, G. & Mayor, M., 2003, *Nature* **422**, 143.

Yee, H. K. C., Mallen-Ornelas, G., Seager, S., et al., 2003, *Proc. SPIE* **4834**, 150.

Young, A. T., 1969, *Appl. Opt.* **8**, 869.

Young, A. T., 1974, in *Methods of Experimental Physics*, Vol. 12, *Astrophysics, Part A, Optical and Infrared*, ed. N. Carlton. New York: Academic Press, 95.

4. From clouds to planet systems: formation and evolution of stars and planets

GÜNTHER WUCHTERL

The discovery of more than one hundred extrasolar planet candidates challenges our understanding of star and planet formation. Do we need to modify theories that were mostly developed for the Solar System in order to understand giant planets orbiting their host stars with periods of a few days? Or do we have to assume particular circumstances for the formation of the Sun to understand the special properties of the Solar System planets? I review the theories of star and planet formation and outline processes that may be responsible for the diversity of planetary systems in general. I discuss two questions raised by extrasolar planets: (1) the formation of Pegasi planets and (2) the relation between discovered extrasolar planets and the metallicity of their host stars. Finally, I discuss the role of migration in planet formation and describe three tests to distinguish whether planets migrated long distances or formed near their final orbits.

4.1. Witnessing the discovery

What happened to the theory of star and planet formation when almost ten years ago, in October 1995, Mayor and Queloz (1995) announced that they had found a planet, in a four day orbit around the fifth magnitude star 51 Pegasi? Theory at this time was preparing for the discovery of extrasolar planets in orbits around common main-sequence stars. Yet the first discoveries seemed to lie well in the future, not to be expected before the start of the new, the third, millennium.

4.1.1. *Observations: 12 years, 21 stars – no planets (yet)*

In August of that year, planet searchers from the University of British Columbia, led by Gordon Walker, had published their 12 year long effort to monitor 21 solar-like stars for planetary companions. It was widely considered to be the first search having the required sensitivity to detect planets. And, as everybody considered self-evident, the easiest one to detect would be a Jupiter-like planet in a Jupiter-like orbit with an approximately 12 year period. After all, that would be the case for the Solar System place at a distance of 10 parsec, the typical distance of a nearby star.

When Walker *et al.* (1995) published that they had not found any planets[1] and gave an upper limit of less then 1 in 10 stars having a 'jupiter', the formation theorists started to wonder about the result. It was generally considered that Solar System formation theory was not yet complete. But should the Solar System be more special in any case when the Sun was compared to other stars (i.e. a typical member of the galactic minority of single stars that is outnumbered by a factor of about three by stars that are in multiple systems)?

---

[1] However, γ Cep was a special case where the planetary interpretation could not be rejected with certainty, cf. Walker *et al.* (1992). In the meantime two of the stars in Gordon Walker's list have been found to host planets: ε Eridani (Hatzes *et al.* 2000) and γ Cepheï A (Hatzes *et al.* 2003). Twelve years of data were not enough.

Extrasolar Planets, eds. Hans Deeg, Juan Antonio Belmonte and Antonio Aparicio.
Published by Cambridge University Press.
© Cambridge University Press 2007.

The finding by Wolszczan and Frail (1992) of planets in orbit around the pulsar PSR B1257+12 had been confirmed by Wolszczan (1994) using modifications of the planetary orbits by mutual perturbations of the inferred planets themselves. Hence, planet formation was apparently a rather ubiquitous process also occurring in one of the very exotic environments that the millisecond pulsar had provided during its long history.

4.1.2. *Theory not quite in place*

That October, theorists had assembled on the island of Hawaii to discuss extrasolar planets and their formation. The discovery seemed like a reality shock. I still remember the faces of other theorists after the message of the discovery of 51 Pegasi was put on an overhead projector at the end of the extrasolar planets session. We realized that we were on the wrong side of the globe. The discovery was announced in Florence, close to Galileo's last home. Theory was taken by surprise. A 'jupiter' at 1/100 of Jupiter's orbit, a few stellar radii from its host star, was bewildering. It was not that anyone had proved, or even tried to prove, that giant planets could not form close to their stars. Apparently, nobody had even thought about it. Everybody wondered how this could have happened and there was at the same time concentrated thinking: how could it form? I am sure that, as we travelled home, we all had an idea of how to make it. When I stepped out of the plane in Frankfurt, I had convinced myself that it actually would be easy to form 51 Peg's planet and planned the calculations to prove it. Similar events must have happened for many people because within a year many ideas for the formation of 51 Pegasi b were published.

I am telling this because it is exciting and also because it helps us to understand the diversity of ideas that developed after the discovery of 51 Pegasi b. Before that, there was a fairly detailed framework for the understanding of the formation of the Solar System. It was a step-wise approach, tied into the data from the history of the Solar System. Supposedly, it could be generalized to other systems. But it had and has its open questions. Hence research focused on filling the holes for the Solar System where the data were rich. An understanding of a general theory of planet formation was only just beginning. As the pioneering work for understanding planet formation around stars of various masses, I recommend a series of three papers: Nakano (1987, 1988a,b). Nakano discusses the limitations for planet formation based on growth timescales and nebular and stellar lifetimes in the framework of the *Kyoto*-model.

Just before the discovery, Jack Lissauer gave a talk on the *Diversity of Plausible Planetary Systems* (Lissauer 1995). It is a careful assessment and weighting of the uncertainties trying to focus on what can be expected for other systems from what we have learned for the Solar System. I think every person interested in what could be expected from theory will find the overview there.

4.2. Planet formation theory

4.2.1. *The 'original' Solar System perspective*

So there is the classical picture based on the formation of the Solar System until 1995 and the diversity of ideas thereafter, with manifold scenarios for planet formation, many of them probably inconsistent with Solar System formation or of unknown relation to the Solar System.

To get an overview of the classical picture of Solar System formation, I recommend the review article of Hayashi's Kyoto group (Hayashi *et al.* 1985), where the stage is set for modern theories[2] – planets forming in a circumstellar protoplanetary disc of gas and dust that originates from an interstellar cloud together with the host (proto)star. The article discusses the key physical processes and outlines a complete picture of Solar System formation. The overall picture is basically unchanged up to now, but some of the numbers have been corrected and gaps in the argumentation have been closed. To get a more quantitative view, especially on terrestrial planet formation, and a concept to resolve the issue of the long growth time of the cores of giant planets, I recommend Jack Lissauers' (1993) article on planet formation in *Annual Reviews of Astronomy and Astrophysics*. It is written after a half-year long get-together of essentially all researchers on planet formation in 1992 at Santa Barbara. It is a superb review and covers the situation of planet formation theory at that pre-51 Peg time.

4.2.2. After 51 Pegasi – forget the Solar System?

With the 'discovery shock' many researchers apparently felt that the Solar System approach was 'dead' and invented numerous schemes to rapidly explain the new object. A year after the discovery it was still a close race between the number of discovered planets and the number of new theories to explain them.

New processes were invoked – often rather ad hoc – to explain the unexpected properties of 51 Peg, as well as the growing diversity in the observed exoplanet population. Some of the new processes apparently played no (or only a minor) role in the Solar System. The unforeseen properties of most extrasolar planets – giant planets in orbits with periods of a few days, planets with eccentricities much larger then any of the Sun's planets, and many giant planets with orbital radii much smaller than Jupiter's – prompted strong comments.

'With these discoveries, theorists have lost their understanding of the formation of the solar system,'

as Pavel Artymovic put it in IAU Symposium 202 at the Manchester General Assembly in 2000, or more drastically,

'Forget the solar system!,'

as a well-known astrophysicist recommended off the record in 2002. Are we witnessing the final failure of the Copernican principle in cosmogony? Is the formation of our own home system fundamentally different from the typical planet formation process in the Galaxy? Are special processes or unlikely circumstances required to explain the properties of the Solar System?

4.3. Introduction – trying to solve a big problem

The approach I will follow to look at the problem and possibly contribute to answering these questions is not to describe the diversity of theories and ideas but focus on a simplified picture that is closely linked to basic physical principles (essentially to the equations that describe the fluid dynamics of a mixture of gas and dust supplemented by the theory of planetesimal growth).

---

[2] Weizsäcker (1943) distinguishes three major groups of planet formation theories: planets formed from (A) a uniformly rotating mass that filled the space of the present planetary system (Kant, Laplace); (B) a tidal wave (*Flutwelle*) that was excited by a star passing the Sun (Chamberlin, Moulton), and (C) an irregularly shaped nebula filament (Nölke). I consider work following the *nebular hypothesis* (A) supplemented by the *planetesimal hypothesis* as modern formation theories.

The idea is to provide a theoretical backbone that is unlikely to suffer refutation unless one of its key assumptions turns out to be invalid.

Those assumptions are:
  (i) there exists an angular momentum transfer process to separate mass and angular momentum during protostellar collapse;
  (ii) dust growth in the protoplanetary nebula leads to km-size planetesimals on a timescale that is fast compared to nebula evolution;
  (iii) the basic properties of stars and planets can be described in terms of spherical symmetry.

I refrain from discussing many results in the literature based on my judgement of how close they are to a deductive, more theoretical physical approach. This is not because they may not be relevant but because they are usually snapshot studies of a particular process without a clear justification of previous or later phases of star and/or planet formation. Often unknown physical processes are parameterized to make the respective models solvable.

So I emphasize parts of the problem that can be solved with relatively high reliability but neglect the addition of processes that will occur in reality but cannot be reliably quantified. The picture given here will not address many aspects of the problem that are needed to understand all the properties of star–planet systems, but what is discussed should hold, although it may turn out to be of minor relevance for the big picture.

The analogy for the Sun would be to just discuss global properties like mass, luminosity, temperature and age but ignore surface effects like spots and other magnetic activity and global effects like differential rotation and globally relevant circulation patterns inside the convection zone and granulation.

4.4. The plan

We will start our considerations with interstellar clouds and proceed to planets in three steps:
  (i) star formation as the collapse of a gravitationally unstable cloud fragment;
  (ii) the early evolution of the star and a plausible circumstellar protoplanetary nebula;
  (iii) planet formation in the protoplanetary nebula.

I will discuss protostellar collapse and early stellar evolution with detailed solutions for the spherical problem for masses down to the brown dwarf regime. This is analogous to describing stellar evolution to and from the main sequence, with the important difference that fluid dynamical processes are important due to the collapse origin of stars. Accretion flows determine the luminosity, structure and evolution during the earliest phases that last approximately 1 Ma (1 Ma is 1 million years).

A similar first principles approach is not feasible at present for the formation of the protoplanetary nebula. Hence, I will briefly describe the ongoing efforts and then fall back on a pragmatic approach to constructing plausible protoplanetary nebulae for planet formation studies. The resulting models then give estimates for the nebular conditions that can then be used as a starting point for the construction of planet formation models.

In the third step, i.e. planet formation, I return to a more basic principle-orientated approach. The theory of planetesimal accretion will be combined with spherical models of protoplanets that consist of a core and a gaseous envelope that are embedded in a plausible protoplanetary nebula. These models will provide masses, luminosities and accretion-histories for planets. Finally, I discuss some applications, including observational tests of planet formation theory.

For convenience and consistency I will mostly use my own models, which are applicable to stars, brown dwarfs and planets and are based on the same set of basic physical equations and identical descriptions of the microphysics. They all use the same *constitutive relations*, i.e. equations of state and opacities. The system of equations is calibrated to the Sun and observationally tested by the solar convection zone and RR Lyrae lightcurves (Wuchterl and Feuchtinger 1998); Wuchterl and Tscharnuter 2003).

4.5. From clouds to stars

4.5.1. Clouds

Stars form from molecular clouds. Most of the clouds' mass is in the form of molecular hydrogen. Some of them can be seen as nearby dark clouds because the dust they contain obscures the light of the Milky Way stars behind them. In some dark clouds faint stars can be seen that have ages of a few million years (Ma). These *T Tauri* stars have not yet commenced hydrogen burning and will evolve towards the main sequence in a few tens of Ma. Hence this evolutionary phase is often referred to as the pre-main sequence (PMS) phase. Only when their nuclear reactions produce energy at a sufficient rate to balance the energy radiated from their surfaces will the process of gravitational contraction stop. Once that balance is reached, typical stars will stay on the main sequence with approximately constant luminosity and surface temperature for $\sim 10^9$ years.

Star formation preferentially occurs in the spiral arms of the Galaxy. Giant molecular clouds have masses of 1 million solar masses and fragment into substructures that finally lead to the sub-collapse of sub-fragments that typically result in a cluster of stars. Star formation is a multiple-scale process ranging in size from 100 pc clouds with subunits referred to as clumps down to the smallest structures, termed cores, of size 0.1 pc and masses comparable to stellar ones, i.e. typically one solar mass. Galactic tides, magnetic fields and irregular motion (turbulence) play a role on larger scales, but ultimately star formation is a competition between support by thermal pressure and the inward pull of gravity. Once gravitational collapse starts and the cloud cores shrink, rotation becomes more and more important. As the size shrinks, conservation of angular momentum leads to spin up, even for initially slowly rotating structures. The common outcomes are clusters of stars, multiple systems and star-disc systems. That way the classical angular momentum problem (even a slowly rotating cloud would lead to a hypothetical stellar embryo rotating much faster than breakup speed) is most probably being solved – by redistribution of angular momentum between the components.

I focus on the physics of star formation that is most relevant for planet formation. That is most importantly the properties of the newborn protostar and the protoplanetary nebula, as well as the timescales for final collapse and early stellar evolution. A general review of the *physics of star formation* has recently been given by Larson (2003). Two stages of cloud collapse may be discerned:
(i) fragmentation of the cloud;
(ii) collapse and accretion of cloud fragments.
Instead of discussing the fragmentation process of molecular clouds I will assume that it leads to fragments of all masses down to the *opacity limit* of fragmentation, which is estimated to be at approximately 0.01–0.007 $M_\odot$ or 7–10 M_{Jup}.[3] Below the opacity limit, further fragmentation into still smaller mass units is probably made more difficult or impossible because of the temperature increase, which is a consequence of the dense

---

[3] 1 $M_\odot = 1.989 \times 10^{30}$ kg, 1 $M_{\text{Jupiter}} = 1.899 \times 10^{27}$ kg, 1 $M_\oplus = 5.974 \times 10^{24}$ kg, 1 $M_{\text{Jupiter}} = 0.95 \; 10^{-3} \; M_\odot$, 1 $M_{\text{Jupiter}} = 317.8 \; M_\oplus$. I will use 0.001 $M_\odot$ and M_{Jupiter} as approximately equal.

fragments becoming opaque for light of all wavelengths capable of transporting energy and hence cooling the collapsing clouds heated by gravitational self-compression. Indeed, stars may have their masses determined directly by those of the prestellar cloud cores. This is suggested by the fact that the distribution of masses, or the 'initial mass function' (IMF), with which stars are formed appears to resemble the distribution of masses of the prestellar cores in molecular clouds (see Larson 2003 for discussion).

To determine the physical properties of the final fragments that have become gravitationally unstable, we consider the balance of gravity and thermal pressure in a cloud-core of given temperature T and mass-density (or density for short) ϱ. Such a fragment is unstable to perturbations, i.e. density fluctuations, if its volume contains a mass that is larger than the *Jeans-mass*:

$$M_{\text{Jeans}} = \varrho \lambda_{\text{Jeans}}^3, \qquad (4.1)$$

with the *Jeans-length*,

$$\lambda_{\text{Jeans}}^3 = c_T \sqrt{\frac{\pi}{G\varrho}}, \qquad (4.2)$$

where G is the gravitational constant, $c_T = \sqrt{P/\varrho} = \sqrt{kT/m}$, the isothermal sound speed for an ideal gas of particles with mean mass m, with Boltzmann constant k. The other constant factors depend somewhat on the details of the assumed geometry of the cloud and the perturbations respectively (i.e. density fluctuations) assumed in the cloud.[4]

We restrict our discussion of star formation to star formation from gravitationally unstable – to be more specific, Jeans-unstable – cloud cores as an idealization of the smallest cloud fragments. To be more precise, we will mostly discuss star formation from cloud fragments that are actual rigorous solutions for the equilibria of self-gravitating isothermal clouds with spherical symmetry that are embedded in a medium of finite, constant pressure (the ambient larger cloud superstructure). Such self-gravitating equilibrium gas spheres are called Bonnor–Ebert spheres (Bonnor 1956; Ebert 1957). Such theoretical constructs do actually exist in the sky: the structure of a relatively nearby dark cloud, the Bok globule Barnard 68, has been found closely to match a Bonnor–Ebert sphere (Alves *et al.* 2001).

The neglect of rotation, magnetic fields and turbulence, and the use of spherical symmetry that lead to the classical Jeans picture with thermal pressure and gravity as the only players, may seem somewhat restrictive, but we follow Larson (2003) in noting that

The Jeans length and mass are still approximately valid, even for configurations that are partly supported by rotation or magnetic fields, as long as instability is not completely suppressed by these effects. Thus, if gravity is strong enough to cause collapse to occur, the minimum scale on which it can occur is always approximately the Jeans scale, and structure is predicted to grow most rapidly on scales about twice the Jeans scale.

4.5.2. *Clouds and stars*

We have looked at the clouds in the beginning. Let us now look at the outcome of the formation process, the products of the cloud collapse (stars, brown dwarfs and planets)

---

[4] Planar waves are used in the case given. A discussion of variants of the Jeans instability and more rigorous treatments that avoid the *Jeans swindle* of assuming an unperturbed state that is not an equilibrium solution is given in Larson (2003) or text-books such as Kippenhahn and Weigert (1990).

TABLE 4.1 Basic properties of stars, brown dwarfs and planets

Stars	Brown dwarfs	Planets
gas spheres	gas spheres	often gas spheres
self-gravitating	self-gravitating	self-gravitating
self-luminous	self-luminous	often self-luminous
often nuclear fusion	often nuclear fusion	heavy element enriched
main-sequence: luminosity balanced by nuclear burning	initial nuclear burning; luminosity never balanced	no nuclear burning according to historical practice and IAU working definition

that we will refer to as *celestial bodies* for short. Basic properties that emphasize similarities are summarized in Table 4.1. What are the common physical principles of these celestial bodies? Arguably the most important one that distinguishes them from other common objects in the Universe like rocks, plants, animals and cars is the importance of gravity. Newtonian gravity is described by the Poisson equation, which in spherical symmetry can be integrated once and written in the form:

$$\frac{dM_r}{dr} = 4\pi r^2 \varrho, \qquad (4.3)$$

where ϱ is the density; and the mass interior to radius r for a radial density distribution $\varrho(r)$ is obtained by the integration mentioned above as

$$M_r = \int_0^r 4\pi r'^2 \varrho \, dr'. \qquad (4.4)$$

The corresponding gravitational force that acts on the structure at distance r is:

$$F_{\text{grav}} = -\frac{GM_r}{r^2} \varrho. \qquad (4.5)$$

This leads us to one of the most important principles of astrophysics, namely the force balance between gravity and pressure (gradients) that governs the structure of gravitating gas spheres throughout most of their lives, namely hydrostatic equilibrium:

$$\frac{dP}{dr} = F_{\text{grav}} = -\frac{GM_r}{r^2} \varrho. \qquad (4.6)$$

Pressure has entered our considerations, so we have to specify an equation of state that relates pressure and density. For fully degenerate objects, such as white dwarfs and neutron stars, and to some extent for planets, when they behave as liquids, the dominating pressure dependence is on density. Our equations for the structure would then be completed by specifying the pressure as a function of density alone. In general, however, and in any case for high temperatures, the pressure depends on density *and* temperature, as, for example, in an ideal gas with particles of mean mass m:

$$P = \frac{\varrho}{m} kT, \text{ often written as } P = \frac{R_{\text{gas}}}{\mu} \varrho T. \qquad (4.7)$$

The first version uses the mean particle mass directly and hence the Boltzmann constant appears explicitly, the second version uses quantities for a mole of an Avogadro number, N_A, of particles, with the molar gas constant $R_{\text{gas}} = kN_A$ and their mean molar mass, μ. Two important new quantities have entered our considerations via the pressure equation of state (Eq. (4.7)).

(i) The elemental and chemical composition, i.e. its elemental abundances and the respective chemical state: whether they are in molecules, neutral or ionized, and the number of free electrons in the latter case. The mean particle mass $m(\varrho, T)$ changes accordingly.
(ii) The temperature, to be more explicit, the radial temperature distribution $T(r)$ of a celestial body.

The chemical composition (number of particles as H_2, H, e^-, ...) and hence $m(\varrho, T)$ can be determined by thermodynamic calculations for given elemental composition, density and temperature. Such calculations also give the pressure equation of state, as well as other equations of state, such as specific heats or the adiabatic gradients. What remains – apart from solving the whole thing – is to determine the temperature.

Here the self-luminous nature of the objects comes into play. Because celestial bodies usually radiate heat into space, their temperatures – the surface value and the interior temperature distribution $T(r)$ – can be calculated from an energy budget; that is, the difference between the amount of energy that is radiated into space per unit time – luminosity – and whatever amount of thermal energy is generated or available as an internal heat-reservoir.

The ratio between the amount of energy in the interior and the luminosity of a celestial body can be used to calculate a timescale for the change of the energy content. This results in the thermal, cooling, or Kelvin–Helmholtz timescale:

$$\tau_{\mathrm{KH}} = \frac{E_{\mathrm{therm}}}{L} = \frac{M \bar{c}_{\mathrm{V}} \bar{T}}{L} \sim \frac{1}{2} \frac{GM^2}{RL}, \quad (4.8)$$

where we have ignored thermodynamic details and estimated the thermal energy of the celestial body by first using a mean specific heat $\bar{c}_{\mathrm{V}}$ and a mean temperature $\bar{T}$, and then halving the gravitational energy of a sphere of mass M and radius R. This is a valid approach for gas spheres in hydrostatic equilibrium.

Because of their self-luminosity, celestial bodies change their thermal energy content – in the absence of sufficient energy sources – and hence must evolve on the Kelvin–Helmholtz timescale. That timescale is of the order of 1 Ma for the objects under consideration.

The determination of the luminosity and the temperature leads to the introduction of an energy equation that contains the important energy transfer processes. For celestial bodies these are:

(i) *radiative transfer*, which not only transfers heat through the interior but also in the end brings information from the uppermost photosphere to the observer's telescope;
(ii) *heat conduction*, which is usually treated formally together with radiative transfer; and
(iii) *convection*, i.e. energy transfer by small and medium-scale gas motion, turbulent or otherwise, that does not lead to large scale motion, restructuring of the object or disobeying the overall hydrostatic equilibrium.

To summarize energy transfer, the luminosity of an object is determined by its surface area and the amount of radiation that can be emitted by unit surface area in unit time; it is well approximated by and equal to $\sigma T_{\mathrm{eff}}^4$ in the black body case, where σ is the Stefan–Boltzmann constant. For the *effective temperature*, T_{eff}, which holds exactly, we can derive the luminosity $L = 4\pi R_\tau^2 \sigma T_{\mathrm{eff}}^4$ for appropriately defined surface radius R_τ from which the photons typically can travel directly to the observer. The internal temperature structure, $T(r)$, is then determined by the surface temperature, and the temperature increase or gradient, which is controlled by the efficiency of the dominating transfer process. Generally, efficient energy transfer leads to a small temperature gradient for given luminosity and hence a moderate temperature increase towards the centre. Even almost constant temperature is a good approximation in the case of very efficient radiative transfer or heat conduction or almost zero luminosity. In the opposite extreme, if all other energy transfer processes are inefficient, convection takes over and essentially limits the temperature gradients to the adiabatic values. Or, more precisely, to an isentropic[5] structure with the temperature gradients being such that the specific entropy is constant throughout a convective region. In that case the temperature gradient with respect to pressure[6] takes a particularly simple form:

$$\frac{\mathrm{d}\ln T}{\mathrm{d}\ln P} =: \nabla = \nabla_s := \left(\frac{\partial \ln T}{\partial \ln P}\right)_s. \tag{4.9}$$

Following the notation of classical stellar structure theory, ∇ denotes the gradient along the structure $T(r), P(r)$ of a celestial body, whereas ∇_s is the respective slope along an adiabat (or more precisely isentrope), i.e. a thermodynamical property of the particular material under consideration. A simpler way to put this (Eq. (4.9)) is

$$\frac{\mathrm{d}s}{\mathrm{d}r} = 0, \tag{4.10}$$

where s is the specific entropy. The simplicity is paid for by the introduction of another equation of state, e.g. $s(T, P)$.

With the equations outlined above and setting more technical boundary conditions aside, we can calculate the structure of a celestial body – be it a star, brown dwarf or planet – for a given time when we know its structure at a previous time. The evolution is driven by the luminosity that changes the thermal content on the Kelvin–Helmholtz timescale.

This simplified discussion must suffice for the present purpose. We note that the ideal gas is only a very rough approximation that already needs significant corrections in the solar interior. The interactions between particles in the gas that are responsible for those corrections become more and more important as typical densities increase and masses and temperatures decrease towards the brown dwarf and planetary domain. Jupiter and the Earth are much better approximated by a liquid than by a gas, but a similar argument holds involving more elaborate equations of state.

We are left with the fact that the determination of the temperature for self-luminous objects – which stars, brown dwarfs and planets are – has led us towards an evolutionary picture. The evolution is driven by the fact that these objects change their heat content through radiating into space. Hence following their evolution means following how they transfer heat from their interior into the surrounding, cold Universe.

That very fact led Lord Kelvin to estimate the age of the Earth from the temperature increase observed in deep mines[7] to be between 20 and 400 Ma and prompted Eddington to look for an energy source to sustain the luminosities of the Sun and the stars – subatomic energy as he called it.

---

[5] Isentropic: without change in entropy.
[6] I suggest reading pressure in a hydrostatic object as a coordinate. It is then monotonically decreasing with radius and hence runs from the outside in.
[7] Because the temperature on Earth increases as we dig deeper there is apparently a temperature gradient. Because there is a gradient there must be a heat flux. That flux must transfer heat from a reservoir. Given the size of the reservoir – the Earth – and the flux inferred from the gradient we can estimate the time that is needed to reduce the reservoir. That is the Kelvin–Helmholtz timescale, τ_{KH}.

For us it will prompt an important question: what was the initial thermal energy that the formation processes put into stars, brown dwarfs and planets? That input would determine their evolution: for stars until they reach the balance of nuclear energy generation and luminosity at the main sequence, for brown dwarfs and planets until they have 'forgotten' the details of their formation.

Given sufficient energy sources, the losses due to luminosity are balanced and the structure of a celestial body does not change unless its elemental composition changes owing to nuclear reactions. This is the case for a star on the main sequence.

4.5.3. *How to begin – stellar evolution as initial value problem*

Star formation involves phases that follow the gravitational instability of the cloud cores, that is, an instability in the hydrostatic equilibrium. It can be shown that it continues to grow rapidly in the non-linear regime, departing further and further from the initial (close-to) equilibrium conditions. The initial perturbations are rapidly forgotten. This leads to the diminishing role of the gas pressure (e.g. Kippenhahn and Weigert 1990). Hence the subsequent evolution is dominated by gravity alone – a collapse with the cloud falling freely towards its centre. For the prevailing isothermal cloud these conditions hold for about a ten order of magnitude growth in density. In the absence of gas pressure the cloud would collapse to a point in a *free-fall time*[8]

$$\tau_{\rm ff} = \sqrt{\frac{3\pi}{32G\varrho_0}}, \qquad (4.11)$$

where ϱ_0 is the constant (or mean) initial density of the cloud when it becomes unstable.

Because the free-fall time is much smaller than the cooling time, $\tau_{\rm KH}$, (Eq. (4.8)) of a typical newborn celestial body, the collapse leaves a thermal imprint. Energy transfer processes are too slow to erase the $T(r)$ structure that the collapse builds up. Hence it is unlikely that newborn celestial bodies, in particular stars, have adjusted their thermal structure to that required on the main sequence to balance the luminosity needs by appropriate nuclear energy production. Therefore, we have an evolution from an initial thermal structure to the one on the main sequence – *that is the pre-main sequence evolution.*

Early stellar evolution onto the main sequence is a gravothermal relaxation process from the thermal structure produced by the star formation process to the thermal structure determined by the balance of energy radiated into space from the stellar surface and energy generated by nuclear reactions in the stellar interior. If star formation produced young stars with thermal structures closely resembling those of main sequence stars, there would be no, or only a negligible, pre-main sequence phase. Embedded objects would then start shining through their cocoons with main sequence stellar properties. If star formation resulted in a thermal structure of stars that is considerably different from the main sequence, a significant phase of gravothermal relaxation would be expected. Observations of young stars high above the main sequence prove the latter to be the case. Those *classical T Tauri stars* still show accretion-signatures (e.g. H_α-emission, 'veiling' of the spectral lines and IR excess) and cloud remnants in their vicinity (IR and mm emission of discs, as well as residual circumstellar envelopes, at least in many cases).

---

[8] This corresponds to half an orbit with a semi-major axis of twice the radius of a constant density cloud with mass M. Imagine a very elongated ellipse with apastron at the cloud radius and the periastron approaching the cloud centre. Kepler's third law for a semi-major axis of half the cloud radius and the cloud mass as the primary mass then gives the free-fall time as half the orbital period.

Since evolution towards the stellar main sequence depends on the thermal structure provided by the stellar formation process, star formation has to be considered to determine the pre-main sequence evolution. An uncertainty in the stellar structure derived from a study of protostellar collapse causes an uncertainty in pre-main sequence evolution. A quantitative theory of star formation is therefore needed to provide the correct 'initial' thermal structure of stars to derive the pre-main sequence stellar structure and evolution, as well as the stellar properties during that time.

4.5.4. Early stellar evolution theory

Present stellar evolution theory mostly deals with stars that are in hydrostatic equilibrium, where gas pressure balances gravity. The motion and inertia of stellar gas are usually neglected. The starting point of stellar evolution calculations is the early pre-main sequence phase. The mechanical structure there is determined by solving for hydrostatic equilibrium. However, to obtain the thermal structure requires knowledge of still earlier evolutionary stages. Trying to obtain thermal information from the evolution in those embedded stages of star formation complicates the question even more because then the hydrostatic equilibrium cannot be used to determine the mechanical structure.

One important aim of star formation theory, which is still an unsolved problem, is therefore to provide the initial conditions for stellar evolution, i.e. the masses, radii and the internal structure of young stars, as soon as they can be considered to be in hydrostatic equilibrium for the first time.

The calculation of appropriate starting conditions for stellar evolution is complicated by the fact that, in general, young stars still accrete mass. Therefore, the hydrostatic parts of young stars and their photospheres are more or less directly connected to circumstellar material being in motion due to mass-inflow and/or outflow. Moving circumstellar matter and the accretion process are by nature non-hydrostatic. Flows that contain both the hydrostatic protostellar core and the hydrodynamic accretion flow must be calculated by using at least the equations of radiation hydrodynamics, including convection, and cause a wealth of technical difficulties.

Modellers of early stellar evolution and the pre-main sequence phase have therefore relied on simplified concepts to make the problem tractable. Originally, Hayashi *et al.* (1962) argued that the cloud collapse should be so fast that the fragment would evolve adiabatically to stellar conditions and the hydrostatic young star would appear as an isentropic sphere radiating at high luminosities, thus causing a fully convective structure. For a given stellar mass they would appear in the Hertzsprung–Russell diagram on the almost vertical line defined for fully convective models of all luminosities – the Hayashi track. Because Hayashi's estimate for the collapse led to large radii, they should appear near the top of that line.

Larson (1969) showed that radiative losses during collapse are substantial, and that early collapse would proceed isothermally. Detailed models showed the necessity of a careful budgeting of energy losses in the framework of radiation hydrodynamics (RHD) and demonstrated the high accuracy requirements for a direct calculation of the collapse (Appenzeller and Tscharnuter 1974, 1975; Bertout 1976; Tscharnuter and Winkler 1979; Winkler and Newman 1980a,b; Tscharnuter 1987; Morfill *et al.* 1985; Tscharnuter and Boss 1993; Balluch 1991a,b; Kuerschner 1994; Wuchterl and Tscharnuter 2003). Modellers then looked for simplified, sometimes semi-analytical concepts to characterize the collapse and accretion flow with key parameters chosen in accord with values from detailed RHD models of protostellar collapse or from properties of specific collapse solutions like the constant mass accretion rate for the self-similar singular isothermal sphere or those resulting from accretion disc models.

4.5.5. Strategies to determine 'initial' stellar structure

The studies of star formation have not resulted in an easy way to calculate star formation before beginning a pre-main sequence stellar evolution calculation. To obtain the initial stellar structure for such calculations the thermal structure produced by star formation has been approximated using different concepts to separate the early stellar evolution and the dynamics of protostellar collapse. The simplification strategies differ in how the complete problem is split into a quasi-hydrostatic 'stellar' and a hydrodynamic 'accretion' part, in both space and time.

Quasi-hydrostatic, constant mass stellar evolution

Quasi-hydrostatic calculations use high luminosity initial conditions for a given stellar mass, i.e. with accretion assumed to have terminated or only causing negligible effects on the pre-main sequence. Once the mass is chosen, the initial entropy and radial entropy structure has to be specified before the initial model can be constructed. The choice of entropy essentially results in a value for the stellar radius. To be sure that the details of the star formation process, or more precisely the specific initial conditions, do not influence such studies, the initial luminosities are chosen to be very high. If they are sufficiently high, the later evolution rapidly becomes independent of the earlier evolution.

The internal thermal structure of the star (temperature or entropy profile) is usually assumed at a moment when dynamical infall motion from the cloud on to the young star is argued to have faded and contraction of the star is sufficiently slow (very subsonic) so that the balance of gravity and pressure forces, i.e. hydrostatic equilibrium (Eq. (4.6)), accurately approximates the mechanical structure (pressure profile) of the star. The absolute ages associated with these states are obtained by the homological back-extrapolation of the so-obtained initial hydrostatic structure to infinite radius. This leads to typical initial ages of $\sim 10^5$ a, i.e. of the order of a free-fall time for a solar mass isothermal equilibrium cloud. At $t > 10^6$ a the initially set up thermal structures are assumed to have decayed away sufficiently – as is the case in a familiar, non-gravitating thermal relaxation process after a few relaxation times – and the calculated stellar properties would then well approximate the properties of young stars, which do form by dynamical cloud collapse in reality. This relaxation issue is somewhat complicated by the thermodynamic behaviour of self-gravitating non-equilibrium systems that stars resemble. However, it can be shown that the memory of the initial thermal structure is quickly lost in some cases (Bodenheimer 1966; von Sengbusch 1968; Baraffe et al. 2002), in particular if the star is initially *fully convective*.

A fully convective structure is considered to be the most likely result of the protostellar collapse of a solar mass cloud fragment (Hayashi 1961, 1966; Hayashi et al. 1962; Stahler 1988a) and is thus used as a stellar evolution initial condition.

Following this argument, young star properties are now usually calculated from simple initial thermal structures without considering the gravitational cloud collapse: Chabrier and Baraffe (1997), D'Antona and Mazzitelli (1994) and Forestini (1994) use $n = 3/2$ polytropes to start their evolutionary calculations; Siess et al. (1997, 1999) also use polytropes; Palla and Stahler (1991)[9] found $n = 3/2$ polytropes insufficient and use fully convective initial models.

---

[9] Initial conditions have been kept in use by the authors since then; for example, Palla and Stahler (1992, 1993).

Hydrostatic stellar embryo and parameterized accretion

To arrive at a more realistic description of the transition of cloud collapse to early stellar evolution, the later parts of the accretion process have been modelled. The strategy is to start with a stellar embryo, i.e. with a hydrostatic structure of less than the final stellar mass. The remaining mass growth is described by a separate accretion model. The argument for the central hydrostatic part is the same as the above, but applied to the initial stellar embryo (protostellar core). Stahler (1988a), for example, chose an embryo mass $M_0 = 0.1\,M_\odot$ after trial integrations for embryo masses $>0.01\,M_\odot$. Typically, an embryo of $0.05\,M_\odot$ is chosen to be embedded in a steady accretion flow with a given mass accretion rate. The procedure is thought to reduce the ambiguity in the initial entropy structure by lowering the initial hydrostatic mass. The entropy added to the initial core is calculated self-consistently with the prescribed mass addition from the steady inflow due to disc or spherical accretion. The lowered arbitrariness in the initial entropy structure (mass and initial entropy only have to be chosen for a small fraction of the mass and result in an initial radius for the initial hydrostatic core) is accompanied by the requirement of additionally specifying the mass accretion rate, $\dot{M}$, and the state of the gas at the cloud boundary. The key advantage, however, is that, due to the assumption of the steadiness of the accretion flow, the mathematical complications are reduced considerably by changing a system of partial differential equations into one of ordinary differential equations.

Stahler (1988a) followed this approach to discuss pre-main sequence stellar structure based on a study of steady protostellar accretion in spherical symmetry (Stahler et al. (1980a,b), SST in the following) and argued that the fully convective assumption should be valid for young stars below $2\,M_\odot$. Stahler (1988b) discusses the history of initial stellar structure and the role of convection in young stars and summarizes (p. 1483): 'This nuclear burning, fed by continual accretion onto the core of fresh deuterium, both turns the core convectively unstable and injects enough energy to keep its radius roughly proportional to its mass.' But Winkler and Newman (1980a,b), who did a fully time-dependent study of protostellar collapse, but excluded convection a priori, found a persistence of the thermal profile produced by the collapse and very different young star properties after the accretion had ended.

The fully convective assumption was discussed in Stahler (1988a) for the last time (p. 818) and although it was remarked that it had been shown only by SST for a solar mass it had been widely used for low mass stars of all masses based on a semi-analytical argumentation.

Unlike the competing Winkler and Newman (1980a,b) study that neglected convection, SST did not calculate the evolution before the main accretion phase and the transition towards the first hydrostatic core because 'the detailed behaviour of these processes depends strongly on the assumed initial conditions, and there is little hope of observing this behaviour in a real system.'

Therefore, instead of calculating this transient phase, SST assumed an initial hydrostatic core of $0.01\,M_\odot$ accreting matter in a quasi-steady way (p. 640). The entropy structure was assumed to be linear in mass and convectively stable inside the surface entropy spike due to the shock. Surface temperature was estimated to exceed values for Hayashi's (Hayashi et al. 1962) forbidden zone and SST therefore assumed $T_g = 3000$ (Stahler et al. 1980b, p. 234). Two values for dimensionless entropy gradient were tested (Stahler et al. 1980b, p. 235).

The entropy structure was then assumed to be linear in mass. After a comparison of the effects on later evolution for the different values of the assumed initial entropy gradient it was concluded that the differences had only small effect for the later evolution. The constant gradient assumption, however, was never investigated.

Non-spherical accretion

The effects of non-spherical accretion on early stellar evolution have been taken into account in an analogous way as described in the previous section. The descriptions of accretion accounts for discs and magnetic fields in parameterized ways, but only after an initial stellar or stellar embryo structure has been obtained, as discussed above; see Hartmann *et al.* (1997), Siess and Forestini (1996) and Siess *et al.* (1997) for a discussion.

Summary on initial stellar structure

Our theoretical knowledge about protostellar collapse and the pre-main sequence thus remained separated. On the one hand, modelling based on classical stellar structure theory was able to produce pre-main sequence tracks that could be related to observations if the question of the initial entropy distribution was put aside. On the other hand, models of the protostellar collapse revealed the entropy structure to be a signature of the accretion history, but were unable to provide pre-main sequence observables needed to confront with observations. That led to an *invited debate* at the IAU Symposium 200 – *The Formation of Binary Stars*, which I tried to summarize in dialogue form (Wuchterl 2001a).

Apparently, there is a discrepancy about initial stellar structure between time-dependent radiative studies and time-independent convective studies. Recent advances, both in computational techniques and in the modelling of time-dependent convection, now make it possible to calculate the pre-main sequence evolution directly from cloud initial conditions by monitoring the protostellar collapse until mass accretion fades and the stellar photosphere becomes visible (Wuchterl and Tscharnuter 2003).

4.6. Calculating protostellar collapse

Why is it that protostellar collapse calculations are not routinely used to determine the starting conditions for stellar evolution calculations? The reason lies in the very different physical regimes that govern the original clouds and the resulting stars, as well as the fact that the transition between them is a dynamical one. Let us first compare the physical regimes. Cloud fragments are (1) quasi-homogeneous, i.e. the rim–centre density contrast is less then 100, say, (2) cool, with temperatures of typically 10 K, (3) opaque for visible light but transparent at the wavelengths that are relevant for energy transfer – mostly controlled by dust radiating at the above temperature – and consequently, (4) isothermal.

On the other hand, stars, even the youngest ones (1) are opaque, (2) are compact, (3) are hot, (4) are non-isothermal, with temperature gradients becoming so steep that convection is driven and plays a key role in energy transfer, (5) require non-ideal and often degenerate and generally elaborate equations of state to describe ionization and other processes, and consequently also (6) require detailed calculations of atomic and molecular structure to quantify the corresponding opacities.

Simplifying approximations can be found that are valid in the cloud or the stellar regime, respectively. But to calculate the transition from clouds to stars a comprehensive system of equations has to be formulated. It has to contain both extreme regimes – clouds and stars. In addition, because the transition involves a gravitational instability of the cloud equilibrium, an equation of motion has to be used instead of the hydrostatic equilibrium (Eq. (4.6)). It is the instability of that force equilibrium that initiates the collapse in the first place. Following the instability the clouds collapse and their motion approaches free fall, i.e. gravity and pressure are nowhere near balance.

The equation of motion for a spherical volume V containing matter of density ϱ that is moving with velocity u can be written in the form:

$$\frac{d}{dt}\left[\int_{V(t)} \varrho u \, d\tau\right] + \int_{\partial V} \varrho u (u_{\rm rel} \cdot dA) = -\int_{V(t)} \left(\frac{\partial p}{\partial r} + \varrho \frac{GM_r}{r^2}\right) d\tau + C_{\rm M}, \quad (4.12)$$

where the original force balance of hydrostatic equilibrium (Eq. (4.6)), now reappears as the first term on the right hand side. It has become a generally non-zero 'source' of momentum density ϱu. The left side of Eq. (4.12) is the total change of momentum for the volume V under consideration, i.e. the change inside the volume and what is transferred in and out across its surface ∂V. These changes are due to changes in the mass (density) inside the volume and the acceleration due to the forces on the right hand side. Thus, the above equation is the continuum version of Newton's second law written in reverse:

$$ma = F, \quad (4.13)$$

where the momentum can change due to both a change in mass and an acceleration due to the forces. For the forces we have the pressure gradient, gravity as in the hydrostatic equilibrium and $C_{\rm M}$, the term describing the momentum coupling between matter and radiation. We can loosely speak of 'radiation pressure'. To explain that term we briefly look at the case of an accreting object that produces a radiation field dominated by the energy generation due to the accretion process itself. If the gas-pressure gradient is negligible and if gravity were to balance $C_{\rm M}$, an object would then accrete at the *Eddington limit* with any increased radiation pressure becoming stronger than gravity and reverse the acceleration in an outward direction, and hence render more rapid accretion impossible.

If we set $C_{\rm M}$ at zero and require the explicit and implicit time derivatives on the left side to be zero we recover (leaving the integral aside) Eq. (4.6), i.e. hydrostatic equilibrium.

4.6.1. Inertia governs the collapse

The introduction of the equation of motion has important consequences. It introduces new timescales. Globally, things change on the free-fall time, Eq. (4.11), for the collapse. Locally, the typical timescales for a significant change in a given volume are now comparable to the time a sound wave needs to cross that volume. This is the *dynamical timescale*,

$$\tau_{\rm dyn} = \frac{R}{c_{\rm s}} = \frac{R}{\sqrt{\Gamma_1 \frac{P}{\varrho}}} \approx \frac{R}{\sqrt{\frac{kT}{m}}}, \quad (4.14)$$

where R is the linear size of a typical region under consideration, e.g. the cloud radius or ultimately the stellar radius, and $c_{\rm s}$ is the isentropic (adiabatic) sound speed. Initially the dynamical timescale is equal to the free-fall time for the initial equilibrium cloud. But as the stellar embryo takes shape and heats up, the dynamical timescales drop dramatically. Finally, for mature main-sequence stars we are typically at the solar sound crossing time of 1.5 hours and minutes for oscillations in the upper layers, like the five minute oscillations used in helioseismology. It has to be kept in mind that the dynamical timescale is an estimate based on the sound speed. Hence, events can be and are even faster in the hypersonic flows that appear in stellar collapse.

The physical ingredient preventing the rates of change from being even faster is the inertia of matter. Unlike the *thermal inertia*, which is controlled by the transfer processes and hence by the opaqueness of matter, inertia is universal. Therefore, if a cloud starts to collapse, the initial cloud structure imprints a timescale on to the accretion process. The question is then how much energy the transfer processes can get out of the cloud during the collapse time.

4.6.2. *Isothermal three-dimensional collapse*

The equation of motion significantly complicates the mathematics and numerical analysis compared to stellar evolution calculations that assume the hydrostatic equilibrium to hold. Yet the collapse is calculated almost routinely if additional assumptions are made about the cloud energetics. The simplest assumption is to use the initial cloud temperature. Such 'isothermal' calculations can be carried out without restrictive symmetry assumptions, i.e. in three dimensions, and are accurate for the early stages of cloud collapse. They provide insight into the fragmentation process. However, they also show that important processes happen at the transition to the non-isothermal phases. At this transition the cloud centres become opaque and the temperature starts to rise, signalling the first transient stopping of the collapse process. That happens at some sufficiently high density, typically after an increase of about ten orders of magnitude from the original cloud conditions. One of the most important of these non-isothermal effects is likely to be the opacity limit for the fragmentation process itself that we briefly touched on earlier. Hence, while the isothermal calculation (and calculations with other special 'equation of state assumptions') can show the dynamics without symmetry restrictions, they cannot answer the question of how much energy leaves the cloud during the collapse process and how much heat remains inside the star after accretion is completed.

4.6.3. *The heat in young stars*

To determine the temperature in the clouds during the non-isothermal phases of collapse, energy gains and losses have to be budgeted much in the way we have seen for the pre-main sequence phase above. The key differences result from the fact that now dynamics is under control of the timescales – inertia rules – not the energy transfer processes themselves. Therefore, it becomes important how and how fast a transfer process can react on a change of the situation that is imposed by the dynamics. The collapsing clouds are mostly non-hydrostatic, sooner rather than later changing faster than the energy transfer processes can respond.[10] That requires a time-dependent treatment of all these processes. This is usually not necessary for later pre-main sequence phases because the transfer processes control the changes leading to a quasi-equilibrium situation, as described in Section 4.5.2.

In addition *radiative transfer* is complicated by the fact that the problem cannot be separated into a hydrostatic, very opaque part – the stellar interior – and a geometrically and optically thin part that emits the photons into the ambient space – the stellar atmosphere. A complete solution of the radiation transfer equation is needed.

Finally, *convection* is driven in the rapidly changing accretion flows and the rapidly contracting, rapidly heated young stellar embryos. Convection is rapidly switched on or off, requiring a description of how fast convective eddies are generated or vanish. This

---

[10] So fast that they play no role in parts of the flow for significant timespans, leading to adiabatic phases.

is unlike stellar (pre-) main sequence evolution, where changes are driven by chemical evolution or slow contraction and hence the convection pattern always has enough time to adjust itself to new interior structures so that the time-independent convective energy flux of mixing length theory can be used. It is also unlike stellar pulsations, where convection zones are pre-existing and are modulated by the oscillation of the outer stellar layers. Convection in young objects is initiated in fully radiative structures and hence the rapid creation of a convection zone has to be covered by the theory – surprisingly, a non-trivial requirement (see Wuchterl 1995b; Wuchterl and Feuchtinger 1998). In short, a time-dependent theory of convection is needed.

4.6.4. Beyond the equilibria of stellar evolution theory

Overall, this results in a departure from the three major equilibria of stellar structure: hydrostatic equilibrium, radiative equilibrium and convective equilibrium. Instead of the equilibrium budget, equations have to be used. *Three* hydrodynamic equations to describe the motion of matter, *two* moment equations derived directly from the full radiative transfer equation that constitute the equations of motion for the radiation field, and *one* equation describing the generation and fading of convective eddies to calculate a typical kinetic energy for the eddies from which the convective flux can be obtained. The system includes mutual coupling terms that describe, for example, how moving matter absorbs and emits radiation or dissipating convective eddies create heat. Supplemented with the Poisson equation, we arrive at the full set of the equations of fluid dynamics with radiation and convection, also referred to as convective radiation hydrodynamics: altogether seven partial differential equations instead of four ordinary ones for stellar evolution. To show the budget character of these equations more clearly, and also because of advantages for solving them, we give the integral version of the equations that is valid for spherical volumes V:

$$\Delta M_r = \int_{V(t)} \varrho \, d\tau, \tag{4.15}$$

$$\frac{d}{dt}\left[\int_{V(t)} \varrho \, d\tau\right] + \int_{\partial V} \varrho(u_{\rm rel} \cdot dA) = 0, \tag{4.16}$$

$$\frac{d}{dt}\left[\int_{V(t)} \varrho u \, d\tau\right] + \int_{\partial V} \varrho u(u_{\rm rel} \cdot dA) + \int_{V(t)} \left(\frac{\partial p}{\partial r} + \varrho \frac{GM_r}{r^2}\right) d\tau = C_{\rm M}, \tag{4.17}$$

$$\frac{d}{dt}\left[\int_{V(t)} \varrho(e+\omega) \, d\tau\right] + \int_{\partial V} [\varrho(e+\omega)u_{\rm rel} + j_{\rm w}] \cdot dA + \int_{V(t)} p \, {\rm div}\, u \, d\tau = -C_{\rm E}, \tag{4.18}$$

$$\frac{d}{dt}\left[\int_{V(t)} E \, d\tau\right] + \int_{\partial V} [Eu_{\rm rel} + F] \cdot dA + \int_{V(t)} P \, {\rm div}\, u \, d\tau = C_{\rm E}, \tag{4.19}$$

$$\frac{d}{dt}\left[\int_{V(t)} \frac{F}{c^2} \, d\tau\right] + \int_{\partial V} \frac{F}{c^2}(u_{\rm rel} \cdot dA) + \int_{V(t)} \left(\frac{\partial P}{\partial r} + \frac{F}{c^2}\frac{\partial u}{\partial r}\right) d\tau = -C_{\rm M}, \tag{4.20}$$

$$\frac{d}{dt}\left[\int_{V(t)} \varrho\omega \, d\tau\right] + \int_{\partial V} \varrho\omega u_{\rm rel} \cdot dA = \int_{V(t)} (S_\omega - \tilde{S}_\omega - D_{\rm rad}) d\tau, \tag{4.21}$$

$$C_{\rm M} = \int_V \kappa\varrho \frac{F}{c} \, d\tau, \quad C_{\rm E} = \int_V \kappa\varrho(4\pi S - cE) d\tau, \quad P = \frac{1}{3}E. \tag{4.22}$$

The equations connect *matter*, described by mass density ϱ, gravitating mass M_r, interior to radius r, velocity u, specific internal energy e, gas pressure p and *radiation*, characterized by radiation energy density per unit volume E, radiative flux density per unit surface area F, and radiative pressure P, to convective eddies described by the specific turbulent kinetic energy density per unit mass ω (that is, the square of a mean convective velocity, $u_c = \sqrt{2/3\omega}$) and the convective energy flux density per unit surface area j_w. The connection runs via the matter–radiation coupling terms for momentum exchange C_M, energy exchange (absorbtion and emission of radiation) C_E and radiative cooling of convective elements $D_{\rm rad}$. κ is the frequency average of the mass extinction coefficient per unit mass and S the source function. S_ω and $\tilde{S}_\omega$ are the production and dissipation rates per unit volume of turbulent kinetic energy, ω, due to convective eddy generation by buoyancy forces and eddy dissipation by viscous forces. V is the time-dependent volume under consideration (usually a shell in a celestial body), ∂V its surface, $d\tau$ and dA are volume and surface elements, and $u_{\rm rel}$ is the relative velocity of the flow across the volume surface. Wuchterl and Tscharnuter (2003) discuss how to set up and solve these equations and describe solutions relevant to the formation of stars and brown dwarfs. The authors calculate the collapse of cloud fragments with masses ranging from 0.05 to 10 $M_\odot$ and discuss the consequences for the hydrostatic stellar evolution on the pre-main sequence.

4.7. Early stellar evolution – hydrostatic versus collapse

We now look at the key differences between hydrostatic and collapse calculations of early stellar evolution for the case of one solar mass. For the hydrostatic comparison case we follow Wuchterl and Tscharnuter (2003) and choose the calculations by D'Antona and Mazzitelli (1994), because atmospheric treatment, equations of state, opacities and convection treatment closely match those of Wuchterl and Tscharnuter (2003). Comparison with studies that include more physical processes (e.g. disc accretion or frequency-dependent photospheric radiative transfer) can then be made by using existing intercomparisons of different hydrostatic studies to the D'Antona and Mazzitelli (1994) study. The luminosity as a function of age is shown for the collapse of a Bonnor–Ebert sphere and a hydrostatic, contracting, initially fully convective young star in Figure 4.1. Both studies use close to identical equations of state and calibrate mixing length theory of convection with the Sun. The two important differences are (1) the initial conditions and (2) the model equations.

Initial conditions The starting point for collapse is a solar-mass Bonnor–Ebert sphere as the initial gravitationally unstable cloud fragment. The calculation of pre-main sequence contraction starts with an initially high-luminosity, fully convective structure.

Equations Collapse is calculated using convective radiation fluid dynamics with an equation of motion and time-dependent radiative and convective energy transfer. The calculation of pre-main sequence contraction assumes hydrostatic equilibrium and accounts for time-dependence only due to slow (very subsonic) gravitational contraction via the energy equation. To a good approximation, the equations of the contraction calculation, D'Antona and Mazzitelli (1994), are a hydrostatic limiting case of the collapse equations used by Wuchterl and Tscharnuter (2003).

The collapse calculation starts at zero luminosity (at the beginning of cloud collapse) and stays above the luminosity of the quasi-hydrostatic contraction calculation to beyond 2.5 Ma. The cloud collapse does not lead to a fully convective structure as assumed for the hydrostatic calculation. Even after most of the mass becomes hydrostatic, most of the

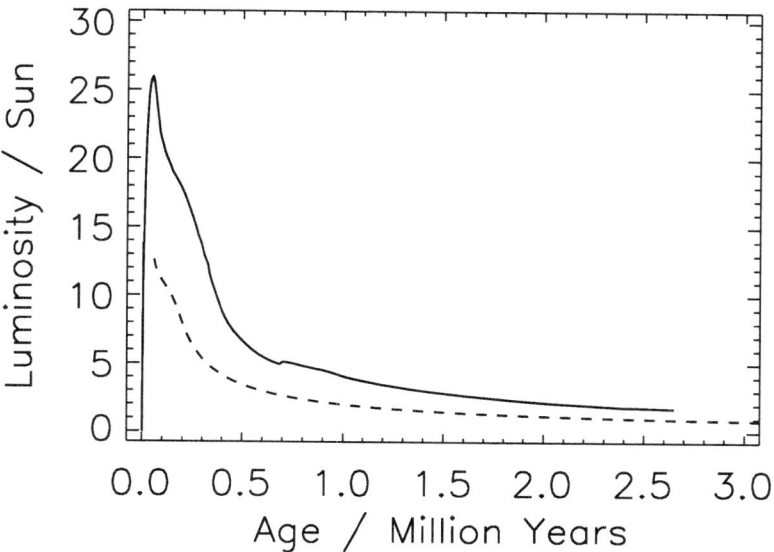

FIGURE 4.1. Early stellar evolution: collapse vs. Hayashi-line contraction. Luminosity as a function of age for a solar mass star. *Full line*: calculated from the protostellar collapse by Wuchterl and Tscharnuter (2003). *Dashed line*: the quasi-hydrostatic contraction of an initially fully convective young star by D'Antona and Mazzitelli (1994).

deuterium has been burnt and accretion effects have ceased to dominate, at approximately 0.7 Ma, the internal structure stays partially radiative. The inner two thirds in radius remain radiative with a convective shell in the outer third of the radius – reminiscent of the present solar interior structure. Wuchterl and Tscharnuter (2003) found similar results for 2 to 0.05 $M_\odot$ Bonnor–Ebert spheres, indicating that at least spherical collapse would not lead to fully convective structures over a considerable mass range. The question then was whether the initial cloud conditions or the cloud environment or non-spherical effects could change the result. Therefore, Wuchterl and Klessen (2001)[11] studied the fragmentation of a large, dense molecular cloud by isothermal hydrodynamics in three dimensions and followed the collapse of one of the resulting fragments, which was closest to a solar mass, throughout the non-isothermal phases to the end of accretion using the spherically symmetric equations of Wuchterl and Tscharnuter (2003). Despite the very different cloud environment – interactions with neighbouring fragments, competitive accretion, varying accretion rate and orders of magnitude higher average accretion rates – the structure of the resulting, young solar mass star after 1 Ma was almost identical to that resulting from the quiet Bonnor–Ebert collapse.

There were large differences during the embedded high-luminosity phases and the earliest pre-main sequence phase (see Figure 4.2), but at 1 Ma, when the accretion effects had become minor the overall structure resulting from the Bonnor–Ebert collapse was confirmed: a convective shell on top of a radiative interior. This has consequences for the observables of very young stars. A solar precursor at 1 Ma should have twice the luminosity and an effective temperature that is 500 K higher than that of the respective fully convective structure resulting from a hydrostatic, high-luminosity start.

---

[11] The Wuchterl and Tscharnuter (2003) article was submitted in 1999 and subject to four years of peer reviewing by numerous reviewers.

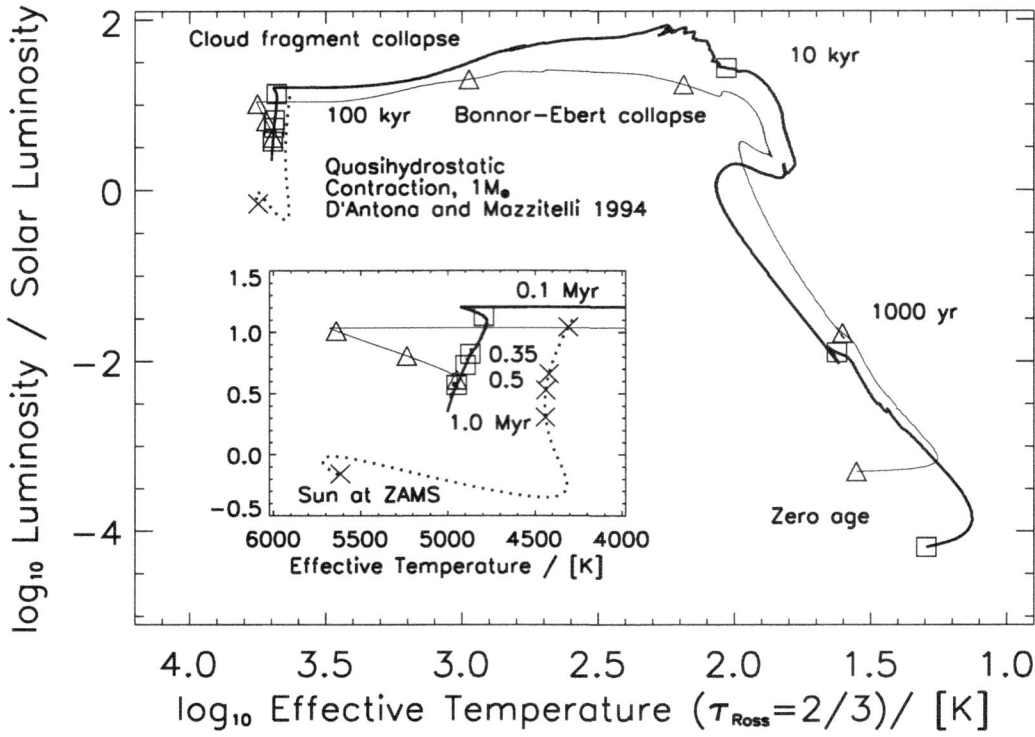

FIGURE 4.2. Protostellar collapse and early stellar evolution for a solar mass in a large theoretical Hertzsprung–Russell diagram. The collapse resulting from a fragmenting cloud (thick line) and a Bonner–Ebert sphere (thin line) are compared to a quasi-hydrostatic pre-main sequence calculation (D'Antona and Mazzitelli 1994). The insert shows an amplification of the evolution beyond 0.1 Myr. After Wuchterl and Klessen (2001).

The differences of the young Sun's properties at 1 Ma are the result of a number of differences that result when early stellar evolution is calculated directly from the cloud collapse instead of the hydrostatic evolution of initially high luminosity, fully convective structures. The collapse calculations predict a series of changes to the classical picture. In summary (Wuchterl and Tscharnuter 2003):
  (i) young solar-mass stars are not fully convective when they have first settled into hydrostatic equilibrium;
  (ii) their interior structure with an outer convective shell and an inner radiative core extending across 2/3 of the radius rather resembles the present Sun than a fully convective structure;
  (iii) in the Hertzsprung–Russell diagram they do not appear on the Hayashi line but to the left of it;
  (iv) most of their deuterium is burned during the accretion phases;
  (v) deuterium burning starts and proceeds off-centre in a shell;
  (vi) therefore, there is no thermostatic effect of deuterium during pre-main sequence contraction and hence no physical basis for the concept of the stellar birthline, as proposed by Stahler (1988a);
  (vii) these results are independent of the accretion rates during the cloud fragmentation phase and non-spherical effects in the isothermal phase.

These differences have important consequences for surface abundances (of deuterium and lithium), rotational evolution, the stellar dynamo and stellar activity that have to be worked out.

The observational tests are still to be done. What is needed are very young binary systems where masses can be determined independently of the model calculations *and* where the stellar parameters, effective temperature, surface gravity, or ideally the radius can be determined with sufficient accuracy. Furthermore, the binary has to be sufficiently wide to exclude interactions during accretion and the evolution to the observed stage.

But for theoretical reasons alone – namely, the requirement of a physical description of star formation and protostellar collapse – the dynamical models should be used when masses of stars, brown dwarfs and planets are determined from the luminosities, effective temperatures, gravities or radii of those objects, at least for the ages that can be currently covered with such models, i.e. up to 10 Ma.

4.8. Protoplanetary discs

In the previous sections we have shown how the problem of star formation can be solved when angular momentum is neglected. The gravitational collapse of a cloud is then stopped when its central parts become opaque and heat up compressively, and the thermal pressure finally re-balances gravity. Because pressure and gravity act isotropically, the resulting structures – young stars – are spherical. When angular momentum is taken into account there is a second agent that can bring the collapse to a stop: inertia, which gives rise to the centrifugal force. Even a slowly rotating cloud spins up dramatically as material collapses towards the centre reducing the axial distance by many orders of magnitude[12] under conservation of angular momentum. Unlike gas pressure, the centrifugal force is anisotropic and perpendicular to the axis of rotation. Hence collapse parallel to the axis of rotation is not modified by rotation. Gas can fall directly on to the central protostar along the polar axis and parallel to it towards the equatorial plane. The centrifugal force builds up during radial infall until it balances gravity at the *centrifugal radius*,

$$R_{\text{centrifugal}} = \frac{R^4 \omega^2}{GM}, \quad (4.23)$$

for a cloud of mass M, initial radius R, rotating with an initial angular frequency ω. Upon approaching that radius, the gas flow is more and more directed towards a collapse parallel to the axis of rotation. Finally, material arrives in the equatorial plane. There, the vertical component of the central star's gravitational force, i.e. the component parallel to the rotational axis, is zero. The radial component of the primary's gravitational force is balanced by the centrifugal forces. The cloud has collapsed to a flattened structure in the stellar equatorial plane – a circumstellar disc. Because the gas in the disc has a finite pressure, the disc has a finite thickness determined by the force equilibrium of gas pressure and the vertical component of the star's gravity. That is another hydrostatic equilibrium, but only in one direction. The force balance in the radial direction is somewhat more subtle. Because the primary's gravity increases with decreasing orbital distance, parts of the disc that are nearer to the stars are more highly pressurised (more compressed and hence thinner in the vertical direction) than those farther out. Therefore, there is a radial pressure gradient in the disc that contributes to the force balance. Without that pressure,

---

[12] Typically by a factor of 1000 from the initial radius, 50 000 AU, of a solar mass cloud fragment to 50 AU for the disc radius.

the force equilibrium would be gravity vs. centrifugal force, leading to the usual circular Keplerian orbit. With the gas pressure, an orbiting blob of gas is subject to the outward-directed pressure gradient of the disc. This partially reduces the effect of gravity. As a result the centrifugal force required for a balance is slightly less than in vacuum and the disc orbits somewhat below the respective Kepler speeds. While this story is simply told, the details bring an enormous computational task so that there are only very view studies in the literature (e.g. Morfill et al. 1985; Tscharnuter 1987) that can actually calculate the protoplanetary nebula as an outcome of the collapse and follow the evolution from the cloud to stages where planet formation is expected. Again, the problem is the non-isothermal part of the evolution, complicated further by our incomplete knowledge of disc angular momentum transfer processes.

If such processes are assumed, the cloud collapse results in a central pressure-supported protostar and a centrifugally supported accretion disc (Tscharnuter 1987) that may become a protoplanetary nebula.

We can only briefly list here the most important elements of our knowledge, or more appropriately our ignorance of quantitative protoplanetary disc structure. For a discussion we refer to Wuchterl et al. (2000), Wuchterl (2004a) and, with an emphasis on Jupiter, to Lunine et al. (2004). The key problem is that at present neither observation, due to resolution and sensitivity restrictions, nor theory, due to computational difficulties and the problems with the angular momentum transfer process, can provide sufficient information about the physical and chemical state of the protoplanetary nebula to build a satisfactory planet formation theory. However, the following concepts and constraints have been collected for the properties of protoplanetary nebulae.

(i) Theoretically, protoplanetary nebulae may form as a byproduct of the collapse of clouds with properties that are observed in actual clouds on the sky.

(ii) Accretion discs form when an appropriate anomalous viscosity exists that allows the accretion of mass onto the central protostar and outward angular momentum transfer through the discs. Candidate processes for the source of that viscosity are turbulence driven by a shear instability in the Keplerian flow, convective instability due a vertical temperature gradient in the disc, or a magneto–centrifugal instability in a conductive disc with a small seed magnetic field as well as gravitational torques induced by non-axisymmetric structures in the disc.

(iii) Assuming that km-sized planetesimals form, and that the material at present found in the Solar System is converted into such planetesimals, the theory of planetesimal accretion by growth via pairwise collisions can show that the terrestrial planets form within about 100 Ma. A similar statement holds true for the cores of the giant planets, but it is necessary to assume that the original planetesimal disc in the outer Solar System is a factor of a few more massive than the amount of condensible elements currently inferred for the outer Solar System (mostly the condensible element cores of the giant planets). There are two problems for Solar System formation theory here: planetesimal formation and the timescales for giant planet formation. These will be discussed below.

(iv) From meteoritic chronology, disc lifetimes can be inferred as the time span when there was chemistry in the presence of nebula gas and impact driven melting during planetesimal accretion (Wadhwa and Russell 2000). The events can be absolutely dated from the decay of radio nuclides (Lugmair and Shukolyukov 2001; Allègre et al. 1995) and planet formation timescales can be derived by relative chronology that points to a few hundred Ma. Recent studies obtain a formation timescale of the Earth mantle of approximately 30–60 Ma (Yin et al. 2002), depending on the accretion scenario that determines the size and frequency of reservoirs, which in turn contribute to the mixing process producing the isotopic evidence that can be analyzed today.

(v) Observations at mm wavelengths have shown that the majority of young stars have circumstellar material of a few to ten per cent of the stellar mass.

(vi) In many cases the inferred discs have been resolved by mm-interferometry, optical and infrared imaging. Sizes range from disc radii of 1000 AU down to the resolution limits.

(vii) The properties in these discs are only weakly constrained by observation. Mostly because of insufficient resolution, mm interferometry is able to provide information about the temperature, density and turbulent motion for scales down to 50 AU in the best cases, i.e. for bright, nearby systems.

(viii) The disc structure below 50 AU is spatially unresolved. Spectral energy distributions can be used to infer properties via disc modelling. From near and mid-IR data, the surface properties of discs can be inferred for distances of a few to a few tens of stellar radii.

(ix) Those observations show a decay of IR emission from inner disc regions on timescales of a few to a few tens of Ma. These observations give statistical information for the frequency of detected IR emission for clusters that can be dated via the stellar pre-main sequence evolution. Because a cluster sample is observed the obtained timescale may indicate various evolutionary events in the systems (binaries) and their environment, with planet formation or disc dispersal being possible underlying processes.

In summary, the theoretical prediction of the frequent occurrence of circumstellar discs is well corroborated by observations. The observationally inferred disc masses are in the expected range of a few to ten per cent of the stellar masses. Evolution of disc indicators is seen on timescales expected for planet formation, but the information about *local* disc properties that planet formation theory could use as input is presently beyond observational capabilities for orbital distances below 50 AU. This unfortunately corresponds to the size of the Solar System. Even less accessible are the orbital distances of less then 5 AU of extrasolar planets detected by radial velocity and transit techniques. In search of more information about the planet formation era, with hardly any options left, we turn now to our well studied home system.

Our approach is to look at the Solar System first and then to try to generalize protoplanetary nebula structure by assuming that planet formation might occur in any gravitationally stable nebula. We will thus obtain a range of plausible protoplanetary nebula discs that are scaled from what we know from reconstructions of the solar nebula from the distribution of matter in the present Solar System.

4.9. The Solar System

More planetary candidates are now known in our galactic neighbourhood than planets orbiting our star. But the Solar System is by far the best studied and most completely known system. The masses, radii and composition of the major and some minor bodies are known. Almost a million orbital elements allow a detailed study of the dynamical and stability properties of the system (Lecar *et al.* 2001). The information about the Sun, the planets and their satellite systems provides a clear picture of the angular momentum distribution. Interior structures and surface properties contain a record of the formation history. The heat budgets of the giant planets show an excess of emitted radiation over absorbed sunlight (except for Uranus where only an upper limit is available). These faint intrinsic luminosities, in the $nL_\odot$ range ($1\,nL_\odot = 10^{-9}$ solar luminosities) are directly related to the formation process that stored the heat 4.5 Ga ago ($1\,\text{Ga} = 10^9$ years). The solid, icy and rocky surfaces of terrestrial planets and many satellites show a record of impacts that samples a time-range reaching from the present, e.g. the surface of Io, back to the formation epoch, e.g. in the lunar highlands. Radioactive dating provides absolute ages for the Earth, Moon and meteorites. The latter provide an accurate age of $4565 \pm 1\text{Ma}$ for the oldest nebula condensates, the calcium–aluminium rich inclusions in primitive meteorites (Allègre *et al.* 1995). Ongoing processes such as orbital evolution, asteroidal collisions and even large impacts, such as those of the fragments of comet Shoemaker–Levy 9 on to Jupiter, can be studied directly and in great detail. Finally, there is abundant information gathered by fly-by spacecraft missions such as the *Voyagers*, orbiters such as *Galileo*, *Mars-Odyssey* and *Mars-Express*, as well as from *in situ* exploration by landers and atmospheric entry probes for the Moon, Venus, Mars and Jupiter, and last but not least the successful atmospheric entry and landing of the *Huyghens* probe on Saturn's moon Titan in January 2005; finally, there is the recent artificial collision with a comet of the impactor of the *Deep-Impact* mission on 4 July 2005 to bring material from the interior of a comet to light.

Our four giant planets contain 99.5% of the angular momentum of the Solar System, but only 0.13% of its mass. Terrestrial planets contribute another 0.16% to the angular momentum. On the other hand, more than 99.5% of the mass and thermal energy of the planetary system is in the four largest bodies, with the remaining 0.5% mostly in the second group of four, the terrestrial planets. Modern models of the interiors and evolution of giant planets in our Solar System account for the high pressure properties of hydrogen, helium and the heavier elements, as well as energy transfer by radiation and convection. When fitted to the observed global properties of Jupiter at an age of 4.5 Ga they show that 10–42 of its 318 Earth masses are due to heavy elements. This corresponds to between 3 and 13%. The respective mass fractions implied for the other Solar System planets are all higher. That points to a bulk enrichment of heavy elements more than a factor of two above solar composition and implies heavy element cores ranging from greater than one Earth mass to a considerable fraction of the total mass (Guillot 1999, 2005; Wuchterl *et al.* 2000). The heavy element enrichment is even more obvious in terrestrial planets, which may also be viewed as cores of failed giant planets. The hydrogen and helium that constitute 98% of the Sun's mass and between 87% and 15% of the giants' is a minor constituent of terrestrial planets. Planet formation enriches heavy elements relative to the central star, which is formed from the same protostellar cloud. Such extensive enrichment is not predicted by any mechanism proposed for the formation of stars and brown dwarfs. That alone already indicates that planet formation is fundamentally different from star formation.

4.10. Solar System formation

The distribution of mass and angular momentum in the Solar System can be understood on the basis of the *nebular hypothesis* (Kant 1755). The nebular hypothesis assumes concurrent formation of a planetary system and a star from a centrifugally supported flattened disc of gas and dust with a pressure-supported central condensation (Laplace 1796; Safronov 1969; Lissauer 1993). Flattened preplanetary nebular discs explain the coplanarity and circularity of planetary orbits by the respective properties of the parent disc. Theoretical models of the collapse of slowly rotating molecular cloud cores have demonstrated that such preplanetary nebulae are the consequence of the observed cloud core conditions and the dynamics of radiating fluids, provided there is a macroscopic angular momentum transfer process (Morfill et al. 1985). Assuming turbulent viscosity to be that process, dynamical models have shown how mass and angular momentum separate by accretion through a viscous disc on to a growing central protostar (Tscharnuter 1987; Tscharnuter and Boss 1993). Such cloud collapse calculations, however, still do not reach to the evolutionary state of the nebula where planet formation is expected. Observationally inferred disc sizes and masses are overlapping theoretical expectations and confirm the nebula hypothesis. High resolution observations at millimetre wavelengths are sensitive to disc conditions at orbital distances > 50 AU. However, observations thus far provide little information about the physical conditions in the respective nebulae on scales of 1 to 40 AU, where planet formation is expected to occur. Planet formation studies therefore obtain plausible values for disc conditions from nebulae that are reconstructed from the present planetary system and disc physics. The so-obtained *minimum reconstituted nebula masses*, defined as

the total mass of solar composition material needed to provide the observed planetary/satellite masses and compositions by condensation and accumulation,

are a few per cent of the central body for the solar nebula *and* the circumplanetary protosatellite nebulae (Kusaka et al. 1970; Hayashi 1980; Stevenson 1982). See Lunine et al. (2004) for a more detailed discussion of reconstructing the preplanetary nebula from observational constraints provided by Solar System data.

4.11. Planet formation – the problem

Giant planet formation requires (1) a compression of the solar nebula gas by about 10 orders of magnitude to form a gaseous condensation held together by its own gravity, at Jupiter's present mean density of 1.33×10^3 kg/m^3, and (2) an enrichment of the heavy elements – which are condensible in the nebula – by typically a factor of at least three above the nebula value, most probably with a substantial fraction contained in a core. Gas in the midplane of a *minimum mass solar* nebula typically has a density of 10^{-8} kg/m^3 at Jupiter's present orbital radius (Hayashi et al. 1985) and a temperature around 100 K. The nebula gas pressure, the young Sun's tides and the radially decreasing orbital velocities in a circumstellar disc, that shows an almost 'Keplerian' shear, counteract the compressing force of nebular gas self-gravity. Accordingly, most circumstellar nebulae – modelled and observed – are gravitationally stable. Unlike in interstellar clouds, larger mass fragments in a circumstellar disc are generally no longer unstable. That is because larger mass fragments at given nebula densities require larger scales that are subject to stabilization against self-gravity by the stellar tidal pull and the Keplerian shear. While in an interstellar cloud sufficiently large scales are always gravitationally unstable, nebulae are stabilized on short *and* long scales.

In addition to the above mechanical barriers against gravitational self-compression of nebular gas, there is a thermal barrier: a Jupiter-mass fragment is optically thick even under unperturbed nebular conditions. The optical depth, τ, for a blob of mass M at nebula densities $\varrho_{\rm neb}$, and for typical (dust) opacity κ is:

$$\tau = 36 \left(\frac{M}{[M_{\rm Jupiter}]}\right)^{1/3} \left(\frac{\kappa}{[0.01 \text{ m}^2/\text{kg}]}\right) \left(\frac{\varrho_{\rm neb}}{[10^{-8} \text{ kg/m}^3]}\right)^{2/3}, \qquad (4.24)$$

where the problem-orientated units in the scaling have been chosen according to the values for the jupiter's position in Hayashi's minimum mass nebula. Any rapid, i.e. dynamical compression under such conditions will result in a temperature increase determined by the efficiency of transfer processes and a much stronger counteracting pressure than in a simple isothermal analogue of protostellar collapse scaled down to planetary masses. In the stellar case with $\tau \ll 1$, compressional heat leaks out as fast as it is produced, keeping the parent cloud isothermal for many orders of magnitude[13] in compression. The rapid compression under optically thick conditions in the planetary case produces an immediate thermal pressure increase, that typically leads to a slowing down of compression from the dynamical timescale of the fragment, a few years, to the thermal (cooling) timescale which is found to be of the order of a million years in detailed models. Compression of the nebula gas – and therefore giant planet formation, their mass-growth and evolution – is then controlled by the heat loss of the fragment or protoplanet, e.g. Safronov and Ruskol (1982). Collapse, i.e. fast, gravity-driven compression, in an essentially free-falling manner, with pressure playing a negligible role, is a very unlikely event in such conditions.

The physical nature of giant planet formation – collapse, thermally controlled quasi-hydrostatic contraction or static accumulation – is decided by the dynamical stability of the nebula and the pressure build-up inside the protoplanets regulated by the thermal budget of the protoplanetary envelopes. The thermal budget contains (1) heating due to contraction of the gaseous envelopes, (2) dissipation of planetesimal kinetic energy at impacts and, (3) 'cooling' due to energy transfer to the ambient nebula by radiation and convection.

4.12. How to compress by 10^{10}?

The transition from dilute, weakly gravitating nebular conditions to compact planets with a spherical shape rounded by self-gravitation involves a compression of ten orders of magnitude. The nebular gas apparently had to be compressed by a macroscopic process from the earliest stages of planetary growth to the final planetary densities of $\sim 1000 \, \text{kg/m}^3$. For stars, the runaway of the Jeans instability easily multiplied the original cloud density by a factor of 10^{10}, only inhibited from proceeding further by the centrifugal force forming the nebula itself. But, unlike the protostellar collapse, the planetary compression process cannot be analogous for the remaining 10^{10} from the nebula to the final planetary densities because it has to enrich the condensable material at the same time.

Since the 1970s two hypotheses have been discussed that try to account for nebular gas compression *and* condensible element enrichment. The *gravitational disc instability hypothesis* tries to find a nebula analogue of the gravitational Jeans instability of star formation, whereas the *nucleated instability hypothesis* aims to explain giant planet formation as a consequence of the formation of solid, condensable element planetary embryos that act as gravitational seeds for nebular gas capture/condensation. The disc instability hypothesis requires nebulae that undergo self-compression in a dynamically unstable situation and lead to a transition from a smooth regular disc to an ensemble of clumps in orbit around the Sun. Such clumps may be regarded as candidate precursors of protoplanets.

The nucleated instability model looks at giant planet formation as a second step in rocky planet formation. A terrestrial planet embryo acts as a gravitating seed to permanently bind nebular gas, thus forming a massive gaseous envelope around a condensable element core.

The key problem for both compression processes is that preplanetary discs are only weakly self-gravitating equilibrium structures supported by centrifugal forces augmented

---

[13] A factor $\sim 10^{10}$ in density for the collapse of a Jeans-critical solar mass.

by gas pressure.[14] Any isolated orbiting object below the Roche density is pulled apart by stellar tides. Typical nebular densities are more than two orders of magnitude below the Roche density, so a finite nebular pressure is needed to confine a condensation of mass M inside its tidal or Hill radius[15] at orbital distance a:

$$R_T = a \left(\frac{M}{3M_\odot}\right)^{1/3}. \qquad (4.25)$$

Mature planets are dense enough for their radii to be much smaller than the Hill radius – hence, their high densities – and, as a consequence, their high surface gravities usually protect them from tidal disruption or noticeable mass loss. Stellar companions in comparison reduce their densities due to evolutionary effects when they become giants. Consequently, their radii may approach the Hill values depending on their orbital radii. The consequence is Roche-lobe overflow when such stars are very close in binary systems. Planet formation requires a somewhat inverse process, where an extra force compresses the nebular material into the Hill-sphere, allowing more material to flow into the Roche lobe, thereby increasing the planetary mass inside the lobe. All theories of planet formation rely on an extra gravity field to perform this compression.

4.13. How to provide the extra gravity field?

Giant planet formation theories may be classified according to how they provide the gravity enhancement:
   (i) the *nucleated instability* model relies on the extra gravity field of a sufficiently large solid core (condensed material represents a gain of ten orders of magnitude in density and therefore self-gravity compared to the nebular gas);
  (ii) a *disc instability* may operate on length scales between short scale pressure support and long scale tidal support; or
 (iii) an *external perturber* could compress an otherwise stable disc on its local dynamical timescales, e.g. by accretion of a clump on to the disc or rendezvous with a stellar companion.

4.14. From dust to planets

Dust growth in the nebula via pairwise collision to centimetre sizes is now fairly well understood theoretically and experimentally (see Lissauer 1993). A key open question is how the transition from ~ 0.1 m dust-agglomerates to km-sized planetesimals can be accomplished. Planetesimals, which are the building blocks of gravitationally controlled planetary accumulation, may form by a gravitational instability of a dust subdisc or by continued growth via pairwise collisions, provided growth is sufficiently large to dominate over losses due to a radially inward drift (see Lissauer 1993). I will follow the *planetesimal hypothesis* here without further discussion and assume that protoplanetary nebulae form km-sized bodies made of condensible elements within a time-frame of about 10 000 a, see, for example, Hueso and Guillot (2003).

The next step, *runaway planetesimal accretion*, proceeds from $\sim 10^{-9} M_\oplus$ planetesimals to $\sim 0.1 M_\oplus$ *planetary embryos*. In the simplest case a single planetesimal grows within a

---

[14] For a review of circumstellar and protoplanetary discs I recommend the review chapters in the *Protostars and Planets IV* book, Hollenbach *et al.* (2000); Calvet *et al.* (2000); and Beckwith *et al.* (2000).
[15] The Hill radius is the radius of a sphere concentric on a planet within which planetary tidal forces on a small body are larger than the tidal forces exerted by the central star.

swarm of other planetesimals of density ϱ_{sw}. Note that ϱ_{sw} is the mass per unit volume in the planetesimal disc. It is a property of the ensemble of planetesimals and depends on the number and masses of the swarm-members and on the statistics of their orbits, which determine the thickness of the planetesimal disc. It has to be distinguished from the density of individual planetesimals. For a relative velocity, v, the planetesimals' mass, M, grows according to the *particle in box accretion-rate*:

$$\frac{dM}{dt} = \varrho_{sw} v \pi R_s^2 \left[1 + \left(\frac{v_e}{v}\right)^2 \right], \qquad (4.26)$$

were R_s is the distance between the planetesimal centres at contact and v_e is the escape speed at contact. Assuming that the mean ratio of horizontal to vertical motions remains fixed, the rate can be rewritten using the surface mass density, Σ_{solid}, of the planetesimal swarm:

$$\frac{dM}{dt} = \frac{\sqrt{3}}{2} \Sigma_{sw} \Omega_{Kepler} \pi R_s^2 F_g, \qquad (4.27)$$

where we have also introduced the gravitational focusing factor, $F_g = 1 + (v_e/v)^2$, and used the Keplerian angular velocity, $\Omega_{Kepler} = \sqrt{GM/r^3}$. Half the ratio of relative speed to escape speed is also known as the Safronov number, $\theta = 1/2(v_e/v)^2$. Depending on the planetesimal swarm properties, the gravitational focusing factor is typically a few 1000 during early runaway growth, and <8 during the late stages of planetary accretion, see Lissauer (1993).

Based on detailed n-body calculations for a number of planetary embryos together with the self-consistent determination of the properties of the planetesimal swarm, Tanaka and Ida (1999) estimate the respective runaway accretion time for protoplanets of mass M_p, at orbital radius a:

$$\frac{\tau_{grow}}{[a]} = 8 \, 10^5 \left(\frac{M_p}{M_\oplus}\right)^{1/3} \left(\frac{a}{[AU]}\right)^{12/13}. \qquad (4.28)$$

Mutual interactions and accretion of planetesimals and embryos are accounted for. Runaway accretion stops at the *isolation mass*. The isolation mass is reached when a planetary embryo has accreted all the planetesimals within its gravitational range – the so called *feeding zone*. The feeding zone of an embryo extends typically to a few Hill radii (~ 5, say) around its orbit. The values of the isolation mass depend on the initial nebula solid surface density, which specifies the amount of condensible material available per unit nebula surface area, and the orbital radius of the embryo. Values for the isolation mass in a minimum mass nebula are typically 1 $M_\oplus$ in the outer Solar System and a Mars-mass, 0.1 $M_\oplus$, at 1 AU. Protoplanets with masses larger than the isolation mass then must enter an *oligarchic growth stage*. In that stage planetary embryos can only continue to grow by collisions with other embryos. Such collisions can only occur after their motions evolve into crossing orbits by mutual 'secular' gravitational perturbations. These perturbations are slow by nature, particularly because the embryo masses are comparatively low and the embryos are isolated, i.e. outside their neighbour's feeding-zones. The eccentricities required for crossing orbits also result in an increase in relative velocities during encounters and reduced gravitational focusing factors in the collision cross-sections. Thus growth times are much larger for protoplanets above the isolation mass. Kokubo and Ida (2002) estimated the total accretion times of planetary cores through runaway accretion and the late phases of oligarchic growth in the Jovian planet region to be:

$$\frac{T_{\text{grow}}}{[a]} \sim 9\,10^4 \left(\frac{e}{h_M}\right)^2 \left(\frac{M}{[10^{26}\,\text{g}]}\right)^{1/3} \left(\frac{\Sigma}{[4\,\text{g/cm}^2]}\right)^{-1} \left(\frac{a}{[5\,\text{AU}]}\right)^{1/2} \quad (4.29)$$

for an eccentricity in Hill units, e/h_M, $h_M \equiv R_{\text{Hill}}$, solid surface density Σ, final protoplanetary mass M and semi-major axis a. They estimate that the final condensible element mass of a protoplanet at 5 AU would be 5 $M_\oplus$. Accretion would be completed in 40 Ma. A 9 $M_\oplus$ core at the distance of Saturn would require 300 Ma.

These growth times probably have to be shortened due to the enhancement of collision cross-sections of planetary embryos by their envelopes (Inaba and Ikoma 2003). The effect is especially strong when it is considered that some collisions of smaller embryos lead to fragments that are more affected by gas drag in the atmospheres of the larger embryos. Inaba et al. (2003) find that their largest planetary embryo at 5.2 AU, with a mass of $21 M_\oplus$, formed in 3.8 Ma!

4.15. Solar System formation modelling

Understanding the formation of the Solar System currently means the reconstruction of its history. That approach is necessary because, given our incomplete knowledge of important physical processes, it is necessary to include parameterized descriptions of uncertainties. The most famous parameterizations is that of anomalous, turbulent α-viscosity, which is assumed to allow angular momentum redistribution and accretion of mass onto the star. Next, an initial structure model of the preplanetary nebula has to be assumed. Two major classes of nebula models may be distinguished: (1) *active* viscous α-discs (e.g. Ruden and Pollack 1991; Drouart et al. 1999; Hueso and Guillot 2003) and (2) *passive* discs that are heated by absorbed stellar radiation (e.g. the *Kyoto* minimum mass nebula, Hayashi et al. 1985). Once the class and parameterization of the nebula model are chosen (passive or active) the planet formation processes have to be specified and parameterized. The key nebula processes and parameters are:

(i) Distribution of temperature and density as a function of orbital radius. This follows for a given class of nebula models from a chosen mass and mass distribution. In practice the discussion is parameterized by the local surface densities, Σ_{dust} and Σ_{gas}, of nebula condensates and nebula gas, respectively.

(ii) Planetesimal properties and size distribution.

(iii) Planetesimal collision properties, i.e. coefficients of restitution and outcome of collisions – merging into a larger planetesimal or fragmentation into smaller pieces.

(iv) Energy transfer properties of the nebula gas:
   (a) Dust properties (size distribution, composition, mineralogy) to determine the dust opacities and the efficiency of radiative transfer. Nebula dust differs considerably in size and composition from the interstellar dust due to growth and condensation processes in the nebula.
   (b) A prescription and parameterization for convective energy transfer.

Solar System data are used at two stages: (1) in the construction of the nebula surface densities and (2) in the adjustment of parameters by comparing the final outcome of modelling to the empirical data from Solar System planets. Because the uncertainties in the initial nebula structure are very large, the respective structure parameters are the prime ones that are adjusted. A typical procedure is as follows: in the first step a nebula is constructed, for example, by assigning a volume to every Solar System planet. Hydrogen and helium is then added until presolar abundances are reached. The resulting mass of solids and gas is smeared throughout the volume and fitted to the chosen class of nebula models. The result is, for example, a minimum reconstituted nebula with solid and gas surface densities described by parameterized power laws (e.g. Hayashi *et al.* 1985). For the nebula so constructed the outcome of planet formation is deduced in a multi-step process: (1) planetesimal formation, (2) planetesimal accretion, (3) formation of planetary envelopes, (4) nebula gas capture by large envelopes and (5) termination of planetary accretion and dissipation of remnant nebula gas. A typical result for the minimum-mass nebula is that predicted accretion times turn out to be much longer than plausible nebula lifetimes. In consequence, the original assumptions going into the construction of the nebula are reconsidered. Lissauer (1987), for example, described how a solid surface density increased by a factor of less than ten could account for a jovian planet within the time constraints. Wuchterl (1993) showed how an increase in gas surface density of less than a factor of ten would lead to a new class of protoplanets with massive envelopes that dynamically could grow to a few hundred Earth masses (Wuchterl 1995a). Pollack *et al.* (1996) adjusted nebular and planetesimal parameters to account for the accretion of Jupiter and Saturn with detailed models of planetesimal accretion and gaseous envelope capture. When coupled with evolutionary models, the observed properties of gravitational fields, radii and present excess luminosities can be reproduced when interior structures are fitted by detailed planetary structure and evolution models with three compositional layers (Guillot 1999, 2005).

4.15.1. *Gaseous envelopes – giant planets*

Planetesimals in the solar nebula are small bodies surrounded by gas. A rarefied equilibrium atmosphere forms around such objects. The question is then how massive the planetesimal or planetary embryo has to become to capture large amounts of gas. In particular, at what mass it could bind more gas from the nebula than its own mass. It could then become a Jupiter or Saturn precursor object. A *proto-giant planet* would then form. The respective mass values are referred to as the *critical mass*. Some care has to be taken because the usage of the term is not homogeneous in the literature and physically differing variants are often used synonymously.

Because planetary masses are optically thick in the nebula (see Eq. (4.24)), such objects are much hotter in the interior than the ambient nebula. Consequently, the energy budget of the envelope has been modelled more and more carefully. Mizuno (1980) calculated the first realistic protoplanetary structures that could be related to the Solar System planets. Mizuno found that the required mass for gas capture would be similar anywhere in the nebula. That could explain the similarity of the Solar System giants' cores, despite their widely differing envelopes and total masses. Bodenheimer and Pollack (1986)[16] accounted for heat generated by gravitational contraction of the envelopes by building quasi-hydrostatic models. Pollack *et al.* (1996) showed that planetesimal accretion would control the timing and onset of envelope accretion. Dynamic effects and possible accretion flows were added by Wuchterl (1989, 1990, 1991a,b). The hydrodynamical calculations showed that accretion was not the only pathway of planetary evolution and envelope ejection, hence mass loss can also occur at the critical mass. Wuchterl (1993)

---

[16] The group is continuing to refine their models in the quasi-static approximation. Pollack *et al.* (1996) included detailed planetesimal accretion rates, Bodenheimer *et al.* (2000) applied it to extrasolar planets and Hubickyj *et al.* (2004) accounted for dust depletion effects due to planetesimal growth.

showed that protoplanetary structure and hence the critical mass can vary a lot once the outer envelopes become convective, and Wuchterl (1995a) showed that largely convective protoplanets would allow the onset of accretion at much lower core masses than the dominating population of protoplanets with radiative outer parts. For further discussion see Wuchterl et al. (2000).

Most aspects of early envelope growth, up to $\sim 10\,M_\oplus$, can be understood on the basis of a simplified analytical model given by Stevenson (1982) for a protoplanet with constant opacity, κ_0, core-mass accretion-rate, $\dot{M}_\text{core}$, and core-density, ρ_core, inside the Hill radius, r_Hill. The key properties of Stevenson's model come from the 'radiative zero solution' for spherical protoplanets with static, fully radiative envelopes, i.e. in hydrostatic and thermal equilibrium. Wuchterl et al. (2000) presented an extended solution relevant to the structure of an envelope in the gravitational potential of a constant mass for zero external temperature and pressure and using a generalized opacity law of the form $\kappa = \kappa_0 P^a T^b$.

The critical mass, defined as the largest mass a core can grow to with the envelope kept static, is then given by:

$$M_\text{core}^\text{crit} = \left[\frac{3^3}{4^4} \left(\frac{R_\text{gas}}{\mu}\right)^4 \frac{1}{4\pi G} \frac{4-b}{1+a} \frac{3\kappa_0}{\pi\sigma} \left(\frac{4\pi}{3}\rho_\text{core}\right)^{\frac{1}{3}} \frac{\dot{M}_\text{core}}{\ln(r_\text{Hill}/r_\text{core})} \right]^{\frac{3}{7}}, \qquad (4.30)$$

and $M_\text{core}^\text{crit}/M_\text{tot}^\text{crit} = 3/4$; R_gas, G and σ denote the gas constant, the gravitational constant, and the Stefan–Boltzmann constant respectively. Neither does the critical mass depend on the midplane density, ϱ_neb, or on the temperature T_neb of the nebula in which the core is embedded. The outer radius, r_Hill, enters only logarithmically weakly. The strong dependence of the analytic solution on molecular weight, μ, led Stevenson (1984) to propose 'superganymedean puffballs' with atmospheres assumed to be enriched in heavy elements and a resulting low critical mass as a way to form giant planets rapidly (see also Lissauer et al. 1996). Except for the weak dependencies discussed above, a proto-giant planet essentially has the same global properties for a given core wherever it is embedded in a nebula. Even the dependence on $\dot{M}_\text{core}$ is relatively weak: detailed radiative/convective envelope models show that a variation of a factor of 100 in $\dot{M}_\text{core}$ leads only to a 2.6 variation in the critical core mass (Wuchterl 1995a).

However, other static solutions are found for protoplanets with *convective* outer envelope, which occur for somewhat larger midplane densities than in minimum mass nebulae (Wuchterl 1993; Ikoma et al. 2001). These largely convective proto-giant planets have larger envelopes for a given core and a reduced critical core mass. Their properties can be illustrated by a simplified analytical solution for fully convective, adiabatic envelopes with constant first adiabatic exponent, Γ_1:

$$M_\text{core}^\text{crit} = \frac{1}{\sqrt{4\pi}} \frac{\sqrt{\Gamma_1 - \frac{4}{3}}}{(\Gamma_1 - 1)^2} \left(\frac{\Gamma_1 \mathcal{R}}{G\,\mu}\right)^{\frac{3}{2}} T_\text{neb}^{\frac{3}{2}} \rho_\text{Neb}^{-\frac{1}{2}} \qquad (4.31)$$

and $M_\text{core}^\text{crit}/M_\text{tot}^\text{crit} = 2/3$. Γ_1 is a density-exponent of the pressure equation of state. It is relevant for adiabatic processes that leave the specific entropy s constant:

$$\Gamma_1 = \left(\frac{\partial \ln P}{\partial \ln \varrho}\right)_s. \qquad (4.32)$$

In this case, the critical mass depends on the nebular gas properties and therefore the location in the nebula, but it is independent of the core accretion rate. Of course, both the radiative zero and fully convective solutions are approximate because they only roughly estimate envelope gravity, and all detailed calculations show radiative *and* convective regions in proto-giant planets. The critical mass can be as low as $1\,M_\oplus$, and subcritical static envelopes can grow to $48\,M_\oplus$. See Wuchterl et al. (2000) and Wuchterl (1993) for more details. Ikoma et al. (2001) study largely convective protoplanets for a wide range of nebula conditions and show the limiting role of gravitational instability.

The early phases of giant planet formation discussed above are dominated by the growth of the core. The envelopes adjust much faster to the changing size and gravity of the core than the core grows. As a result the envelopes of proto-giant planets remain very close to static and in equilibrium below the critical mass (Mizuno 1980; Wuchterl 1993). This has to change when the envelopes become more massive and cannot re-equilibrate as fast as the cores grow. The nucleated instability was assumed to set in at the critical mass, originally as a hydrodynamic instability analogous to the Jeans instability. With the recognition that energy losses from the proto-giant planet envelopes control the further accretion of gas, it followed that quasi-hydrostatic contraction of the envelopes would play a key role.

4.15.2. *Hydrodynamic accretion beyond the critical mass*

Static and quasi-hydrostatic models rely on the assumption that gas accretion from the nebula onto the core is very subsonic, and the inertia of the gas and dynamical effects as dissipation of kinetic energy do not play a role. To check whether hydrostatic equilibrium is achieved and whether it holds, especially beyond the critical mass, hydrodynamical investigations are necessary. Two types of hydrodynamical investigations of protoplanetary structure have been undertaken in the last decade: (1) linear adiabatic dynamical stability analysis of envelopes evolving quasi-hydrostatically (Tajima and Nakagawa 1997) and (2) non-linear, convective radiation hydrodynamical calculations of core-envelope proto-giant planets (see Wuchterl et al. 2000). In the linear studies it was found that the hydrostatic equilibrium was stable in the case they investigated. The non-linear dynamical studies follow the evolution of a proto-giant planet without a priori assuming hydrostatic equilibrium, and they *determine* whether envelopes are hydrostatic, pulsate or collapse, and at which rates mass flows onto the planet assuming the mass is available in the planet's feeding zone. Hydrodynamical calculations that determine the flow from the nebula into the protoplanet's feeding zone are discussed in Section 4.23. The first hydrodynamical calculation of the nucleated instability (Wuchterl 1989, 1991a,b) started at the static critical mass and produced a surprise: instead of collapsing, the proto-giant planet envelope started to pulsate after a very short contraction phase; see Wuchterl (1990) for a simple discussion of the driving κ-mechanism. The pulsations of the inner protoplanetary envelope expanded the outer envelope, and the outward travelling waves caused by the pulsations resulted in mass loss from the envelope into the nebula. The process can be described as a pulsation-driven wind. After a large fraction of the envelope mass has been pushed back into the nebula, the dynamical activity fades and a new quasi-equilibrium state is found that resembles Uranus and Neptune in core and envelope mass (see Figure 4.3, full line).

The main question concerning the hydrodynamics was then to ask for conditions that allow gas accretion, i.e. damp envelope pulsations. Wuchterl (1993) derived conditions for the breakdown of the radiative zero solution by determining nebula conditions that would make the outer envelope of a 'radiative' critical mass proto-giant planet convectively

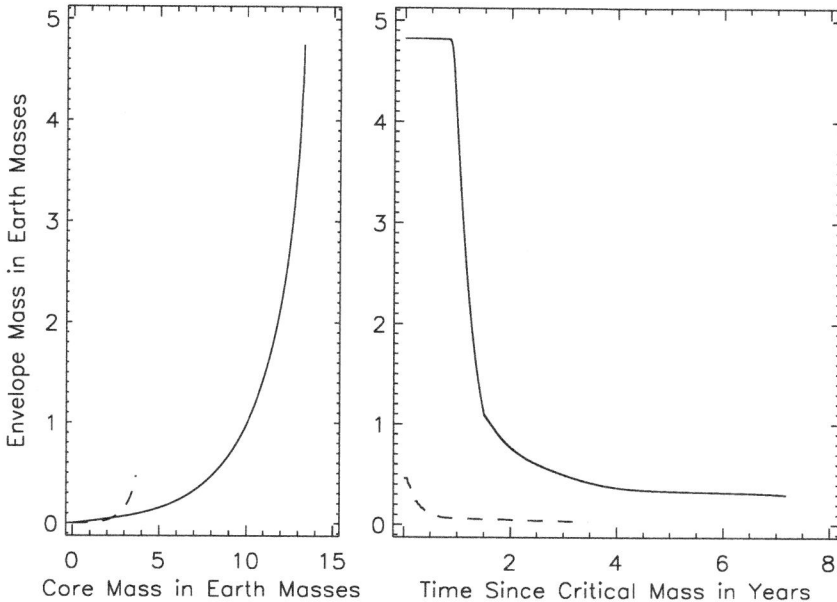

FIGURE 4.3. Hydrostatic envelope accretion due to a core growing by accretion of planetesimals (left) and hydrodynamical ejection of protoplanetary envelope gas due to pulsation-driven mass loss (right). The evolution is shown for Mizuno (1980)'s Neptune conditions and a planetesimal accretion rate of the core $\dot{M}_{\rm core} = 10^{-6}\,M_{\rm Earth}\,{\rm a}^{-1}$ (solid line). The dashed line is the same but with time-dependent MLT-convection, updated molecular opacities, and a particle-in-a-box core accretion rate, with a planetesimal surface density of $10\,{\rm kg\,m^{-2}}$ and a gravitational focusing factor of 2000 (Wuchterl 1997).

unstable. The resulting criterion gives a minimum nebula density that is necessary for a convective outer envelope:

$$\frac{\varrho_{\rm neb}^{\rm crit}}{[10^{-10}{\rm g\,cm^{-3}}]} = 2.2 \left(\frac{T}{[100{\rm K}]}\right)^3 \left(\frac{\nabla_{\rm s}}{[2/7]}\right) \left(\frac{\mu}{[2.2]}\right) \left(\frac{\kappa}{[{\rm cm^2\,g^{-1}}]}\right)^{-1}$$
$$\times \left(\frac{\dot{M}_{\rm core}}{[10^{-6}\,M_\oplus\,{\rm a}^{-1}]}\right)^{-1} \left(\frac{M_{\rm core}}{[10\,M_\oplus]}\right)^{\frac{1}{3}} \left(\frac{\varrho_{\rm core}}{[5.5\,{\rm g\,cm^{-3}}]}\right)^{-\frac{1}{3}}, \quad (4.33)$$

where $\nabla_{\rm s}$ and μ are the logarithmic isentropic temperature gradient (with respect to pressure) and the mean molecular mass, cf. Eqs. (4.7) and (4.9). Protoplanets that grow under nebula conditions above that density have larger envelopes for a given core and a reduced critical mass. For sufficiently large nebula densities, Wuchterl (1995a) found that the pulsations were damped and rapid accretion of gas set in and proceeded to $300\,M_\oplus$. The critical core masses required for the formation of this class of proto-giant planets are significantly smaller than for the Uranus/Neptune-type (see Wuchterl 1993, 1995a; Ikoma et al. 2001).

4.16. The importance of convection

Convection plays an important role in determining the mass of protoplanets by controlling energy transfer in the outer layers under specific nebula conditions. It also controls the dynamical behaviour of the protoplanetary envelopes beyond the critical mass, as described in the last section. Most giant planet formation studies use zero-entropy gradient convection, i.e. set the temperature gradient to the adiabatic value in convectively unstable layers of the envelope, or use time-independent mixing length theory. That is done for simplicity but can be inaccurate, especially when the evolution is rapid and hydrodynamical waves are present (Wuchterl 1991b). Furthermore, convection in the outer layers of a protoplanet occurs under weak gravities and relatively low optical depths. Hence, departures from adiabatic behaviour might be expected. It was therefore important to develop a time-dependent theory of convection that can be solved together with the equations of radiation hydrodynamics in the entire protoplanetary flow regime. Such a time-dependent convection model (Kuhfuß 1987) has been reformulated for self-adaptive grid radiation hydrodynamics (Wuchterl 1995b) and applied to giant planet formation (Götz 1989; Wuchterl 1996, 1997). In a reformulation by Wuchterl and Feuchtinger (1998), it closely approximates standard mixing length theory in a static local limit and accurately describes the solar convection zone and RR Lyrae lightcurves.

The heart of this convection model is a dynamical equation for the specific kinetic energy density, ω, of convective elements. The following equation accounts for creation of eddies by buoyancy, the dissipation of eddies due to viscous effects, and eddy advection and radiative losses:

$$\frac{d}{dt}\left[\int_{V(t)} \varrho\omega\,d\tau\right] + \int_{\partial V} \varrho\omega u_{\text{rel}} \cdot dA = \int_{V(t)} (S_\omega - \tilde{S}_\omega - D_{\text{rad}})\,d\tau, \qquad (4.34)$$

where the eddy kinetic energy generation rate, the eddy dissipation rate, the convective enthalpy flux, the reciprocal value of the mixing-length, Λ, and the timescale for radiative eddy losses, respectively, are: (For a description of the variables, see text below Eq. (4.22).)

$$S_\omega = -\nabla_{\text{s}} \frac{T}{P}\frac{\partial P}{\partial r}\Pi, \qquad (4.35)$$

$$\tilde{S}_\omega = \frac{c_{\text{D}}}{\Lambda}\omega^{3/2}, \qquad (4.36)$$

$$j_{\text{w}} = \varrho T \Pi, \quad \Pi = \frac{w}{T} u_{\text{c}} F_{\text{L}}\left[-\sqrt{3/2}\,\alpha_{\text{S}}\Lambda \frac{T}{w}\frac{\partial s}{\partial r}\right], \qquad (4.37)$$

$$\frac{1}{\Lambda} = \frac{1}{\alpha_{\text{ML}} H_{\text{p}}^{\text{stat}}} + \frac{1}{\beta_r r}, \quad H_{\text{p}}^{\text{stat}} = \frac{p}{\varrho}\frac{r^2}{GM_r}, \qquad (4.38)$$

$$\tau_{\text{rad}} = \frac{c_p\,\kappa\,\rho^2\,\Lambda^2}{4\sigma\,T^3\,\gamma_{\text{R}}^2}, \quad D_{\text{rad}} = \frac{\omega}{\tau_{\text{rad}}}. \qquad (4.39)$$

In the time-independent and static limit this is essentially mixing-length theory and accuracy is assured by fitting the prescription to the Sun via a solar model. The difference is that the parameterization is now brought into a fluid-dynamical framework and basic physical plausibility constraints that are required in the time-dependent regime are fulfilled (see Wuchterl and Feuchtinger 1998). The Schwarzschild–Ledoux criterion for convective stability, $\nabla < \nabla_{\text{ad}}$, is contained in the formulation via $-\partial s/\partial r = c_{\text{p}}/H_{\text{p}}(\nabla - \nabla_{\text{s}})$ and $\nabla_{\text{s}} = \nabla_{\text{ad}}$ in the absence of energy sources and sinks inside eddies. Convectively unstable stratifications occur in this model when pressure and temperature gradients have the same sign and produce a positive value of S_ω that then contributes a source of turbulent kinetic energy, $\omega = 3/2u_{\text{c}}^2$, to the balance equation of turbulent kinetic energy, Eq. (4.34), u_{c} being the convective velocity corresponding to mixing length theory. A general problem of mixing-length theory – the violation of a convective flux limit – has been corrected, as described by Wuchterl and Feuchtinger, by introducing a flux-limiting function (see Wuchterl and Tscharnuter 2003). The great advantage of this approach is that a general prescription can be used for the Sun, stellar evolution, pulsating stars, brown dwarfs, planets and protoplanets. Any calibration of parts of the convection model obtained in one astrophysical system – the mixing-length parameter calibrated by the Sun, the time-dependent behaviour tested by RR Lyrae stars – will decrease the uncertainties in applications to not-so-easy-to-observe systems such as protoplanets.

4.17. The fluid dynamics of protoplanets

The time-dependent convection model allows the formulation of a fully time-dependent set of equations (see Eqs. (4.15)–(4.20) discussed earlier in Section 4.6.4). These equations describe the radiative and convective envelopes of protoplanets as well as the protostellar collapse and pre-main sequence evolution (see Wuchterl and Tscharnuter 2003).

The equations are applied to the volume taken by the protoplanetary envelope, assuming spherical symmetry. They determine the motion of gas in the protoplanetary envelope or determine hydrostatic equilibrium if the forces balance out. As a consequence of the structure and motion in the envelope, mass exchange with the nebula results and determines whether the planet gains or loses mass. Material belongs to the planet when it is inside the planet's gravitational sphere of influence. The sphere of influence is approximated by a spherical volume of radius R_{Hill} around the condensible element planetary embryo at the centre. At the outer boundary of the volume, i.e. at the Hill sphere, the protoplanet radiates into the ambient nebula and may exchange mass with it. Planetesimals enter the sphere of influence and collide with the core. The core's surface is the inner boundary. The surface changes its radius as the core grows due to planetesimal accretion. The planetesimals add mass to the core and dissipate their kinetic energy at the core surface. The core radiates the planetesimal's energy into the adjacent planetary envelope gas. That heats the inner parts of the protoplanetary envelope. The resulting temperature increase relative to the nebula induces temperature gradients that drive energy transport through the envelope towards the nebula. In general, transfer occurs by both radiation and convection.

4.18. Dynamic diversity

Quasi-hydrostatic models of giant planet formation always encounter envelope growth once the critical mass has been reached. When departures from hydrostatic equilibrium are allowed and the dynamics of the envelopes are calculated the situation is more diverse: the occurrence of accretion depends on the nebula properties and the properties of the protoplanet at the critical mass, see Wuchterl *et al.* (2000). Furthermore, the onset of planetary envelope mass loss depends on the planetesimal accretion rate of the core and the treatment of energy transfer. The dependence is quantitatively significant on the scale of a few Earth masses, which is comparable to the masses of terrestrial planets, the cores of giant planets and the envelopes of planets like Uranus and Neptune. In Figure 4.3 two calculations are compared for the position of Neptune in the Kyoto-nebula.[17] Hence nebula properties and orbital radius effects (orbital dynamic timescale, solar tides, size of

---

[17] Mizuno's Neptune (17.2 AU, 45 K, $3.0\,10^{-13}$ g/cm^3), for orbital radius, nebula temperature and midplane nebula density respectively is located inside Neptune's present orbital radius (semi-major axis 30.06 AU) to allow for outward migration after formation (see Hayashi *et al.* 1985).

the Hill-sphere) are identical for both calculations. The difference is in the energy input and content of the envelopes, i.e. the thermal structure. The first calculation (full line in Figure 4.3) is for a constant mass accretion rate and simple instantaneous zero-entropy gradient convection. The second calculation (dashed line) is physically more refined, with a particle-in-box planetesimal accretion rate and time-dependent convection as described above. The gravitational focusing factor is chosen appropriately for the runaway phase up to the isolation mass. The outcome is qualitatively very similar to that employing simpler physics: with growing core mass, the envelope mass increases until the slope becomes almost vertical in the vicinity of the critical mass (Figure 4.3, left panel). But the values of the critical core mass and the envelope mass at given core mass are significantly different (critical core masses, 13 and 4 $M_\oplus$ with envelopes of 5 and 0.5 $M_\oplus$ respectively). The evolution beyond the critical core mass is shown as a function of time in the right panel. Note that the evolution is now on the short dynamical timescale (a few years) of the envelopes. Mass loss is driven in both cases, and both calculations approach a new quasi-equilibrium state with smaller envelope mass. But the envelope masses ultimately differ by approximately a factor of ten. Even with the relatively well-known properties of giant planets at the critical mass, no general conclusion is possible about the dynamical processes that happen thereafter and the expected envelope mass of, for example, a Uranus-type planet. It is obvious that a more general understanding is needed to predict the outcome of planet formation when realistic physics, such as runaway planetesimal accretion, dynamical effects and plausible convection, are included.

Following the usual approach for Solar System planet formation, we might try to adjust the parameters of planetesimal accretion to account for the observed properties of, say, Neptune, but that will not lead to a predictive theory or a general understanding of planet formation. I will outline an alternative approach below.

4.19. A few problems of Solar System theory

To conclude the discussion of Solar System planet formation theory I will describe open problems that were known before the discovery of the first extrasolar planet. These problems might help to understand what parts of the theory might need modification for general application to planet formation in the galactic neighbourhood. With dust growth to cm size now increasingly well understood by theoretical and experimental work (Blum and Wurm 2000), the most important remaining problems are:

 (i) planetesimal formation,
 (ii) the total growth times in the outermost Solar System, and
 (iii) the final planetary eccentricities.

4.19.1. *Planetesimal formation*

Planetesimal formation by the coagulation and agglomeration of dust grains may stall at dm to m size, where loss processes by radial drift may halt the planet formation process. Planetesimal formation through the gravitational instability of a dust subdisc may require special nebula conditions that are incompletely explored to decide in how wide a range the instability will operate and whether the non-linear outcome is the consolidated condensible element bodies that are envisaged and assumed in the planetesimal hypothesis. A related key question is how wide a diversity of nebulae will lead to instabilities that produce appropriate planetesimals. 'Appropriate' mostly means a size large enough to decouple from the head wind of the nebula gas; an event that typically occurs at km size. The production of non-standard planetesimals does not automatically mean that planet formation will not proceed as currently imagined, but new pathways in a theoretically essentially unexplored regime have to be worked out in that case.

4.19.2. Late accretion: total planetary growth times

The standard model is centred around the planetesimal hypothesis, which has been successful in helping us to understand a wide range of Solar System bodies, mainly in a quantitative way. But observational results obtained for nearby star-forming and young star regions quantitatively challenge the standard model because indicators of the presence of circumstellar discs, Haisch et al. (2001), suggest disc depletion timescales comparable to or shorter than calculated formation times for Solar System giant planets of $\approx 10^8$ years (Safronov 1969). Moreover, unless the eccentricities of the growing embryos are substantially damped, embryos will eject one another from the star's orbit (Levison et al. 1998). Runaway growth, possibly aided by migration (Tanaka and Ida 1999), appears to be the way by which solid planets can become sufficiently massive to accumulate substantial amounts of gas while the gaseous component of the protoplanetary disc is still present (Lissauer 1987; Kokubo and Ida 2002). The theoretical estimates for planetary growth times have been known to be idealized because the size distribution of planets, embryos and planetesimals, and the interaction with the residual nebular gas can only be accounted for incompletely in the n-body calculations necessary for reliably calculating the final orbital outcome, at least for an idealized situation, to allow a quantitative discussion and theoretical progress. Inaba and Ikoma (2003) and Inaba et al. (2003) looked at the collisional cross-sections of planetesimals with gaseous envelopes and found a significant increase for their accretion rates, thereby considerably reducing planet growth times. This is especially important for the giant planet regime, where the envelopes may become comparable in mass to the condensible element cores during the runaway phase. Hence, total growth times can be expected to decrease further when the nebular gas is not neglected in determining the collision cross-sections of planetary embryos.

4.19.3. Late accretion and final eccentricities

Late accretion and hence the evolution to the final orbital parameters of a planet are governed by interactions with other planets, the remaining planetary embryos and planetesimals (Levison et al. 1998; Thommes and Lissauer 2003; Levison and Agnor 2003). The relevant overall masses in all components may or may not be comparable to the mass of the largest planet. There is probably still a large and locally dominant number of bodies around that in the case of the Solar System are responsible, e.g. for the formation of the Moon and the late heavy cratering bombardment. Late accretion effects are apparently important in the asteroidal region of the Solar System, where it is possible that Jupiter's perturbations precluded the accretion of embryos into a planet. The important remaining dynamical process is then the orbital evolution of planets and embryos due to secular mutual perturbations. In models of final planetary growth they typically lead to eccentricities larger than those observed in the Solar System (Wetherill 1990; Chambers and Wetherill 1998) or found in long-range backward integrations of the planetary system (Lecar et al. 2001). Studies with an increased number of planetary embryos reduce the discrepancy (Chambers 2001) but still do not reproduce the low time-averages of the planetary eccentricities in the Solar System. Eccentricity damping by residual nebular gas or by a remnant population of planetesimals or small planetary embryos might resolve that problem.

4.20. Theory – planet mania

In addition to Solar System problems, the planetary properties of the first exoplanet harvest were unpredicted by theory and were surprising because of the detection of:
   (i) giant planets with orbital periods of a few days, corresponding to 0.01 of Jupiter's orbital radius,
   (ii) planet candidates with $M \sin i$ up to 13 M_{Jupiter},[18]
   (iii) a broad range of eccentricities higher than those known for the Solar System planets,[19]
   (iv) planets in binaries.[20]

The 1995 discovery of a planetary companion to 51 Peg electrified the theorists. Very rapidly, Guillot et al. (1996) showed that planets like 51 Peg b could indeed survive for the estimated ages of their host stars. Within a year it was shown that 51 Peg b could form at its present location when existing fluid dynamical models of giant planet formation were applied to orbital distances of 0.05 AU, provided there was sufficient building material in the planet's feeding zone (Wuchterl 1996).

But very rapidly alternative theories emerged. They held the often communicated view that giant planets would only be able to form beyond the ice line, typically beyond a few AU from their parent star. If that remained true the planets had to move from their formation place to a position much closer to the star, like 51 Peg b's. How could a massive planet like Jupiter move from 5 to 0.05 AU, say? Mechanisms to change the orbital elements were proposed:
   (i) violent dynamical relaxation of multi-giant planet systems – the so-called *jumping jupiters*, and
   (ii) a gradual decrease in the planetary orbital radius due to interaction with the disc of gas and planetesimals: *orbital migration*.

Violent dynamical relaxation (Weidenschilling and Marzari 1996; Marzari et al. 2005) needs synchronizing of planet formation to provide a number of giant planets within a narrow time span. They would subsequently very rapidly interact via mutual perturbations that typically destroy the system on a dynamical timescale, leaving a close-in giant planet in some cases. While it is unlikely that the assumed very unstable initial state would be reached as the final state of the preceding planet formation process, there are additional problems. The close orbits typically produced would not be as small as those observed and hence would require further orbital evolution, the final systems being somewhat disturbed with one planet closer in and the other farther out. That is unlike a system such as υ Andromedae with relatively close orbiting giant planets in addition to the *Pegasi planet*[21] at 0.05 AU.

The other alternative, involving migration caused by disc–planet interactions, is favoured by many researchers (e.g. Lin et al. 1996; Trilling et al. 1998). It starts out with a standard situation of planet formation: a planetary embryo or proto-giant planet orbiting at a conventional giant planet orbital distance. The change in orbital radius is continuous to very small values and there is no requirement for other planets to be

---

[18] 1 $M_{\text{Jupiter}} = 317.71\, M_{\oplus} = 1.898\, 10^{27}$ kg $= 0.95\, 10^{-3}\, M_{\odot}$.

[19] This might have been expected because of the difficulties in explaining the low eccentricities, but it was not predicted, most likely because missing elements in late accretion were obvious (see above).

[20] In spite of the fact that dynamicists had shown planetary orbits to be stable in binary systems as well as having already classified them as P- and S-type (Dvorak 1986; Dvorak et al. 1989) in analogy to planets and satellites in the Solar System.

[21] By 'Pegasi planet' I mean a Jupiter-sized planet orbiting a main sequence star with a period of a few days. Other common expressions are 'hot jupiter' and 'hot giant planet'.

present at the same time. I will discuss migration as the dominating theory of Pegasi-planet formation below.

An even more radical rethinking of planet formation has been proposed. It has been proposed that giant planets might form directly via a disc instability (see Wuchterl *et al.* 2000 for a review). While this was more directed towards the timescale problem of planet formation in the outer Solar System, it also might offer a way to explain the diversity in the detected extrasolar planets. Maybe some of the systems, in particular the very massive planetary candidates, with minimum masses $M \sin i \sim 10\,M_{\rm Jupiter}$, were formed by a disc instability and others by the nucleated instability.

With the formation process reconsidered, the relatively large minimum masses of many of the early exoplanet discoveries, the large eccentricities, which are hardly distinguishable from those of binary stars, and the Pegasi planets, an old question resurfaced: what is a planet?

4.21. What is a planet?

Given the unexpected properties of extrasolar planet candidates and claims of discoveries of so-called *free-floating planets*, the IAU's Working Group on Extrasolar Planets[22] issued a preliminary working definition[23] based on the following principles:
- objects with true masses below the limiting mass for thermonuclear fusion of deuterium that orbit stars or stellar remnants are planets,
- substellar objects above that mass are brown dwarfs,
- free-floating objects in young star clusters with masses below the deuterium limit are not planets.

With these guidelines for a definition, most of the discovered extrasolar planets are candidates because their true masses are not yet sufficiently well known. Because of the unknown orbital inclination only the minimum masses, $M \sin i$, are known. These values also probably approximate the true masses well, but a few per cent may turn out not to be below the 13 $M_{\rm Jupiter}$ mass limit. Only the masses of the pulsar planets (from mutual perturbations) and for the few known transiting planets (orbital inclination determined from transit lightcurves) are true masses in the sense of the above definition.

The IAU working definition explicitly excludes the way of formation from the definition. But by referring to the deuterium limit, the formation history implicitly enters the definition through the back door.

The minimum mass for thermonuclear fusion of deuterium is a concept that is shaped in some analogy with the minimum mass for hydrogen burning that defines the lower end of the stellar main sequence. But the main sequence is defined by stellar thermal equilibrium, in which nuclear burning fully balances the surface energy losses. The stellar luminosity is balanced by nuclear energy production of the same magnitude. Objects with masses below the lower end of the main sequence also burn hydrogen, but insufficiently to maintain the luminosity (Kumar 1963) – they are called brown dwarfs. Because of this, stellar thermal equilibrium and a phase of constant radius are never reached. The objects have to contract forever to (at least partly) supply their luminosity need from contracting in their own gravitational field. Because they never reach an equilibrium state (such as the main sequence) their evolution always depends on their history. Ultimately, that means it depends on their formation. Recent calculations of the spherical collapse of stars and brown dwarfs show that the young objects after the end of significant mass

---

[22] www.dtm.ciw.edu/boss/IAU/div3/wgesp/.
[23] www.dtm.ciw.edu/boss/IAU/div3/wgesp/definition.html.

accretion (on the pre-main sequence for young stars) contain the thermal profile shaped by the collapse (Wuchterl and Tscharnuter 2003). But it is exactly that thermal profile which controls the ignition of thermonuclear fusion processes. This is particularly true for deuterium, whose burning is in all models (hydrostatic and dynamic) a very episodic event in the first few million years for stars *and* brown dwarfs. This means that for masses below the main sequence the question of whether deuterium (or hydrogen) will just start to burn – and this defines the borderline in the IAU definition – will depend on the history of the respective low mass object. For deuterium, which burns early on, it will depend on the formation process.

In summary, using *hydrogen burning* to define the lower end of the main sequence means using a major, physically dominating process that defines a long-lasting equilibrium state that contains no memory of the formation history. On the contrary, using *deuterium burning* for characterizing a planet means using an essentially irrelevant process for the evolution of low mass objects that is history-dependent in a way that will be very hard to predict.

I think we should rethink the definition of a planet along the following major characteristics:

(i) heavy element enrichment,
(ii) orbital stability properties,
(iii) mass.

The first item is straightforward with the large enrichments (bulk and atmospheric) of planets relative to their parent star. A factor of three or more should also be a working basis that is empirically very probably much less challenging in terms of future determinations in exoplanets than trying to observe the presence of deuterium in a few Jupiter-mass companions in a 10 AU orbit, even around a nearby star.

The second point is still hard to characterize quantitatively, but great progress has been made in the understanding of the stability properties of the Solar System (Lecar *et al.* 2001). The basis could be volume exclusion principles based on the non-overlap of planet domains with a width of multiples of the Hill radii. Laskar (2000) has recently shown that they are the consequence of simple assumptions about planetary growth via pairwise embryo collisions. The low Solar System planetary eccentricities could be a special case of that. Certainly, low eccentricity planets can orbit closer together in terms of Hill exclusion.

The third point is the most uncertain. Observationally, the characteristic mass of the detected planetary population seems to decrease as more discoveries are made. Currently, it may be around three Jupiter masses, with the estimated true distribution still peaked towards the detection limit. Theoretically, the planetary masses are at present essentially unconstrained at the upper end. The best hypothesis for the moment is that planetary masses are limited by the amount of material that is within the respective feeding zone in the nebula. This definition contains a considerable degree of circularity in general but has been consistently worked out, at least for planetary embryos.

In summary, I think we will see the definition of a planet remaining a *morphological type*, i.e. without an explicit, constructive definition for some time. But I think that condensible elements should play the major role, not a hydrogen trace-isotope.

4.22. Why not abandon Solar System formation theory?

If the extrasolar planet properties are so different, and the theories developed for the Solar System did not predict their properties, why not look for a completely new theory of extrasolar planet formation? Should this new theory be more along the lines of binary star formation? Thus one might conclude, given the fact that the period–eccentricity distribution of extrasolar planets is indistinguishable from that of binary stars.

I think we should not throw away the Solar System formation theory as a general theory of planet formation too hastily. It not only provides a fairly consistent picture of Solar System bodies ranging in size from interplanetary dust particles to Jupiter, but has also led to the development of predictive elements that led, for example, to the correct prediction of many orbital properties of trans-neptunian objects.

On the other hand, scenarios like the *jumping jupiters* to explain Pegasi planets raise more questions than they answer. Instead of one planet at 0.05 AU the simultaneous formation of many massive planets is required as a presumption.

Moving a planet into place by migration requires, in addition to the planet, a mechanism that counteracts the migration process to *park* the planet once it has arrived at the intended final orbit. This is by no means trivial because of the large migration rates (on the local disc evolution timescale and shorter) that increase as the star is approached, resulting in acceleration rather then slow-down for small orbital radii. Numerous *parking* processes have been suggested, but ultimately the way out of the dilemma might be only the dissipation of the nebula. Planets would then continue to form, drift inward and disappear into the star until the exhaustion of nebular material finally ends this road of destruction. Observationally, no metal trend versus effective temperature is found on the main sequence, reflecting the different sizes of convection zones that would play the role as a planetary graveyard and that hence might be expected to be heavy-element enhanced (Santos *et al.* 2003).

The disc instability model, if it works and does indeed form planets, in order to be a general alternative would have to be augmented by a separate way of terrestrial planet formation. For the giant planets the disc instability would probably require a separate heavy element enrichment process. Even if Jupiter formed by a disc instability, the craters on a Galilean moon would recall the planetesimal picture.

The basic feature of non-standard planet formation theories is that they quickly provided scenarios for newly discovered objects. But typically they would fail the Solar System test. Let us look at planetary migration as an example of the new pathways of planet formation and discuss it in more detail.

4.23. Planet–disc interaction

4.23.1. *What is planetary migration?*

Planetary migration seems at present to denote any systematic change of the orbital semi-major axis of a planet that does not change direction. Historically, the outward migration of Uranus and Neptune as a consequence of 'passing comets down to Jupiter' seems to be the first large-scale post-formation reshaping process of planetary orbits to be considered. It had been noticed by Fernandez and Ip (1984) and has been considered by Hayashi *et al.* (1985) as a process that would allow shorter growth times for Uranus and Neptune. In the late stages of outer Solar System formation these planets would move outwards as a consequence of angular momentum exchange when perturbing comets into a Jupiter-controlled orbit with subsequent ejection to the Oort cloud.

After the discovery of 51 Peg b it has become a custom in planet formation theory to denote many kinds of changes in the planetary semi-major axis or orbital distance as planetary migration. This is usually independent of the physical process underlying the respective orbital change. With processes proposed and a terminology of types I and II, suggested by Ward (1997), and a type III added later by analogy we have in particular:

Type I migration: an embedded planetesimal or planetary embryo that interacts with its own disc density waves;
Type II migration: a protoplanet that has opened a gap in the nebula – i.e. produced a region of reduced nebula density in its feeding-zone – is locked in that gap and follows the gradual inward motion of viscous disc gas together with the gap;
Type III migration: an instability of the planet–disc interaction that leads to orbital decay within a few orbital periods.

We distinguish here between migration processes that modify the orbit by less than a factor of e^2 (or ~ 10) and those that may lead to larger changes up to orders of magnitude in the orbital radius, and may ultimately result in the loss of the planet. The latter processes we will call *violent migration* in the following. They may dominate the planet formation processes if they operate in many and diverse nebulae.

After planetesimal formation violent migration is the second key problem of planet formation. Like an inefficient planetesimal formation mechanism it has the potential to make the formation of solar-type systems very unlikely. It is expected by many investigators to become important in the mass range resulting from the early fast *runaway* mode of planetesimal growth. The runaway phase ends when all planetesimals within the gravitational range of the locally largest body have been accreted and hence its feeding zone has been emptied. Planetary embryos gravitationally interact with the ambient gas disc, planetesimal disc and other planetary embryos or planets. As a result *planetary migration* can come about (see Thommes and Lissauer 2005 for a review).

The migration effects become severe at larger sizes because they are proportional to the planetesimal mass, for type I (after Thommes and Lissauer 2005, cf. Ward 1997):

$$v_{\rm I} = k_1 \frac{M}{M_*} \frac{\Sigma_{\rm d} r^2}{M_*} \left(\frac{r\Omega}{c_{\rm T}}\right)^3 r\Omega, \qquad (4.40)$$

where k_1 is a measure of the torque asymmetry, M, M_* the masses of the planet and the primary respectively, r is the orbital radius, Ω is the disc angular velocity, that is approximately Keplerian with $\Omega_{\rm Kepler} = \sqrt{GM_*/r^3}$, $\Sigma_{\rm d}$ the disc surface density and $c_{\rm T}$ the isothermal sound speed. For a planet that has opened a gap and is locked to the disc the rate of change in orbital radius (type II migration) is (Ward 1997):

$$v_{\rm II} = k_2 \frac{\nu}{r} = k_3 \alpha \left(\frac{c_{\rm T}}{r\Omega}\right)^2 r\Omega, \qquad (4.41)$$

where a nebula viscosity $\nu \sim \alpha c_{\rm T}^2/\Omega$ has been assumed, and k_2 and k_3 are further constants. For easier reading it is worth noting that the vertical disc scale height $h \sim c_{\rm T}/\Omega$ and h/r is roughly constant and ~ 0.1 in some nebulae. Note that both rates are proportional to the Keplerian orbital velocity $r\Omega$.

4.23.2. *Violent migration*

Violent migration is a back-reaction of the planetary embryo's 'bow wave' in the nebula on to the embryo itself. As the embryo orbits the star, its gravitational potential adds a bump to the stellar one. At the embryo's orbit – at the corotation resonance, in the linear terminology of migration theory – the embryo and its potential move at the same, almost Keplerian, velocity. That is co-orbital motion, as in the case of Jupiter and the Trojan asteroids. Inside the embryo's orbit, the gas in a quasi-Keplerian disc orbits faster and hence the embryo's potential and gravitational acceleration travels at a different speed relative to the gas. This accelerates the disc gas and excites a pressure and density wave that travels with the embryo. Because matter deeper in the primary's potential must orbit faster the waves are dragged forward inside, and backwards, relative to the embryo, outside the embryo's orbit. These rather particular protoplanetary bow waves include density enhancements that gravitationally back-react onto the planet. Due to the inherently asymmetric nature of the situation (Keplerian orbital velocities changing $\propto r^{-1/2}$) and the particular wave pattern, the forces (and in particular the torques) on to the planet may not cancel out. This leads to a net exchange of angular momentum between the planetary embryo and the gaseous disc if the waves dissipate or break in the disc, in the neighbourhood of the planet's gravitational sphere of influence. The result is the familiar reaction of orbiting matter if angular momentum exchange is allowed: most matter (the embryo) moves in and a small amount (some gas) moves out carrying away the angular momentum. The very growth of the embryo would lead to orbital decay and gradual movement towards the star on timescales of disc evolution or much less. Many studies are currently being devoted to determining the strength of the effect and evaluating the rates of orbital decay, and hence the possible survival times, for planets of given mass in a given disc. If migration dominates planet formation, it has the potential to wipe out any and many generations of planets. In that case, and because the basic effect originates from a relatively small difference in a delicate torque balance in a significantly perturbed non-Keplerian disc, I doubt that we will be able to reliably predict much about planet formation any time soon.

4.23.3. *A closer look*

Modern planetary migration theory originated from the study of planetary rings (see Ward 1997). While the basic physical processes – density waves in quasi-Keplerian discs – are well studied, the application to the problem of forming planets in the nebula disc is not straightforward.

The basic problem that has to be solved to determine migration rates for a proto-giant planet orbiting in a nebula disc is the fluid dynamical analogue to the restricted three-body problem of celestial mechanics. In the classical problem of celestial mechanics the motion of a test particle is considered in the combined gravitational field of the Sun and a planet. For a proto-giant planet two modifications have to be made:

(i) a protoplanet, unlike a mature planet, is not well approximated by a point mass,
(ii) the test particles are replaced by a fluid with a finite pressure.

A protoplanet fills its Hill sphere and a considerable fraction of its mass is located at significant fractions of the Hill sphere (e.g. Mizuno 1980; Pečnik and Wuchterl 2005). Furthermore, the protoplanet builds up a significant contribution to the gas pressure at the Hill sphere. Typically, planet and nebula are in a mechanical equilibrium. This may only change when and if the planet collapses into the Hill sphere and does not rebound. Fluid dynamical calculations show that this is a non-trivial question that depends on the structure of the outer protoplanetary layers near the Hill sphere (Wuchterl 1995a). In consequence, the problem of a protoplanet in a nebula disc is not only a problem of gas motion in the gravitational potential of two centres but is controlled by the nebula gas flow and the largely hydrostatic equilibrium of the protoplanets themselves. The Hill spheres are filled by hydrostatic protoplanets at least up to the critical mass – 20 M_{Earth}, say – and by quasi-hydrostatic structures, typically up to 50–100 M_{Earth}. In fact, strictly static solutions for protoplanets are published up to masses that closely approach that of Saturn (Wuchterl 1993). Static isothermal protoplanets may be found with masses comparable to Jupiter's (Pečnik and Wuchterl 2005). As a consequence, the protoplanetary migration problem is very far from the idealizations of essentially free gas motion in the potential of two point masses.

Because the problem is basically three-dimensional and the density structure of a protoplanet covers many orders of magnitude, additional approximations have to be made to solve the problem – either numerically or analytically. The basic analytical results (see Ward 1997) stem from solving the linearized fluid dynamical equations for power-law nebula surface densities and an approximate gravitational potential of the problem. The starting point is an unperturbed, quasi-Keplerian disc. The planet is approximated by an expansion of the perturbations induced by a point mass. The linear effect (spiral density waves launched in the disc) is deduced and the resulting torques of the waves on the planet are calculated, assuming that the waves dissipate by a break in the disc. If the waves (and the angular momentum carried) were to be reflected and return there would be no effect. This approach has at least two potential problems.

(i) The dense parts of the protoplanets, approximately in the inner half of the Hill sphere, that potentially carry a large fraction of momentum are treated as if there were no protoplanet – the density structure of the disc is assumed to be unperturbed by the protoplanetary structure even at the position of the planet's core, certainly throughout the Hill sphere. In that way the pressure inside the Hill sphere is dramatically underestimated. The Hill sphere effectively behaves like a hole in the idealized studies of the problem: the gravity of the protoplanet is introduced into the calculations, but the counteracting gas pressure of the static envelopes is omitted.

(ii) The unperturbed state needed for the linear analysis is an unperturbed Keplerian disc. But if a planet with finite mass is present, the unperturbed state is certainly not an axially symmetric disc and corrections have to be made at all azimuthal angles along the planet's orbit. The Keplerian disc in the presence of a protoplanet or an embryo is an artificial state that is found to decay in any non-linear calculation. It is certainly not a steady state, as would be required for a rigorous linear analysis. Therefore, the approach is not mathematically correct. It may turn out that the corrections are minor, as in the case of the *Jeans swindle*, and the basic results hold despite considerable mathematical violence. But unlike in the Jeans case, where Bonnor–Ebert spheres show that there are indeed nearby static solutions, nothing similar is available for the planet-in-disc problem. In fact the respective steady flows are essentially unknown and it is questionable whether they exist at all in the fluid dynamical problem – they might always be a non-steady planetary wake trailing the planet. High resolution calculations (Koller and Li 2003; Koller *et al.* 2003) indeed show considerable vorticity and important effects on the torques at the corotation resonances with potentially important consequences for the migration rates.

Non-linear three-dimensional (and two-dimensional) hydrodynamic calculations of the problem are very challenging, both in terms of timescales and spatial scales. The 'atmospheric' structure of the protoplanet inside the Hill sphere can barely be resolved even in the highest resolution calculations (D'Angelo *et al.* 2002, 2003, 2005) and the dynamics has to be done for simplified assumptions about the thermal structure and dynamical response of the nebula (usually locally isothermal or locally isentropic). The great value of these calculations is that they provide information about the complicated interaction of the planet with a nebula disc that can only be incompletely addressed by models with spherical symmetry that calculate the structure and energy budget of the protoplanet in great detail. The interaction regime between the outer protoplanetary envelope, inside half a Hill radius, and the unperturbed nebula disc, at five Hill radii, say, is only accessible by two- or three-dimensional calculations. Its nature is unknown owing to the lack of any reference flows, be they experimental or theoretical. This situation in my opinion is similar to that of the restricted three-body problem at the time when numerical integrations had just started.

Hence migration rates calculated numerically or analytically have to be considered preliminary and await confirmation by more complete studies of the problem. Agreement found between different investigations is within the very considerable assumptions outlined above and does not preclude considerable uncertainties in the migration rates of many orders of magnitude.

4.23.4. Is a planet a hole or not a hole?

To illustrate the progress that has been made with great effort in state-of-the-art high resolution calculations, I want to discuss briefly a set of new calculations (D'Angelo et al. 2002, 2003) that have brought considerable insight into the problem of how accretion into the Hill sphere of a protoplanet may occur if the protoplanet is assumed to accrete the gas into a small area, essentially on to a mature planet (D'Angelo et al. 2003). For the first time the planet has not been assumed to be a point mass that accretes everything that approaches the limiting resolution of the calculation; instead, alternative assumptions were made about central smoothing of the protoplanetary gravitational potential, guided by analytical structure models (Stevenson 1982; Wuchterl 1993) in order to look at the pressure feedback of the growing protoplanet onto the accretion flow. The flow in the Hill sphere turned out to be qualitatively and quantitatively very different for the two assumptions: with and without the pressure build up by the protoplanetary envelope – or, in short, with or without a hole. The results demonstrate that the planetary structure feeds back on the flow, both inside and *outside* the Hill sphere. Migration rates derived from the three-dimensional calculations were, depending on planetary mass, reduced down to 1/30 of the respective analytical values (Ward 1997).

It is important to add a note of caution to the interpretation of the planetary masses or mass-scalings used in the two- and three-dimensional calculations of planet–disc interaction and planetary migration. The scale of the critical mass for isothermal protoplanets with typical nebula temperatures (100 K) is $\sim 0.1\,M_\oplus$, i.e. about a factor of 100 below the 'realistic' values of 7–10 $M_\oplus$, which are typically found in detailed planetary structure calculations. Hence, the typical regimes calculated in higher-dimensional isothermal studies are a factor of 100 supercritical! Take the isothermal case of a 10 $M_\oplus$ planet, for example. The results of planet–disc interaction calculations roughly correspond to the accretion of a protoplanet of 3 Jupiter masses, i.e. approximately 1000 $M_\oplus$! The respective Jupiter mass case ends up in a dynamical regime that would occur at 0.1 $M_\oplus$ for a realistic protoplanetary structure! The effective mass-scale of isothermal two- and three-dimensional studies is *10 to 3000 times* supercritical. That scaling relates to all parts of the calculations within the gravitational range of the planet, of a few Hill radii, with the most severe effects located inside the Hill sphere, in the protoplanetary envelope. Hence published studies of disc–planet interactions are at present in a much more violent regime than detailed one-dimensional planet growth models require! An overlap is technically challenging but needed: studies at the same effective physical scale which is set by the critical mass for the appropriate thermal planetary structure.

Calculations that treat the protoplanet *and* the planet–nebula interaction in detail, i.e. by accounting for the heating and cooling processes, as well as realistic thermodynamics at the required resolution and over a significant fraction of the planetary growth time, are still in the future. But a first study of the coupled problem seems to be within reach for the idealized isothermal case.

4.24. Towards a general theory: in search of the planetary main sequence

The isothermal case is well studied in higher-dimensional calculations of planet–nebula interaction but comparatively little attention has been paid to isothermal models of the structure of protoplanets (e.g. Sasaki 1989). The reason is that the isothermal assumption for protoplanets is physically unrealistic: the atmospheres of giant planetary embryos become optically thick very early in their growth (Mizuno 1980), even for extreme assumptions about the metallicity (Wuchterl et al. 2000). The advantage of isothermal models is that they are comparatively easy to understand.

Motivated by unsolved problems of detailed statical and dynamical models – e.g. the physical nature of the critical mass and unpredicted equilibrium structures found in dynamical models (Wuchterl 1991a,b, 1993; Wuchterl et al. 2000) – Pečnik and Wuchterl (2005) classified all isothermal protoplanets to identify possible start and end states of dynamical calculations.

4.24.1. All isothermal protoplanets

The construction of all possible isothermal hydrostatic protoplanets is analogous to the construction of the main sequence for stars – both are defined by equilibria. Because of the nature of the equilibria they are independent of the history that leads to them. For the main sequence the stellar equilibria are long-lived and hence describe most of the observed stars. For planets, no similar survey to look for all equilibria has been performed. Our current knowledge covers only the end-states of planetary evolution – the compact cooled planets. Their best stellar analogue may be white dwarfs. But in their youth, planets had rich and long-lasting equilibria that have not been explored from a global point of view. First steps in the construction of a *planetary 'main sequence'* – the most probable planetary states in the nebula – have now been taken. For the isothermal case, Pečnik and Wuchterl (2005) not only found the end states (mature planets and the planetary embryo states) but a large number of previously unknown planetary equilibria. They found multiple solutions to exist in the same nebula and for the same protoplanetary embryo's core mass. Those calculations can now be used to constrain isothermal two- and three-dimensional calculations such that consistent overall solutions of the planet formation problem may be found.

For the first time, analogously to the stellar main sequence, all possible protoplanets are known for the isothermal case and a statistical discussion in a diversity of nebulae as well as a classification of the pathways of planet formation are now possible.

4.25. Formation of Pegasi planets

To finish the theoretical considerations, I will give an example of a complete formation history of a planet from planetesimal size to its final mass. I will briefly discuss the formation and early evolution of a Pegasi planet, from 0 to 100 Ma.

4.25.1. A Pegasi planet: formation and properties

The first extrasolar planet discovered in orbit around a main sequence star was 51 Peg b (Mayor and Queloz 1995). With a minimum mass $M \sin i = 0.46\, M_{\text{Jupiter}} \sim 146\, M_{\text{Earth}}$, a semi-major axis of 0.0512 AU, a period of 4.23 days and an eccentricity of 0.013, it is the prototype of short-period giant planets, the *Pegasi planets*.

To model the formation of such an object I assume the midplane properties of a standard minimum reconstitutive mass nebula (Hayashi et al. 1985) at 0.052 AU, i.e.

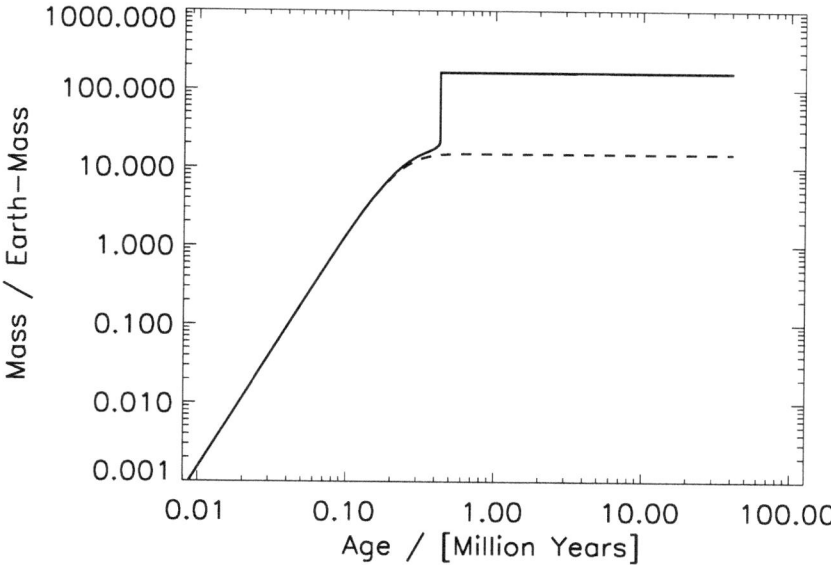

FIGURE 4.4. Total mass (full line) and core mass (dashed line) of a Pegasi planet forming at 0.05 AU from a solar-mass star. The accumulation of a gaseous envelope surrounding a condensible element core that grows by planetesimal accretion is shown for the first 100 million years. The structure of the envelope is calculated dynamically, including time-dependent theory of convection that is calibrated to the Sun and including detailed equations of state and opacities.

0.01 of Jupiter's semi-major axis, and feeding zone masses of 15 $M_\oplus$ of solids (sufficient to easily reach the critical mass) and 150 $M_\oplus$ of nebula gas. This is motivated by the fact that the $M \sin i$ value is only a lower limit for the mass, and accretion may not be 100% efficient.

With these assumptions, the equations of self-gravitating radiation fluid dynamics for the gas (Eqs. (4.15) to (4.20)) with time-dependent convection (Eq. (4.21)), calibrated at the Sun, are solved for a Hill sphere embedded in the standard nebula at 0.05 AU. The condensible element core at the centre of the sphere grows by planetesimal accretion according to the particle-in-box rate. The minimum-mass solid surface density (a safe lower bound) and a gravitational focusing factor of 3 (Safronov number of 1, also an assumed slow growth) are used.

4.25.2. Mass accretion history – the first 100 million years

The resulting mass accretion history from zero to 100 Ma is shown in Figure 4.4 with a logarithmic time axis. Age zero is chosen, following Wuchterl and Tscharnuter (2003), at the moment when the envelope becomes optically thick for the first time and hence a thermal reservoir is formed. The calculation starts at a core of ~10 km size and a mass of ~10^{15} kg. The displayed evolution in Figure 4.4 starts 10 ka after the embryo's envelope has become optically thick, at roughly a tenth of a lunar mass. At that time the total mass (full line) and the core mass (dashed line) are essentially the same because the envelope mass is negligible. At 200 ka and somewhat below 10 $M_\oplus$ the two curves separate, due to the development of a gaseous envelope of significant mass. The planetesimal accretion rate at that time has already dropped due to depletion of the solids, and the core growth curve, $M_{\text{core}}(t)$, starts to flatten out, with the total mass following. As the critical mass is approached, the total mass curve turns upwards with the core mass flattening further. This shows the onset of efficient envelope accretion. The contraction of the envelope is still quasi-static and the gas is practically at rest. The step in the total mass reflects a period of efficient envelope accretion that rapidly increases the total mass until the feeding zone is essentially emptied. The Mach numbers are finite during this stage, but the hydrodynamical part of the flow is basically a transition flow from the nebula onto the contracting inner parts of the protoplanetary envelope that are quasi-hydrostatic. After the flow from the feeding zone on to the planet has faded, the masses remain constant – a Pegasi planet is born.

4.25.3. *Luminosity of a young Pegasi planet*

The luminosity of the Pegasi planet corresponding to the mass accretion history above is shown in Figure 4.5. The luminosity increases during mass growth, passes through a double maximum and then decays roughly exponentially. The two luminosity maxima reflect the maximum accretion of solids and gas respectively. Initially, the growth rate and luminosity rise due to the planetary embryo's increasing gravity-enhanced cross-section for planetesimal accretion. As the planetesimals are removed from the feeding zone and incorporated into the embryo, the surface density of the remaining condensible population fades and the luminosity turns over. Planetesimal accretion passes thermal control to the contracting gaseous envelope. As the envelope mass becomes comparable to the core, its contraction controls the luminosity of the planet. On approaching the critical mass, the luminosity again turns upwards due to the rapid growth of the envelope, reaching the sharp peak at maximum accretion. Once the final mass has been reached, no further material is added and the only luminosity supply is contraction of the envelope, which slows down as larger parts of the planet degenerate. Thereafter, the planet cools into its present state, its luminosity being inversely proportional to age.

Most of the planetary evolution turns out to be quasi-hydrostatic with a brief dynamical period around maximum accretion: the step-like increase in Figure 4.4 and the narrow luminosity peak in Figure 4.5. During this brief period, most of the mass is brought into its final position and acquires its initial temperature. The rapid, dynamical phase is so fast that there is essentially no thermal evolution occurring. Hence, it sets the initial thermal state and determines the bulk starting properties of the planet's evolution at its final mass.

The luminosity of the planet during this period lasting a few hundred years at an age of a few hundred thousand years is shown in Figure 4.6. The entire evolution shown in this figure is present in Figure 4.5 but unresolved in the luminosity spike. The peak accretion phase starts at the turnover of the luminosity, see Figure 4.6. Because the contraction is rapid, the outer parts of the envelope are adiabatically cooled and the nebula gas starts radiating *into* the protoplanetary envelope. The luminosity becomes increasingly negative as the inner, most massive parts of the protoplanet contract further. The trend is reversed when the contraction of the central parts is slowed down again and the heat produced in the process reaches the outer boundary of the planet for the first time (the brief spike at 180 years in Figure 4.6). The overall contraction and accretion of the planet then takes over again and the luminosity rises to positive values, reaching its peak at maximum gas accretion. As accretion fades, the luminosity turns over and decreases with the vanishing amount of nebula gas remaining in the gas-feeding zone. At the end of significant gas accretion, the luminosity slope changes sharply, as the luminosity becomes determined by contraction alone (beyond 220 a in Figure 4.6). The initial luminosity of the planet at its final mass is $\sim 1\,\mathrm{m}L_\odot$ but fades rapidly. The total width of the spike in Figure 4.5, at a tenth of the peak value, i.e. at $\sim 10^{-4}\,L_\odot$, is approximately 100 000 a.

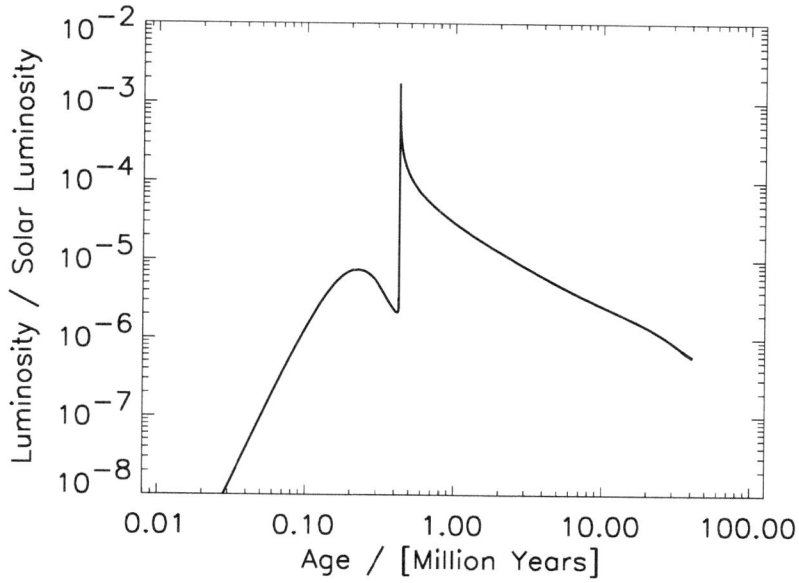

FIGURE 4.5. Luminosity of a Pegasi planet during the first 100 million years.

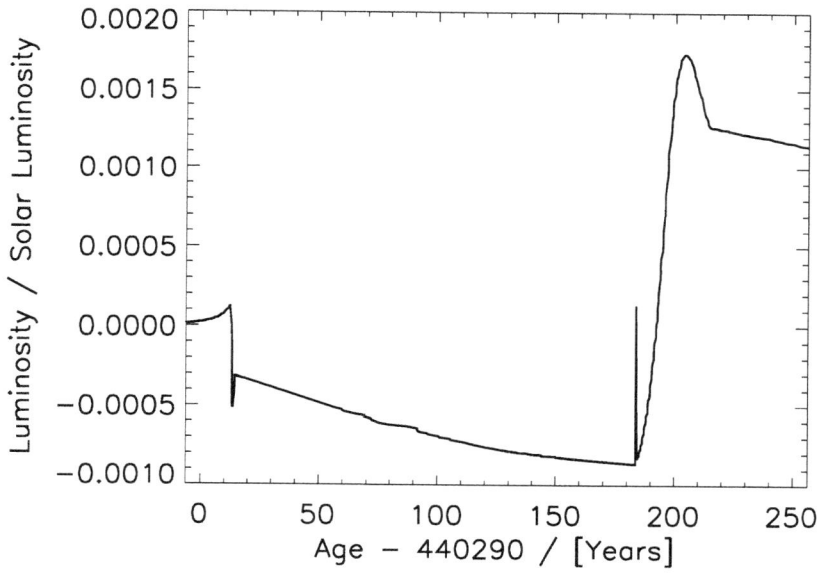

FIGURE 4.6. Luminosity of a Pegasi planet: detail around maximum luminosity.

The final mass of the planet is approximately equal to $M \sin i$ for 51 Pegasi b. Being below a Jupiter mass (see Wuchterl et al. 2000), there is a chance that spherical symmetry will give fairly correct results not only up to 100 $M_\oplus$, or $\sim$0.317 $M_{\rm Jup}$, but even for the gas accretion rates towards peak accretion and stagnation when the feeding zone is emptied. The rapid transition from 20 to 150 $M_\oplus$ also may provide a way to escape violent migration by quickly removing any relevant, migration-driving mass from the nebula disc before significant orbital decay occurs in type II migration mode.

The critical assumption concerning the above calculations is whether feeding zones with the assumed masses are plausible. In a minimum-mass nebula, the mass integrated over plausible feeding zones is only a few Earth masses. But to understand extrasolar planets and planet formation in general the minimum-mass nebula that is reconstructed from the Solar System is a much too narrow constraint for the plausibility of the available mass. Most probably, the nebula will be gravitationally stable during the planet formation epoch. That requirement allows us to derive a more general constraint on the feeding zone. The gravitational stability of discs can be roughly estimated by the Toomre criterion for axially symmetric gravitational stability,

$$Q := \frac{\pi G \Sigma}{\kappa c_\mathrm{T}} < 1, \qquad (4.42)$$

where Σ is the surface mass density of the disc, c_T the isothermal sound speed and κ the epicyclic frequency, which approaches the Keplerian angular frequency, $\Omega_\mathrm{Kepler} = \sqrt{(GM/r^3)}$, for Keplerian discs. I suggest using the marginally Toomre-stable nebula as the limit for the variety of nebula conditions that are allowed for consideration. The maximum midplane density values so obtained for planet-forming nebulae are a factor of 200 above the minimum mass at 0.05 AU. In a very conservative feeding zone of one Hill radius around the orbit, a Jupiter mass (318 $M_\oplus$) can easily be accounted for in such an enhanced nebula. Typical feeding zones would have more than five times the radius and hence provide a volume around the orbit that is 25 times larger. Hence many Toomre-stable nebulae can provide sufficient mass in plausible feeding zones at 0.05 AU to form a giant planet. The above calculations therefore show that giant planets form in less than a million years at 0.05 AU if their orbits are stable, i.e. migration rates remain small.

Relying on a standard minimum mass nebula and planet formation fluid dynamics that are physically improved but following a simple model setup, it is possible to explain the formation of a Pegasi planet *in situ* provided there is sufficient mass of gas and solids in the feeding zone. This is not the case for a minimum-mass nebula. But, assuming the diversity in extrasolar planets originates from a diversity in disc properties, we may vary the global nebula parameters. Gravitationally stable nebulae that have less total angular momentum and hence more mass closer in may provide sufficient mass to grow giant planets dynamically at 0.05 AU.

4.26. Is conventional formation theory misleading or not?

I have shown that one of the surprising discoveries – that Pegasi planets can be reconciled with standard Solar System formation theory – is explicable if a diversity of nebulae is accepted. I now return to the empirical bases for the doubts raised about the general validity of Solar System-based understanding of planet formation: the exoplanet discoveries.

4.26.1. Is there no bias . . . ?

One of the first questions that immediately occurs is whether the discoveries are biased, as is very usual in astronomy because faint things are harder to see. Possible sources of bias introduced by the dominating radial velocity technique are:
  (i) sensitivity of the radial velocity measurements (highest for large masses and short periods);
  (ii) the planet hunting grounds and hunting tactics;

(iii) the discovery race and issues of secrecy and selective publishing – in particular the focus on planetary detections and the postponing of the very difficult issue of determining upper limits in case of null-results that lead to an unclear picture of the *de facto* sensitivity of the searches;
(iv) the binary issue, i.e. that the binary fraction in the Galaxy is much larger than in the typical samples of radial velocity planet searches;
(v) the selection of suitable host-stars to avoid variability, activity, youth, giants, etc.;
(vi) the extraction of reliable planetary signals from the data – two massive planets with widely separated orbits are easier to identify then two relatively low mass planets with comparable masses and relatively close orbits (this is especially true when the orbital periods are close to being multiples of each other, as in the case of Jupiter and Saturn).

These questions can only be answered by very well-defined and complete samples and to some extent by other discovery methods. It is interesting to look at the planetary 'yield' of ongoing transit searches. They have so far not detected as many planets as would be (naively) expected from an extrapolation of the radial velocity discoveries. It remains to be seen whether this is due to difficulties in the transit method or simply to different biases in radial velocity and transit searches. Clearly, an overlap of methods is important and seems to be possible for astrometry and direct imaging within a few years.

4.26.2. *Towards normality*

With the radial velocity method providing most of the information by far about extrasolar planets, it is interesting to look at how the typical properties of the discovered exoplanets change as more and more are discovered. It is notable that:
- the 'outskirts' of the eccentricity distribution approaches that of the Solar System planets with an overlap in all parameters expected soon;
- the periods of the discovered planets increase with time, now starting to overlap with those of the Solar System giant planets, and waiting seems to make the Solar System more typical;
- the median of the distribution of minimum masses seems to continue to decrease; it seems that the characteristic mass has changed from about $M \sin i = 7$ $M_{\rm Jup}$ in 1996, to 4 $M_{\rm Jup}$ in 2000, and I understand from obervers' talks that $M \sin i = 2$ or 1 $M_{\rm Jup}$ may be a possible final outcome.

The next important step in the discoveries is an extrasolar planet that overlaps with a Solar System giant planet in all its properties, i.e. mass and eccentricity less than Jupiter's, orbital period larger than Jupiter's.

4.26.3. *Brave-hearted searches – Icarus!*

To close the gap, searches are necessary where they are most difficult:
(i) avoiding 'hunting bias', i.e. without a priori input to select stars for planetary yield or assigning higher observation priority to 'good' stars, i.e. stars with low radial velocity 'noise';
(ii) volume-complete samples (see next section);
(iii) searches for planets in binaries (e.g. α-Cen, see Endl *et al.* 2001);
(iv) searches for planets of stars with a type earlier than late F – most RV samples focus on stars later than that, but Setiawan *et al.* (2003) detected a planet around a K1 III giant (an A-star on the main sequence), expanding the mass range of known stars with planets;
(v) searches for planets in clusters (as opposed to *cluster planets*), which have the advantage of more homogeneous and coeval stellar populations;

(vi) searches for young planets to determine the earliest time at which planets exist;
(vii) the host star mass range is also considerably widened by studying M-stars: Kürster et al. (2003) showed that planets with $M \sin i$ of a few $M_\oplus$ could be detected around M-stars;
(viii) Guenther and Wuchterl (2003) went to the extreme and searched for planets around brown dwarfs: while significant RV variation was detected, only an upper limit could be set for the presence of Jupiter-mass planets;
(ix) direct imaging searches to look for planets in long-period orbits and the start of their direct characterization (Neuhäuser et al. 2000).

4.26.4. Looking at stars near you – metallicity

To show the present state of the discussion and possible problems with biases I will briefly discuss the planet–metallicity relation (Santos et al. 2003; Fischer and Valenti 2005).[24] The authors find the frequency of planets to increase with metallicity (i.e. the [Fe/H] metallicity indicator). Fuhrmann (2002, 2004) studied a volume-complete sample of nearby F-, G- and K-stars that overlaps with the planet-hunting samples. When compared to the Fisher et al. result and assuming the use of the Nidever et al. (2002) volume-*limited* sample, the following is noticed (Fuhrmann, personal communication) when comparing it to the volume-*complete* Fuhrmann-sample: of the 166 Fuhrmann stars only 90 (54%) are in the Fisher et al. study despite the smaller-volume but complete Fuhrmann sample. Missing in the Fischer and Valenti (2005) study are:

(i) a few subgiants,
(ii) fast-rotating stars with $v \sin i > 10$ km/s,
(iii) a few young and chromospherically active stars,
(iv) stars with no precise luminosity class,
(v) binaries and multiple systems.

These are all properties that make planet hunting more difficult. From the findings of Fuhrmann, I can only conclude that stars that are unfavourable for planet hunting are under-represented. The average, [Fe/H] = −0.02, of the 90 Fisher et al. stars that fall into the intersection with the Fuhrmann sample is only 0.01 dex higher than Fuhrmann's respective value. However, for [Fe/H] ≥ 0.2 only five stars are missing, whereas for [Fe/H]≤ −0.2, 13 stars are missing. Hence, there is a slight trend indicating that metal-poor stars are preferentially missing. Considering the still small numbers that are available for comparison and the average metallicity of [Fe/H] = 0.00 for thin-disc stars, there could be a metallicity effect of +0.10 to +0.15 (contrary to the +0.25 favoured by Santos et al. 2003). With the Sun at the thin-disc average of [Fe/H] for our distance to the Galactic Centre (Fuhrmann 2004), the role of metallicity may well be a slight increase overall due to a significant increase for the Pegasi planets' host stars. That would render the Solar System normal as far as metallicity and planet hosting are concerned.

It is just this kind of bias that I think might still be present in planetary discoveries. This should be considered before abandoning what we know about the Solar System.

4.27. Has the Solar System been misleading us?

All things considered, has the Solar System been misleading us? The Sun led to much of modern physics and astrophysics. It is the calibrator of stellar evolution and the age of the Universe. The Solar System offers the best traces for planet formation studies. But there are stars other than the Sun and there are planetary systems other than ours. After

---

[24] See also Chapter 5.

the dust has settled, I think the Solar System will still provide the basis for understanding planetary diversity. It doesn't do to blame the path when getting lost!

The dominance of *strange* planets and the increased metallicity of planet hosts may be a result of hunting biases in the exoplanet sample. The close-in giant planets have not been explicitly excluded by most investigators of planet formation. To a large extent, the question of such planets has not been considered in sufficient detail. The high eccentricities in the presently known exoplanets may be due to two effects:
  (i) they could represent the high-mass end of planet formation (dynamically larger masses result in larger planet–planet interactions that typically increase the eccentricities);
  (ii) eccentric planets might be easier to detect because an eccentric planet excludes a wider range of neighbouring planets according to Hill exclusion stability criteria: as a result, for a sufficiently eccentric planet, there is no neighbouring, competing signal that causes confusion, thus not adding 'planetary noise' due to the unidentified additional radial velocity signals of neighbouring (and smaller) planets.

We need careful studies of the observation biases; most importantly, we need the study of predefined complete samples and we need an analysis of what planets may be extracted first if every planet host has a planetary system that is as dynamically filled as ours.

4.28. Observational tests of formation theory

There are currently two scenarios of close-in giant planet formation theory:

the migration scenario, in which planets first form in a special giant-planet formation region and subsequently move into orbits much closer to the star;

the *in situ* scenario, where planets form near their present orbits.

All migration scenarios use violent migration as defined above with more than a factor of e^2, or roughly an order of magnitude, change in orbital radius from the birth region to the final orbit. Violent migration processes such as those due to disc–planet interaction are thought to impose a severe limit on planetary lifetime in the presence of the protoplanetary nebula. Yet there is apparently no evidence for such a process in the Solar System. Three observational tests have been proposed to check whether violent migration is operating in general planetary systems and to distinguish between the two planet formation scenarios (Wuchterl 2001b, 2004b):
  (i) the existence of hot neptunes,
  (ii) birthplace exclusion in multiple stellar systems, and
  (iii) Hill sphere compression in volume exclusion stability criteria for planetary systems.

Among these tests, (i) uses the fact that migrating planets with about 10 Earth masses usually accrete gas, (ii) uses the fact that there is no 'appropriate' birthplace, e.g. beyond the ice-line, from where the planet might initiate its migration, and (iii) is based on the fact that the Hill spheres, and stability criteria derived from them, face a compression effect during inward migration.

Before we discuss the tests, let us look at the basic assumptions used in the two scenarios. The *in situ* picture assumes (1) that planets form via core growth by planetesimal accretion, (2) that there is only minor orbital evolution and (3) that there is sufficient building material in a feeding zone near the final orbit. Migration studies assume that (1) planets form in a special formation region, (2) there is sufficient building material in the formation region, (3) significant orbital evolution is necessary to arrive at the final planet location and (4) migration operates and stops in time.

In situ theory predicts that giant planets can form whenever sufficient mass is available and hence there is in general no characteristic distance for their occurrence. Nebulae that provide sufficient mass in planets' feeding zones are needed. That requires a diversity of protoplanetary nebulae because in the minimum-mass nebula there is insufficient mass close to the star. The dynamics of envelope accretion controls what happens at the critical mass; for example, only sufficiently convective envelopes allow accretion to Jupiter masses, and hence, depending on nebula properties, the mass spectrum as a function of distance is modulated.

Migration theory assumes formation beyond the ice-line, typically at 2–3 AU for a solar-like star (e.g. Sasselov and Lecar 2000; see Hueso and Guillot 2003 for a discussion of the location). Planets essentially form quasi-statically whenever a critical core with a mass above a few $M_\oplus$ is grown or via a disc instability, if that is possible. The latter is favoured by low temperatures and low tides, generally at large orbital distances. Planets then migrate into their present orbits and are subsequently stopped by a parking process, e.g. at the inner edge of the disc. Many of the physical processes in the migration framework are parameterized, even for a given nebula structure, and hence there are few quantitative parameter-independent predictions. But in any case the formation region should be beyond the ice-line, resulting in a minimum orbital radius, $a_{\text{form}} > a_{\text{snow}} = 2\text{–}3$ AU. Furthermore, a critical core mass of a few $M_\oplus$ or a gravitational instability with fragment mass of $>300\ M_\oplus$ is necessary. The migration rates (of all types) increase towards the star, mainly because of the decrease in orbital periods and increase in nebula surface densities, see Eq. (4.40). And finally, to perform the angular momentum transfer efficiently, planets need a disc of comparable mass to migrate.

4.28.1. *Migration test I: hot neptunes*

A hot neptune is a planet with supercritical mass that orbits inside the ice-line; for violent migration $a_{\text{HN}} < a_{\text{snow}}/e^2$. Let us assume a hot neptune formed in the migration scenario. It then had to accrete first beyond the ice-line and then start migration. Migration only operates if the disc is massive, i.e. has a mass at least comparable to that of the planets. A hot neptune is by definition supercritical, hence it will accrete gas if any is present. Therefore, a migrating neptune will continue to accrete. This does not change for type II migration because accretion can continue through gaps that have been opened by the planet in the nebula. Because the disc is still at least of comparable mass to make the planet migrate, it will accrete at least more than its own mass during the inward migration process. Hence, when arriving at the final orbit, near the star, it will have doubled its mass. It then has at least two Neptune masses, or at least twice the critical mass value and is not a hot neptune any more. Therefore, a hot neptune cannot be due to formation far out and subsequent migration.

Because we considered violent migration in the test, the planet has changed its orbital radius by much more then one feeding-zone diameter and hence would have had access to many times the mass that it accreted in its formation zone. It is therefore probably much larger than the factor of two above critical mass that we used in the above argument. Hence, finding a hot neptune refutes the migration scenario because hot neptunes must have formed *in situ*.

The recent detections of Neptune-mass[25] planets in the planetary systems of μ Arae (Santos *et al.* 2004; 0.044 M_{Jupiter}, 9.55 d, 0.0955 AU) and ρ^1 Cancri (McArthur *et al.* 2004; 0.045 M_{Jupiter}, 2.81 d, 0.038 AU) easily qualify as hot neptunes by their orbital

---

[25] $M_{\text{Neptune}} \approx 14.6\ M_\oplus \approx 0.046\ M_{\text{Jupiter}}$, $M_{\text{Uranus}} \approx 17.2\ M_\oplus \approx 0.054\ M_{\text{Jupiter}}$.

radii (being less than 1/10 of the ice-line) and minimum masses ($M \sin i \approx 14\,M_\oplus$). Given the $\sin i$ uncertainty, there is still the unlikely possibility that their true masses could double or triple and hence weaken the test. But keeping in mind that only a few systems have been so thoroughly observed so far – if any except for those two – I consider it unlikely that these $M \sin i$ values will turn out to be outliers due to orbital projection effects and because the detected planetary populations may be compatible with migration theory.

4.28.2. Migration test II: the binary snow-plough

This test uses the fact that the migration scenario adopts a special formation region for giant planets. Let us consider a binary system of two equal mass components for simplicity. The formation region is then defined to lie outside the ice-line, i.e. at orbital distances $a > a_{\text{snow}}$ from the respective component. The gravitational sphere of influence of each stellar component reaches the midpoint in our system and to the respective L_1 point between the components in general. If we consider subsequently closer binary systems there will be a separation where the gravitational sphere of influence of each of the components does not reach beyond the ice-line, because that orbital distance is already within the gravitational reach of the other component, beyond the L_1 point in general. In such a system there is no planet-formation region beyond the ice-line because there exist no more orbits around the respective component. Distances beyond the ice-line belong to the other component. If a planet were to be found in orbit around a binary component where the formation region is taken away by the gravitational reach of the other component, the respective planet cannot have been formed in that special formation region and subsequently migrated to its detected orbit because the formation region does not exist. Hence the migration scenario could then be ruled out.

In practice, the restrictions for the formation regions are more severe because orbits near L_1 are unstable (Dvorak 1986; Holman and Wiegert 1999). While planets with such S-type orbits are known in binary systems (e.g. γ Cep as discussed earlier) the respective binary orbits are not tight enough to perform the test. But this shows that planet searches in binary systems could provide key information about the formation processes. Thus the binary snow-plough might well remove the formation regions beyond the ice-line that are required in migration scenarios.[26]

4.28.3. Migration test III: Hill sphere expansion

A key result for the Solar System is that it is dynamically full (Lecar *et al.* 2001). This means that there is no space to introduce additional planets because there are no additional stable orbits. This can be roughly understood by using a criterion that requires a minimum orbital spacing between planets such that their mutual interactions are limited over a certain time span. These criteria can be expressed in terms of the gravitational sphere of influence of a planet, approximated by their Hill spheres that generate a torus with radius R_{Hill} (Eq. (4.25)) around the planets' orbit. This approximate criterion requires that certain multiples of the Hill spheres of the two planets must never touch; i.e. for any two planets, i and $i+1$, with orbital radii a_i, a_{i+1} and masses M_i, M_{i+1},

$$|a_i - a_{i+1}| > n\left[r_{\text{Hill}}(a_i, M_i) + r_{\text{Hill}}(a_{i+1}, M_{i+1})\right], \qquad (4.43)$$

must hold to ensure the stability of a planetary system.[27] The number n depends on the timescale considered and lies between 4 and 15. The case of two planets is well

---

[26] Konacki (2005) reports a planet in the triple system HD 188753 where a tight binary companion excludes the ice-line.
[27] This can be generalized by requiring it to hold for $a_i(1 - e_i)$ and $a_{i+1}(1 + e_{i+1})$.

investigated now and $n = 2, \ldots, 4$ seems to guarantee permanent stability (see Lecar et al. 2001). For the Solar System $n = 13$ seems to guarantee stability over a few Ga. $n = 15$ might be required in some resonant systems.

If extrasolar planetary systems are dynamically full, like the Solar System, their planetary spacing must obey the criterion with some n. That the systems are full seems to be indicated by the rather dense systems of massive planets in the systems discovered so far, that are often *on the edge* (Barnes and Quinn 2004), meaning that the observed orbital parameters are close to many unstable ones. Hence, the Hill exclusion criteria will be closely matched; in other words, the systems will be dense. On the other hand, migration rates increase with decreasing orbital distance, typically with orbital velocity, i.e. $\propto r\Omega_{\text{Kepler}} = a^{-1/2}$, see Eqs. (4.40), (4.41).

If the systems form in a special formation region and planet formation is prolific, producing full systems, the systems should obey the Hill exclusion criterion in their formation region. But then violent migration comes into play and the innermost planets migrate away from those formed farther out with increasing migration rates. Let us assume a system is formed in the formation region and the planetary spacing, Δa_0, at an orbital radius a_0 is $\Delta a_0 = nR_{\text{Hill}}(a_0, M) = na_0(M/(3M_\odot))^{1/3}$. The system then migrates inwards by a factor of $1/e^2$ in radius, because that is violent migration by definition. Let us look at the system at the moment when a planet from a_0 has arrived at its new orbit with $a = a_0/e^2$. For small spacings $\Delta a_0/a_0 < 1$ we find for the spacing, Δa, after migration to the new orbital radius, a:

$$\frac{\Delta a}{\Delta a_0} = \left(\frac{a_0}{a}\right)^{\frac{1}{2}}, \quad (4.44)$$

which equals e for a violent migration shrinking of an orbit by $1/e^2$. If we keep the masses constant during migration we can express the ratio of orbital separation between two neighbouring planets in multiples n, n_0 of the respective Hill radii, $R_{\text{Hill}} \propto a$, as

$$\frac{n}{n_0} = \left(\frac{a_0}{a}\right)^{\frac{3}{2}}. \quad (4.45)$$

This means that a system satisfying a Hill exclusion criterion with planets spaced by $n_0 R_{\text{Hill}}(M, a_0)$ that migrates by a factor $a_0/a = 10$ to a new orbital radius a will have a spacing of $nR_{\text{Hill}}(M, a_0)$, with $n/n_0 = 10^{3/2} \approx 32$! This means that the planets are a factor of 32 more widely dynamically spaced than the original system.

Since migrating planets must accrete, there is a counteracting effect due to the expansion of the Hill spheres with increasing mass. But this is comparatively weak ($\propto M^{1/3}$) and should not change the conclusion.

With migration invoked for Pegasi planets at 0.05 AU that supposedly migrated by a factor ~100, the effect is even more dramatic. Migration dynamically dilutes planetary systems dramatically, and migrated systems would easily satisfy Hill exclusion principles even when they started out dynamically full, just marginally obeying Hill exclusion.

In summary, if all systems form dynamically full like the Solar System and as the planetesimal accretion process seems to predict, then migrated systems must appear dynamically very underdense. If they form dynamically dense in their formation region, they will appear much more widely spaced than a dynamically full system after they migrated. Hence migrated systems should be widely spaced as counted by Hill radii, whereas *in situ* formed systems should appear closely spaced, just satisfying the respective Hill exclusion principle relevant for their age. *Close-in dynamically dense observed planetary systems therefore refute systems with violent migration.* Of course, this assumes that violent migration processes of types I, II and III dominate more subtle dynamical planetary interactions that govern much of the later evolution, but that is after all the hypothesis of the migration origin of planets.

4.29. Planetendämmerung

To close, let us imagine we could start astronomy and wait for the stars to appear in the evening for the first time. Located at a beach in the Canary Islands an hour or so after sunset, we might only be able to get to a visual limiting magnitude of ∼3, say. Imagine that our entire knowledge of stellar astronomy would have to be derived from, at first, the ten brightest stars, then the 100 brightest stars, which are the first to become visible. What would be the typical mean stellar radius that we would deduce for our nascent stellar astronomy? The answers – taken from the *Catalogue of Apparent Diameters and Absolute Radii of Stars*, are:
- for the ten brightest stars the average radius is 36 $R_\odot$, and
- for the 100 brightest stars the average radius is 32 $R_\odot$.

This is not the astrophysics that we know! Our Galaxy is dominated by low-mass M-dwarfs, with radii smaller than the Sun. They greatly outnumber the giants. It is just that the stellar majority is hard to find and hard to see at first glance. Similarly, I think, we will see more planets 'appear' in the sky! Maybe they will be more like the Solar System planets.

4.30. Towards a broad view

The next steps towards developing a broader picture into which the Solar System and the extrasolar planets fit are:
  (i) the development of a general theory of planet formation for a diversity of protoplanetary nebulae;
  (ii) instruments that are sensitive to the entire mass spectrum of the giant planets and below. The French-led European *COROT* mission with German participation via the DLR should reach that goal for the first time after its launch in Winter 2006;
  (iii) in autumn 2007, the first Jovian year will have passed since the discovery of 51 Peg b and many of the searches should then have the first complete orbits of Jupiter-like planets in their data.

Finally, we will have the required sensitivity for a full Jovian orbit and the theory significantly advanced. Theoretical work is ongoing in preparation for these events. Unlike with 51 Peg b, this time theory will be prepared.

4.31. Conclusion

Have the first 100 exoplanets misled us?

After more than 2500 years of astronomy we have crossed the ocean of space between us and the stars. New worlds become detectable and within the reach of analysis. What has been discovered was unexpected in detail but has shown that planets are abundant in the Galaxy, as expected. Much can be understood on the basis of what has been learned about planet formation from the Solar System, but many problems remain.

After crossing an ocean the first cliff leads to land, but don't try to get too close to it.

REFERENCES

Allègre, C. J., Manhès, G. & Göpel, C., 1995, *Geochim. Cosmochim. Acta* **59**, 1445.

Alves, J. F., Lada, C. J. & Lada, E. A., 2001, *Nature* **409**, 159.

Appenzeller, I. & Tscharnuter, W., 1974, *A&A* **30**, 423.

Appenzeller, I. & Tscharnuter, W., 1975, *A&A* **40**, 397.

Balluch, M., 1991a, *A&A* **243**, 205.

Balluch, M., 1991b, *A&A* **243**, 187.

Baraffe, I., Chabrier, G., Allard, F. & Hauschildt, P. H., 2002, *A&A* **382**, 563.

Barnes, R. & Quinn, T., 2004, *ApJ* **611**, 494.

Beckwith, S. V. W., Henning, T. & Nakagawa, Y., 2000, in *Protostars and Planets IV*, eds. V. Mannings, A. P. Boss, & S. S. Russell, Tucson: University of Arizona Press, pp. 533.

Bertout, C., 1976, *A&A* **51**, 101.

Blum, J. & Wurm, G., 2000, *Icarus* **143**, 138.

Bodenheimer, P., 1966, *ApJ* **144**, 709.

Bodenheimer, P. & Pollack, J. B., 1986, *Icarus* **67**, 391.

Bodenheimer, P., Hubickyj, O. & Lissauer, J. J., 2000, *Icarus* **143**, 2.

Bonnor, W. B., 1956, *MNRAS* **116**, 351.

Calvet, N., Hartmann, L. & Strom, S. E., 2000, in *Protostars and Planets IV*, eds. V. Mannings, A. P. Boss, and S. S. Russell, Tucson: University of Arizona Press, p. 377.

Chabrier, G. & Baraffe, I., 1997, *A&A* **327**, 1039.

Chambers, J. E., 2001, *Icarus* **152**, 205.

Chambers, J. E. & Wetherill, G. W., 1998, *Icarus* **136**, 304.

D'Angelo, G., Henning, T. & Kley, W., 2002, *A&A* **385**, 647.

D'Angelo, G., Kley, W. & Henning, T., 2003, *ApJ* **586**, 540.

D'Angelo, G., Bate, M. R. & Lubow, S. H., 2005, *MNRAS* **358**, 316.

D'Antona, F. & Mazzitelli, I., 1994, *ApJS* **90**, 467.

Drouart, A., Dubrulle, B., Gautier, D. & Robert, F., 1999, *Icarus* **140**, 129.

Dvorak, R., 1986, in *Asteroids, Comets, Meteors II*, Proceedings of the International Meeting, Uppsala, Sweden, June 3–6, 1985. Uppsala: Astronomiska Observatoriet, p. 165.

Dvorak, R., Froeschlé, C. & Froeschlé, C., 1989, *A&A* **226**, 335.

Ebert, R., 1957, *Zeitschrift für Astrophysik* **42**, 263.

Endl, M., Kürster, M., Els, S., Hatzes, A. P. & Cochran, W. D., 2001, *A&A* **374**, 675.

Fernandez, J. A. & Ip, W.-H., 1984, *Icarus* **58**, 109.

Fischer, D. A. & Valenti, J., 2005, *ApJ* **622**, 1102.

Forestini, M., 1994, *A&A* **285**, 473.

Fuhrmann, K., 2002, unpublished, www.xray.mpe.mpg.de/~Fuhrmann.

Fuhrmann, K., 2004, *Astronomische Nachrichten* **325**, 3.

Götz, M., 1989, Ph.D. thesis, University of Heidelberg.

Guenther, E. W. & Wuchterl, G., 2003, *A&A* **401**, 677.

Guillot, T., 1999, *Science* **286**, 72.

Guillot, T., 2005, *Ann. Rev. Earth Planet. Sci.* **33**, 493.

Guillot, T., Burrows, A., Hubbard, W. B., Lunine, J. I. & Saumon, D., 1996, *ApJ* **459**, L35.

Haisch, K. E., Lada, E. A. & Lada, C. J., 2001, *ApJ* **553**, L153.

Hartmann, L., Cassen, P. & Kenyon, S. J., 1997, *ApJ* **475**, 770.

Hatzes, A. P., Cochran, W. D., McArthur, B., *et al.*, 2000, *ApJ* **544**, L145.

HATZES, A. P., COCHRAN, W. D., ENDL, M., et al., 2003, *ApJ* **599**, 1383.
HAYASHI, C., 1961, *PASJ* **13**, 450.
HAYASHI, C., 1966, *ARA&A* **4**, 171.
HAYASHI, C., 1980, *Progr. Theor. Phys.* **70**, 35.
HAYASHI, C., HOSHI, R. & SUGIMOTO, D., 1962, *Prog. Theor. Phys.* **22**, Suppl., 174.
HAYASHI, C., NAKAZAWA, K. & NAKAGAWA, Y., 1985, in *Protostars and Planets II*, Tucson: University of Arizona Press, p. 1100.
HOLLENBACH, D. J., YORKE, H. W. & JOHNSTONE, D., 2000, in *Protostars and Planets IV*, eds. V. Mannings, A. P. Boss, and S. S. Russell, Tucson: University of Arizona Press, p. 401.
HOLMAN, M. J. & WIEGERT, P. A., 1999, *AJ* **117**, 621.
HUBICKYJ, O., BODENHEIMER, P. & LISSAUER, J. J., 2004, in *Revista Mexicana de Astronomia y Astrofisica Conference Series*, **22**, 83.
HUESO, R. & GUILLOT, T., 2003, *Space Sci. Rev.* **106**, 105.
IKOMA, M., EMORI, H. & NAKAZAWA, K., 2001, *ApJ* **553**, 999.
INABA, S. & IKOMA, M., 2003, *A&A* **410**, 711.
INABA, S., WETHERILL, G. W. & IKOMA, M., 2003, *Icarus* **166**, 46.
KÜRSTER, M., ENDL, M., ROUESNEL, F., et al., 2003, *A&A* **403**, 1077.
KANT, I., 1755, *Allgemeine Naturgeschichte und Theorie des Himmels*, Königsberg und Leipzig. Johann Friederich Petersen: English translation: W. Hastie, 1968, *Universal Natural History and Theories of the Heavens in Kant's Cosmology*, New York: Greenwood Publishing.
KIPPENHAHN, R. & WEIGERT, A., 1990, *Stellar Structure and Evolution*, XVI, Berlin, Heidelberg & New York: Springer-Verlag.
KOKUBO, E. & IDA, S., 2002, *ApJ* **581**, 666.
KOLLER, J. & LI, H., 2003, in *Scientific Frontiers in Research on Extrasolar Planets*, ASP Conference Series 294, eds. D. Deming, & S. Seager, p. 339.
KOLLER, J., LI, H. & LIN, D. N. C., 2003, *ApJ* **596**, L91.
KONACKI, M., 2005, *Nature* **436**, 230.
KUERSCHNER, R., 1994, *A&A* **285**, 897.
KUHFUß, R., 1987, Ph.D. thesis, Technical University of Munich.
KUMAR, S. S., 1963, *ApJ* **137**, 1121.
KUSAKA, T., NAKANO, T. & HAYASHI, C., 1970, *Progr. Theor. Phys.* **44**, 1580.
LAPLACE, P. S., 1796, *Exposition du Système du Monde*, Paris: Circle-Sociale. English translation: Pond, J., 1809, *The System of the World*, London: Richard Phillips.
LARSON, R. B., 1969, *MNRAS* **145**, 271.
LARSON, R. B., 2003, *Rep. Progr. Phys.* **66**, 1651.
LASKAR, J., 2000, *Phys. Rev. Lett.* **84**, 3240.
LECAR, M., FRANKLIN, F. A., HOLMAN, M. J. & MURRAY, N. J., 2001, *ARA&A* **39**, 581.
LEVISON, H. F. & AGNOR, C., 2003, *AJ* **125**, 2692.
LEVISON, H. F., LISSAUER, J. J. & DUNCAN, M. J., 1998, *AJ* **116**, 1998.
LIN, D. N. C., BODENHEIMER, P. & RICHARDSON, D. C., 1996, *Nature* **380**, 606.
LISSAUER, J. J., 1987, *Icarus* **69**, 249.
LISSAUER, J. J., 1993, *ARA&A* **31**, 129.
LISSAUER, J. J., 1995, *Icarus* **114**, 217.
LISSAUER, J. J., POLLACK, J. B. WETHERILL, G. W. & STEVENSON, D. J., 1996, in *Neptune and Triton*, ed. D. P. Cruikshank, Tucson, AZ: University of Arizona Press, p. 37.
LUGMAIR, G. W. & SHUKOLYUKOV, A., 2001, *Meteor. Planet. Sci.* **36**, 1017.
LUNINE, J. I., CORADINI, A., GAUTIER, D., OWEN, T. C. & WUCHTERL, G., 2004, in *Jupiter: the Planet, Satellites and Magnetosphere*, eds. F. Bagenal, T. E. Dowling, W. B. McKinnon, Cambridge Planetary Science, 1, Cambridge: Cambridge University Press, p. 19.

MARZARI, F., WEIDENSCHILLING, S. J., BARBIERI, M. & GRANATA, V., 2005, *ApJ* **618**, 502.
MAYOR, M. & QUELOZ, D., 1995, *Nature* **378**, 355.
MCARTHUR, B. E., ENDL, M., COCHRAN, W. D., et al., 2004, *ApJ* **614**, L81.
MIZUNO, H., 1980, *Progr. Theor. Phys.* **64**, 544.
MORFILL, G. E., TSCHARNUTER, W. & VOELK, H. J., 1985, in *Protostars and Planets II*, Tucson: University of Arizona Press, p. 493.
NAKANO, T., 1987, *MNRAS* **224**, 107.
NAKANO, T., 1988a, *MNRAS* **230**, 551.
NAKANO, T., 1988b, *MNRAS* **235**, 193.
NEUHÄUSER, R., GUENTHER, E. W., PETR, M. G., BRANDNER, W., HUÉLAMO, N. & ALVES, J., 2000, *A&A* **360**, L39.
NIDEVER, D. L., MARCY, G. W., BUTLER, R. P., FISCHER, D. A. & VOGT, S. S., 2002, *ApJS* **141**, 503.
PALLA, F. & STAHLER, S. W., 1991, *ApJ* **375**, 288.
PALLA, F. & STAHLER, S. W., 1992, *ApJ* **392**, 667.
PALLA, F. & STAHLER, S. W., 1993, *ApJ* **418**, 414.
PEČNIK, B. & WUCHTERL, G., 2005, *A&A* **440**, 1183.
POLLACK, J. B., HUBICKYJ, O., BODENHEIMER, P., LISSAUER, J. J., PODOLAK, M. & GREENZWEIG, Y., 1996, *Icarus* **124**, 62.
RUDEN, S. P. & POLLACK, J. B., 1991, *ApJ* **375**, 740.
SAFRONOV, V. S., 1969, *Evolution of the Protoplanetary Cloud and Formation of the Earth and Planets*, Moscow: Nauka Press (also NASA–TT–F–677, 1972).
SAFRONOV, V. S. & RUSKOL, E. L., 1982, *Icarus* **49**, 284.
SANTOS, N. C., ISRAELIAN, G., MAYOR, M., REBOLO, R. & UDRY, S., 2003, *A&A* **398**, 363.
SANTOS, N. C., BOUCHY, F., MAYOR, M., et al., 2004, *A&A* **426**, L19.
SASAKI, S., 1989, *A&A* **215**, 177.
SASSELOV, D. D. & LECAR, M., 2000, *ApJ* **528**, 995.
SETIAWAN, J., HATZES, A. P., VON DER LÜHE, O., et al., 2003, *A&A* **398**, L19.
SIESS, L. & FORESTINI, M., 1996, *A&A* **308**, 472.
SIESS, L., FORESTINI, M. & BERTOUT, C., 1997, *A&A* **326**, 1001.
SIESS, L., FORESTINI, M. & BERTOUT, C., 1999, *A&A* **342**, 480.
STAHLER, S. W., 1988a, *ApJ* **332**, 804.
STAHLER, S. W., 1988b, *PASP* **100**, 1474.
STAHLER, S. W., SHU, F. H. & TAAM, R. E., 1980a, *ApJ* **241**, 637.
STAHLER, S. W., SHU, F. H. & TAAM, R. E., 1980b, *ApJ* **242**, 226.
STEVENSON, D. J., 1982, *Planet. Space Sci.* **30**, 755.
STEVENSON, D. J., 1984, in *Lunar and Planetary Science XV*, Houston: Lunar and Planetary Institute, p. 822.
TAJIMA, N. & NAKAGAWA, Y., 1997, *Icarus* **126**, 282.
TANAKA, H. & IDA, S., 1999, *Icarus* **139**, 350.
THOMMES, E. W. & LISSAUER, J. J., 2003, *ApJ* **597**, 566.
THOMMES, E. W. & LISSAUER, J. J., 2005, in *Astrophysics of Life*, eds. M. Livio, I. N. Reid & W. B. Sparks. Space Telescope Science Institute symposium series, Vol. 16., Cambridge: Cambridge University Press, p. 41.
TRILLING, D. E., BENZ, W., GUILLOT, T., LUNINE, J. I., HUBBARD, W. B. & BURROWS, A., 1998, *ApJ* **500**, 428.
TSCHARNUTER, W. M., 1987, *A&A* **188**, 55.

TSCHARNUTER, W. M. & BOSS, A. P., 1993, in *Protostars and Planets III*, Tucson: University of Arizona Press, p. 921.
TSCHARNUTER, W. M. & WINKLER, K.-H., 1979, *Comp. Phys. Comm.* **18**, 171.
VON SENGBUSCH, 1968, *Zeitschrift für Astrophysik* **69**, 79.
WADHWA, M. & RUSSELL, S. S., 2000, *Protostars and Planets IV*, eds. V. Mannings, A. P. Boss, and S. S. Russell, Tucson: University of Arizona Press, p. 995.
WALKER, G. A. H., BOHLENDER, D. A., WALKER, A. R., IRWIN, A. W., YANG, S. L. S. & LARSON, A., 1992, *ApJ* **396**, L91.
WALKER, G. A. H., WALKER, A. R., IRWIN, A. W., LARSON, A. M., YANG, S. L. S. & RICHARDSON, D. C., 1995, *Icarus* **116**, 359.
WARD, W. R., 1997, *ApJ* **482**, L211.
WEIDENSCHILLING, S. J. & MARZARI, F., 1996, *Nature* **384**, 619.
WEIZSÄCKER, C. F. V., 1943, *Zeitschrift für Astrophysik* **22**, 319.
WETHERILL, G. W., 1990, *Ann. Rev. Earth Planet. Sci.* **18**, 205.
WINKLER, K.-H. A. & NEWMAN, M. J., 1980a, *ApJ* **236**, 201.
WINKLER, K.-H. A. & NEWMAN, M. J., 1980b, *ApJ* **238**, 311.
WOLSZCZAN, A., 1994, *Science* **264**, 538.
WOLSZCZAN, A. & FRAIL, D. A., 1992, *Nature* **355**, 145.
WUCHTERL, G., 1989, Ph.D. thesis, Universität Wien.
WUCHTERL, G., 1990, *A&A* **238**, 83.
WUCHTERL, G., 1991a, *Icarus* **91**, 39.
WUCHTERL, G., 1991b, *Icarus* **91**, 53.
WUCHTERL, G., 1993, *Icarus* **106**, 323.
WUCHTERL, G., 1995a, *Earth Moon Planets* **67**, 51.
WUCHTERL, G., 1995b, *Comp. Phys. Comm.* **89**, 119.
WUCHTERL, G., 1996, *Bull. Amer. Astron. Soc.* **28**, 1108.
WUCHTERL, G., 1997, in *Science with the VLT Interferometer*, ESO Astrophysics Symposia, Springer-Verlag, p. 64.
WUCHTERL, G., 2001a, in *The Formation of Binary Stars*, Proceedings of IAU Symp. 200, eds. H. Zinnecker & R. D. Mathieu, p. 492.
WUCHTERL, G., 2001b, in *Abstracts of JENAM 2001*, Astronomische Gesellschaft Abstract Series, Vol. **18**, p. 411.
WUCHTERL, G., 2004a, in *Astrobiology: Future Perspectives*, eds. P. Ehrenfreund, W. M. Irvine, T. Owen & L. Becker, Astrophysics and Space Science Library, Vol. 305, Dordrecht: Kluwer Academic Publishers, p. 67.
WUCHTERL, G., 2004b, *Geophys. Res. Abs.* **6**, EGU04.
WUCHTERL, G. & FEUCHTINGER, M. U., 1998, *A&A* **340**, 419.
WUCHTERL, G. & KLESSEN, R. S., 2001, *ApJ* **560**, L185.
WUCHTERL, G. & TSCHARNUTER, W. M., 2003, *A&A* **398**, 1081.
WUCHTERL, G., GUILLOT, T. & LISSAUER, J. J., 2000, in *Protostars and Planets IV*, eds. V. Mannings, A. P. Boss, and S. S. Russell, Tucson: University of Arizona Press, p. 1081.
YIN, Q., JACOBSEN, S. B., YAMASHITA, K., BLICHERT-TOFT, J., TÉLOUK, P. & ALBARÈDE, F., 2002, *Nature* **418**, 949.

5. Abundances in stars with planetary systems

GARIK ISRAELIAN

Extensive spectroscopic studies of stars with and without planetary systems have concluded that planet host stars are more metal-rich than those without detectable planets. More subtle trends of different chemical elements begin to appear as the number of detected extrasolar planetary systems continues to grow. I review our current knowledge concerning the observed abundance trends of light and heavy elements in planet host stars and their possible implications. These studies may help us to understand the chemical evolution of our Galaxy at supersolar metallicities.

5.1. Introduction

Beginning with the discovery by Mayor & Queloz (1995) of a giant planet, 51 Pegasi b, the number of planets orbiting solar-type stars has now reached 137. Most of the planets have been discovered by the Geneva and California & Carnegie groups using a Doppler technique. This sample size is now sufficient to search for various statistical trends linking the properties of planetary systems and those of their parent stars. It has been suggested that one of the key factors relevant to the mechanisms of planetary system formation is the metallicity of protoplanetary matter (Pollack *et al.* 1996). Note that in the context of this paper we consider as 'metals' all elements except H, He, Li, Be and B.

Chemical abundance studies of planet hosts are based on high signal-to-noise (S/N) and high resolution spectra. Many targets have been observed by more than one group, allowing useful crosschecks of their analyses and spectra. Iron has been used the most often as the reference element in chemical studies (Gonzalez 1997; Laws *et al.* 2003; Murray & Chaboyer 2002; Santos, Israelian & Mayor 2001, 2004a; Santos *et al.* 2003, 2005), whereas some others have discussed the abundance trends of other metals (Gonzalez & Laws 2000; Sadakane *et al.* 2002; Gonzalez *et al.* 2001; Santos, Israelian & Mayor 2000; Bodaghee *et al.* 2003; Ecuvillon *et al.* 2004a,b). The authors in most of these studies have been limited to comparing the results for the planet host sample with other studies in the literature. However, in some articles such a comparison has not been provided, leaving room for any kind of speculation regarding the source of the abundance anomalies. Different authors used different sets of lines, atmospheric parameters, data, etc. These are all potential sources of systematic error. To tackle this problem, Santos *et al.* (2001) prepared a sample of stars with no known planets. To ensure a high degree of consistency between the two samples, these stars were analyzed and observed in the same way as the planet hosts. Further spectroscopic analysis by Santos *et al.* (2003, 2004a), Israelian *et al.* (2004) and Bodaghee *et al.* (2003) were based on this same comparison sample. Recently, Santos *et al.* (2005) have added 54 stars to their original comparison sample and confirmed that stars with planets are metal rich when compared with field stars without planets.

Extrasolar Planets, eds. Hans Deeg, Juan Antonio Belmonte and Antonio Aparicio.
Published by Cambridge University Press.
© Cambridge University Press 2007.

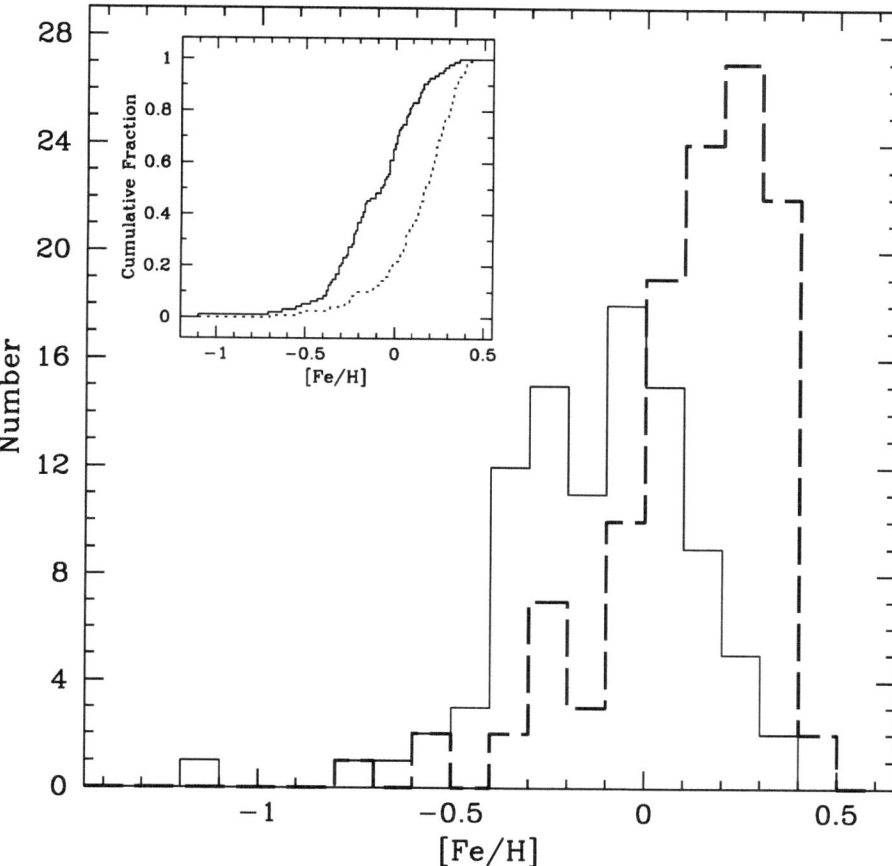

FIGURE 5.1. Metallicity distribution for stars with planets (dashed histogram) compared with the same distribution for the field stars without planets (solid line). The insert shows the same data as a cumulative histogram; the solid line refers again to stars without planets. A Kolmogorov–Smirnov test shows the probability for the two populations being part of the same sample to be 10^{-12}. From Santos et al. (2005). [Fe/H] is the logarithm of the solar Fe/H abundance ratio.

Chemical abundance studies of stars with exoplanets have demonstrated that their metallicities are higher on average than those typically found among solar-type disc stars without known planets (Gonzalez 1997; Laws et al. 2003; Santos et al. 2001, 2003, 2004a, 2005). In Figure 5.1 the metallicity distributions for two volume-limited samples of stars are presented, contrasting 119 planet hosts and 95 comparison stars (Santos et al. 2005). The stars with planets are metal-rich compared with the comparison sample stars by, on average, 0.25 dex,[1] suggesting that the metallicity and the presence of giant planets are linked. The metallicity excess could result from the accretion of planets and/or planetesimals onto the star (Gonzalez 1997). On the other hand, Santos et al. (2000, 2001) proposed that the source of the high metallicity is primordial and the observed abundance trends represent those from the protoplanetary and protostellar molecular

---

[1] 'Dex' indicates the logarithm of the ratio of two values; for example, 0.25 dex corresponds to a ratio of $10^{0.25} \approx 1.78$.

cloud out of which the star and the planets formed. This idea supports the classical core-instability accretion (CIA) model of Pollack et al. (1996) where some 10–15 $M_\oplus$ masses of planetesimals condense into a rocky core. The initial metallicity of the parental cloud is one of the key parameters in this model. On the other hand, Bodaghee et al. (2003) and Ecuvillon et al. (2004a,b) have demonstrated that the metallicity excess observed for planet host stars is not unique to iron. Abundances of chemical elements may provide a clue for checking various planet formation and evolution theories. The self-enrichment scenario of Gonzalez (1997) should lead to a relative overabundance of refractory elements (elements with high ionization temperature, such as iron, α-elements, Si, Ca, Ti, Mg, etc.) compared to volatiles (C, N, O, S, Zn). Volatiles are known to condense into solid grains at relatively low temperatures, and are expected to behave differently compared to the refractories which condense at high temperatures.[2] If the star accreted a considerable amount of planetary material, then high temperatures near the star would favour the addition of refractory elements over volatiles (which are locked in giant planets) and a trend in abundance versus condensation temperatures may appear (see Figure 11 of Smith et al. 2001).

5.2. Abundances of light elements

It is well known that the light elements Li and Be are important tracers of the internal structure and pre-main sequence evolution of solar type stars. These elements provide information regarding the redistribution and mixing of matter within a star. By measuring Li and Be in stars hosting planets we can obtain crucial information about the mixing, diffusion and angular momentum history of the stars. Studies of Be and Li complement each other as Li is depleted at much lower temperatures than Be, where depleted means that the element is being destroyed by nuclear processes in the stellar interior. Accretion of planets and planetesimals, tidal interactions and stellar activity in star–planet systems may considerably modify the surface abundances of the light elements.

5.2.1. Lithium

A first direct comparison of Li abundances among planet-harbouring stars with field stars without planets was presented by Gonzalez & Laws (2000), who proposed that the former have less Li. However, in a critical analysis of this problem, Ryan (2000) concludes that planet hosts and field stars have similar Li abundances. More recently, Israelian et al. (2004) re-investigated the Li problem and looked for various statistical trends. Comparing Li abundances of planet host stars with the 157 field stars of Chen et al. (2001), they found that the Li abundance distributions in the two samples are different (Figure 5.2), albeit with a low statistical significance. A possible excess of Li depletion in planet hosts having effective temperatures in the range 5600–5850 K is observed, whereas there is no significant difference for stars with temperatures in the range 5850–6350 K (Figure 5.3). Given the depth of the surface convection zone, we expect that any effect on the Li abundance will be more apparent in solar-type stars with effective temperatures in the range 5600–5850 K. Cooler low mass stars have deeper convective zones and destroy Li more efficiently, so we can often only set upper limits to the abundance. However, the convective layers of stars more massive than the Sun do not reach the lithium burning layer and therefore these stars generally preserve a large fraction of their original Li. Thus solar-type stars are the best targets for investigating any

---

[2] We note that refractory (such as Si, Ti and Mg) and volatile (H, He, C, N, O, S and Zn) elements have condensation temperatures larger and smaller than 1000 K, respectively.

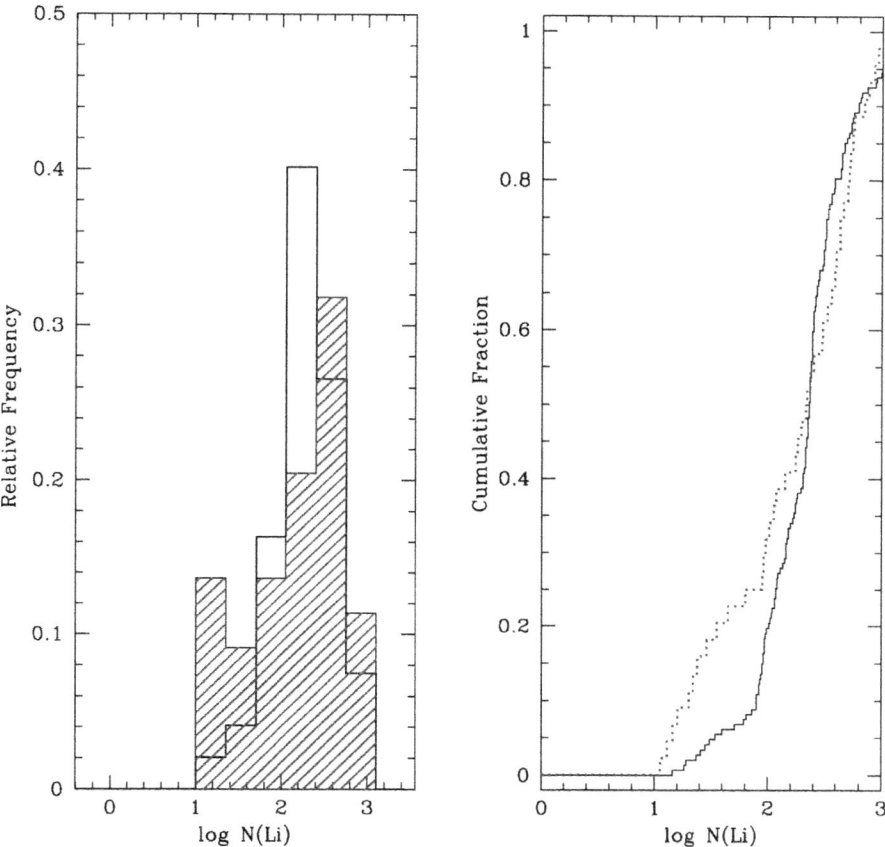

FIGURE 5.2. Lithium distribution for stars with planets (hatched histogram) compared with the same distribution for the field stars from Chen *et al.* 2001 (empty histogram). The horizontal axis is a log Li abundance relative to the solar hydrogen abundance, where log N(H) = 12. A Kolmogorov–Smirnov test shows the probability of the two populations being a part of the same sample is 0.2. From Israelian *et al.* (2004).

possible effects of planets on the stellar surface abundance of Li. According to Israelian *et al.* (2004), extra Li depletion can be associated with a planet migration mechanism at early times in the evolution of the star when the superficial convective layers may have been rotationally decoupled from the interior. Efficient depletion may be caused by a strong mixing due to the migration-triggered tidal forces, which create a shear instability. The mass of the decoupled convection zone in these stars is comparable to the masses of the known exoplanets; therefore, the migration of one or more giant planets could indeed produce an observable effect. The planetary migration may also trigger the accretion of planetesimals, inducing metallicity enhancement. Some fresh Li could also be added in the convective zone. However, if this process takes place in the early evolution of the star, the freshly added Li will be destroyed. Also, the amount of accreted Fe is not expected to be significant and therefore may not account for the observed distribution of [Fe/H]. Finally, let us mention that recently Takeda and Kawanomoto (2005) have confirmed the results of Israelian *et al.* (2004), performing a careful synthesis for the Li line in 160 solar type field stars versus 27 planet hosts.

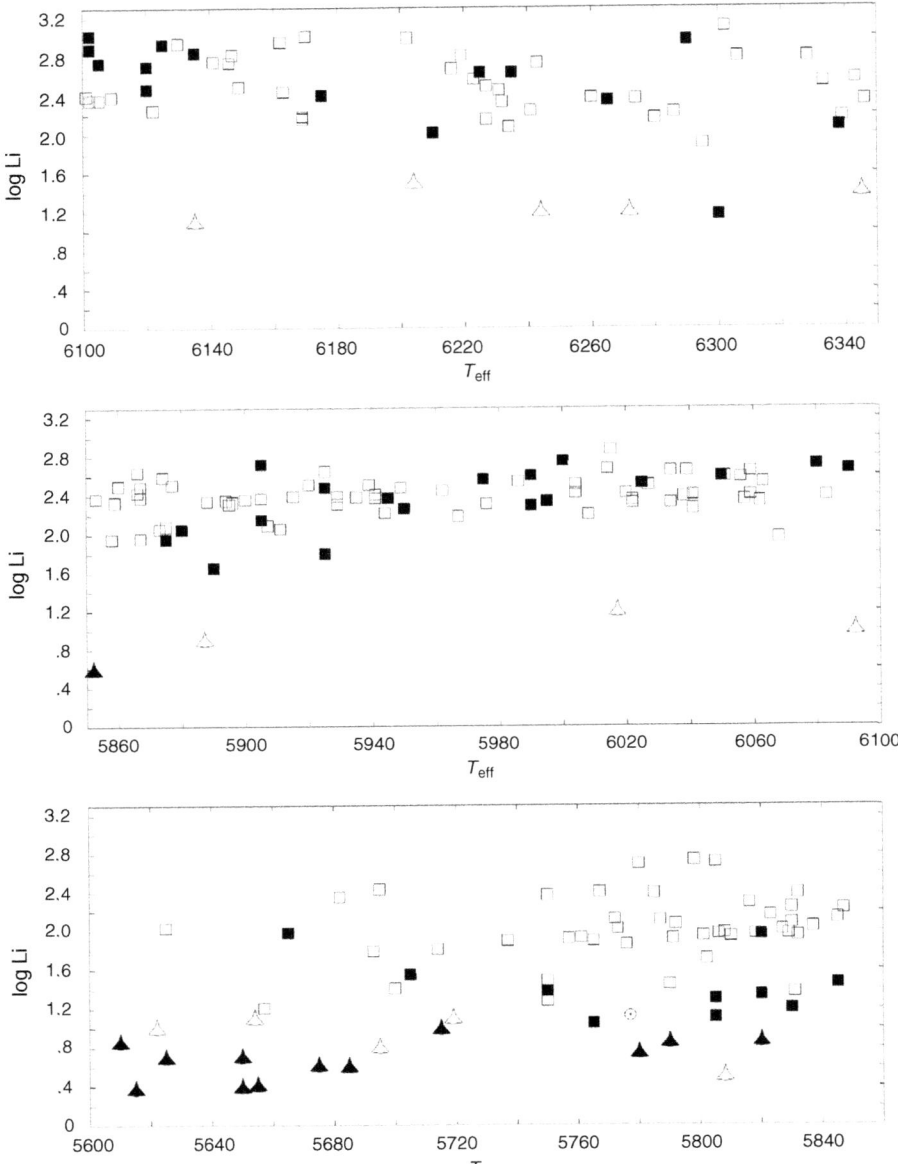

FIGURE 5.3. Lithium versus effective temperature for stars with planets (filled squares) and the comparison sample of Chen *et al.* (2001) (empty squares). Upper limits are filled (planet hosts) and empty (comparison sample) triangles. Meteoritic abundance of Li is assumed, log Li = 3.3. The position of the Sun is indicated. From Israelian *et al.* (2004).

5.2.2. *The ^{6}Li-test*

A unique opportunity for testing the planet and/or planetesimal accretion scenario is offered by a ^{6}Li-test proposed by Israelian *et al.* (2001). The idea is based on looking for an element that should not appear in the atmosphere of a normal solar-type star, but would be present in a star that has accreted planetary matter. According to standard models, nuclear reactions destroy the ^{6}Li and ^{7}Li isotopes in stellar interiors at temperatures of 2×10^6 (^{6}Li) and 2.5×10^6 K (^{7}Li). Furthermore, convection cleans the upper atmosphere of Li nuclei by transporting them to deeper and hotter layers where they are rapidly destroyed. Young pre-MS solar-type stars are fully convective and most of the primordial Li nuclei are destroyed in their interiors in a mere few million years. However, some stars preserve a large fraction of their initial atmospheric ^{7}Li nuclei while completely destroying ^{6}Li. At a given metallicity there is a mass range where ^{6}Li, but not ^{7}Li, is destroyed. Standard models (Forestini 1994) predict that no ^{6}Li can survive pre-MS mixing in metal-rich solar-type stars. The detection of ^{6}Li in HD 82943 (Israelian et al. 2001, 2003) is convincing observational evidence that stars may accrete planetary material, or even entire planets, during their main sequence evolution. Other explanations of this phenomenon such as stellar flares or surface spots have been ruled out (Israelian et al. 2001). It has been also proposed (Sandquist et al. 2002) that ^{6}Li can be used to distinguish between different giant planet formation theories. We also note that Cody & Sasselov (2005) have recently developed a stellar evolution code to model stars with non-uniform metallicity distributions. They found that the primary effects of metal enhancement on stellar structure and evolution are expansion of the convection zone and downward shift of effective temperature.

However, analysis of ^{6}Li is very delicate. Blending of the Li line with some weak absorptions and the placement of the continuum pose problems in metal-rich solar-type stars. Spectra with S/N $\sim$ 1000 and a resolving power of at least $\lambda/\Delta\lambda \sim 100\,000$ are required to analyze ^{6}Li. In metal-rich stars the identification of any weak blends in the region of the Li absorption becomes crucial. For example, Reddy et al. (2002) claimed that a previously noticed weak absorption in the solar spectrum at 6708.025 Å belongs to TiI. With this assumption their study of the Li feature in HD 82943 did not confirm the presence of ^{6}Li. However, recent analysis (Israelian et al. 2003) does not support the identification of a weak absorption feature at 6708.025 Å with the low excitation TiI line. It has been proposed that the unidentified absorption is most probably produced by a high excitation SiI line. The presence of ^{6}Li in HD 82943 was confirmed (Israelian et al. 2003) with the updated value for the isotopic ratio $f(^6\text{Li}) = 0.05 \pm 0.02$. It is worth mentioning that using the Spitzer MIPS photometer, Beichman et al. (2005) observed IR excess in HD 82943, which implies that collisions could have recently showered the star with some solid material. On the other hand, Mandell et al. (2004) reported no ^{6}Li in two lithium-poor stars that host extrasolar planetary systems.

Slow accretion of planetesimals on a timescale longer than 50 Myr was invoked by Murray & Chaboyer (2002) in order to explain the [Fe/H] distribution in planet-harbouring stars. These authors concluded that an average of $6.5 M_\oplus$ of iron must be added to the planet host stars in order to explain the mass–metallicity and age–metallicity relations. Accretion of $6.5 M_\oplus$ of planetesimals of iron during early MS evolution will strongly modify ^{7}Li abundances in these stars. Moreover, given the depth of the convection zone in stars with $T_\text{eff} > 5900$ K, a large amount of the added ^{6}Li may avoid destruction via mixing. Accretion of a chondritic matter with 6.5 $M_\oplus$ of iron by a star with $T_\text{eff} = 6100$ K and with a convection zone mass 10^{-3} $M_\odot$ will raise its ^{7}Li abundance from log (Li) = 2.7 to 3.2, while the isotopic ratio will become $f(^6\text{Li}) = 0.06$. This will create a detectable ^{6}Li absorption feature with an equivalent width (EW) $\sim$4 mÅ. This feature can be measured even if it is blended with the SiI line at 6708.025 Å because the latter is expected to appear with an EW < 2 mÅ in this type of star (Israelian et al. 2003).

5.2.3. Beryllium

The first studies of Be in planet hosts (García López & Pérez de Taoro 1998; Deliyannis et al. 2000) did not arrive at any firm conclusion because of the lack of a comparison sample of stars and the low number of targets. Santos et al. (2002) derived beryllium abundances for a sample of 29 planet host and six 'single' stars, aimed at studying in detail the effects of the presence of planets on the structure and evolution of the associated stars. Their preliminary results (confirmed recently by Santos et al. 2004b) suggest that theoretical models may have to be revised for stars with $T_{\rm eff} < 5500$ K. Santos et al. (2002) found several Be depleted stars at 5200 K which current models cannot explain. A comparison between planet-hosting stars and only six 'single' stars shows no clear difference between either population. More recently, Santos et al. (2004b) have presented Be abundances in a sample of 41 planet hosts and 29 stars without known planets. They confirmed that overall, planet hosts have 'normal' Be abundances. However, small, but maybe insignificant, differences might be present (see Santos et al. 2004b for details). These results support a 'primordial' origin for the metallicity excess observed in the planet hosts.

5.3. Abundances of metals

5.3.1. Volatiles

It was noticed by Santos et al. (2000) and Gonzalez & Laws (2000) that planet hosts tend to show subsolar [C/Fe] values with increasing [Fe/H]. Gonzalez et al. (2001) were less certain about these findings, while Takeda et al. (2001) and Sadakane et al. (2002) found no clear evidence of the constant [C/Fe] in the metallicity range $-0.5 < {\rm [Fe/H]} < 0.4$. All these authors used field stars from the literature to complete their inhomogeneous comparison samples. Sadakane et al. (2002) concluded that [C/Fe] and [O/Fe] ratios in planet hosts closely follow the trends observed in field stars with [C/Fe] = [O/Fe] = 0 at [Fe/H] > 0. Nitrogen abundances were derived by Gonzalez & Laws (2000), Gonzalez et al. (2001) and Takeda & Honda (2005). However, none of these authors discussed the trends of [N/Fe] and/or [N/H].

Recently, Ecuvillon et al. (2004a, 2004b, 2006) used high quality spectra from various telescopes in order to derive the N, C, S, Zn, O abundance in a large number of planet-hosting stars and comparison sample stars from Santos et al. (2001). The near-UV NH band at 3340–3380 Å was employed in their analysis to derive N abundance. Their results indicate a clear difference in [N/H], [C/H], [S/H], [Zn/H], [O/H] distributions for both samples. It has been found that [N/Fe] is flat at [Fe/H] > 0 (Ecuvillon et al. 2004a), while other volatiles, C, S, Zn (Ecuvillon et al. 2004b) and O (Ecuvillon et al. 2006) show [X/Fe] trends decreasing with [Fe/H]. The final abundance ratios of C, N, Zn, O and S for both samples as functions of [Fe/H] are displayed in Figure 5.4.

5.3.2. Refractories

The first abundance studies of several refractory elements (Gonzalez 1997; Gonzalez & Laws 2000; Gonzalez et al. 2001; Santos et al. 2000) revealed a few possible anomalies. Gonzalez et al. (2001) claimed that the stars with planets appear to have smaller [Na/Fe], [Mg/Fe] and [Al/Fe] values than field dwarfs of the same [Fe/H]. These authors did not find any significant differences for the refractories Si, Ca and Ti. However, the abundance trends in the few planet hosts discussed by Takeda et al. (2001) did not show anything peculiar. On the other hand, anomalies were found by Sadakane et al. (2002), who detected a few planet-bearing stars with an interesting abundance pattern in which

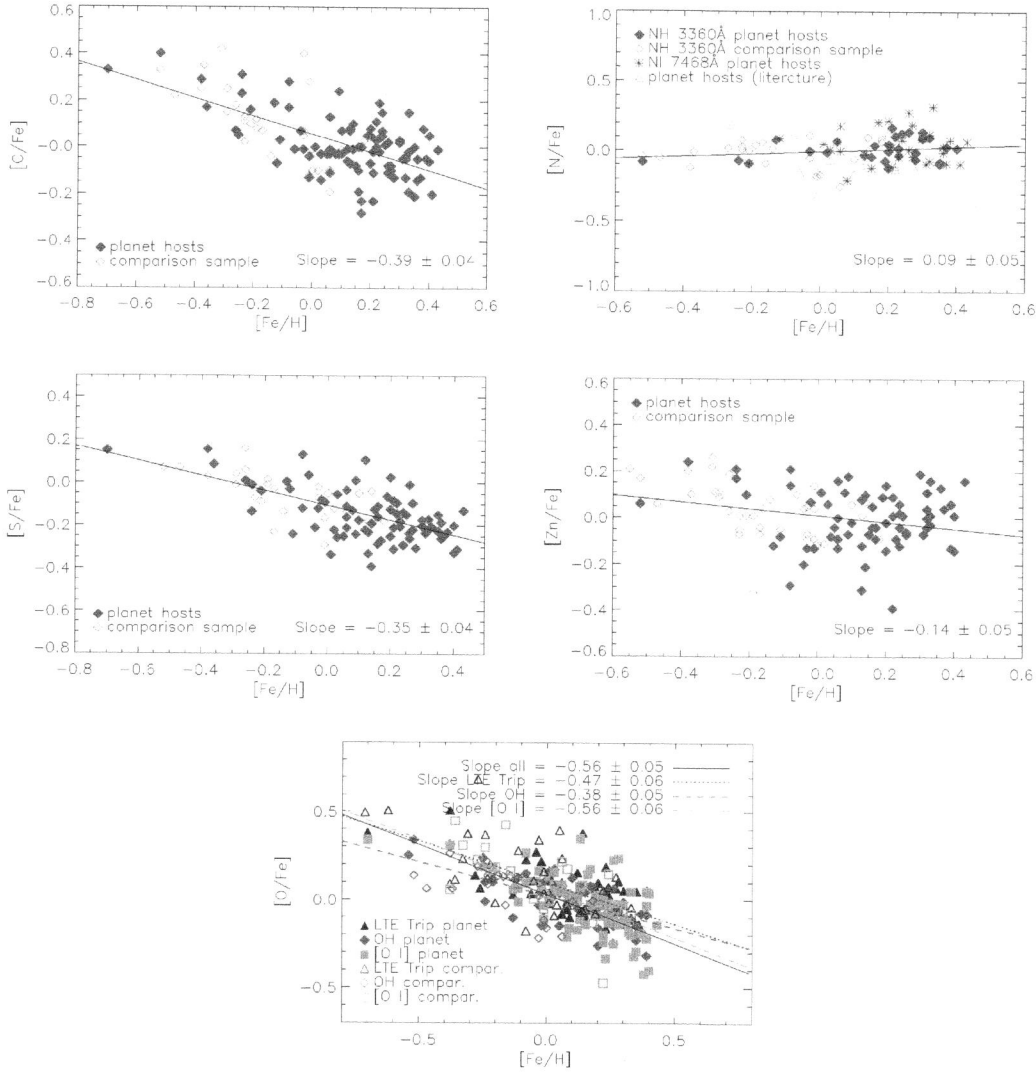

FIGURE 5.4. [X/Fe] vs. [Fe/H] plots for C, N, S, Zn and O. Filled diamonds represent planet host stars, while open symbols denote comparison sample stars. Linear least-square fits to both samples together (*solid lines*) are represented. In the lowest diagram, 'LTE Trip', 'OH' and 'OI' refer to several lines in which oxygen abundances were determined in planet hosts and comparison sample stars. From Ecuvillon et al. (2004a, 2004b, 2006).

the volatile elements C and O are underabundant with respect to refractories Si and Ti. Recently, Beirao et al. (2005) computed Na, Mg and Al abundances in 98 stars with planets and 41 'single' stars. These authors did not find any significant difference in the [X/H] ratios, for a fixed [Fe/H], between the two samples of stars in the region where the samples overlap.

A uniform and unbiased comparison of abundances of some α- (Si, Ca, Ti) and Fe-group (Sc, V, Cr, Mn, Co, Ni) elements in 77 planet host and 42 comparison sample stars without planets was carried out by Bodaghee et al. (2003). These authors

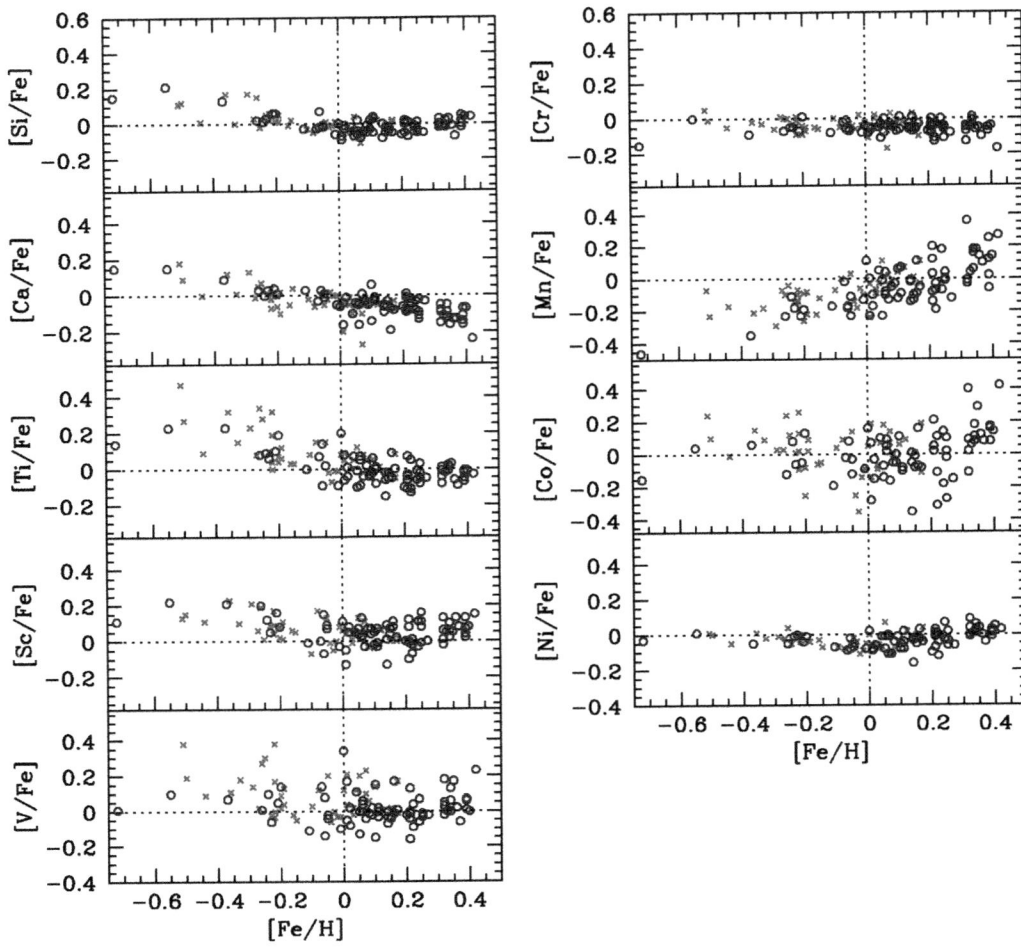

FIGURE 5.5. [X/Fe] vs. [Fe/H] plots for several α- and Fe-group elements. The crosses represent the comparison-sample stars, while the open circles denote the planet-hosting stars. From Bodaghee et al. (2003).

concluded that the abundance trends for the planet hosts are almost identical to those in the field. Slight differences were found for V, Mn and, to a lesser extent, Co and Ti (Figure 5.5). Although the abundance scatter for most of the elements was found to be small, a few elements showed considerable dependence of the derived abundances on the effective temperature. The largest effect was found for Ti, Co and V, where the difference between K- and F-dwarfs has reached 0.2–0.3 dex. These trends might be related to NLTE effects. The work of Bodaghee et al. (2003) was extended recently by Gilli et al. (2006) to include around 100 planet hosts and comparison sample stars from Santos et al. (2005).

In general, the abundance distributions of planet host stars are high [Fe/H] extensions to the curves traced by the field dwarfs without planets. No significant differences are found in the regions of overlap. However, although some differences for certain elements are subtle (and may even be negligible), they are certainly intriguing enough to merit additional studies.

5.4. Implications

5.4.1. Chemical evolution of the Galaxy

One of the byproducts of chemical abundance studies in planet-hosting stars is the possibility of learning about galactic chemical evolution trends at high metallicities. The number of detailed abundance studies at [Fe/H] > 0 is very limited and exoplanet hosts can help to explore this regime. Some of the trends obtained in these studies may be linked with the presence of giant planets. According to Santos et al. (2005), more than 25% of stars with [Fe/H] > 0.3 host planets and the possibility that virtually all metal-rich stars host planetary systems cannot be ruled out. Thus, it is very difficult to compare stars with and without planets in the high [Fe/H] tail of the distribution. The relative frequency of stars with planets increases with [Fe/H], but there is a sharp cutoff once the metallicity reaches about 0.4 dex (Figure 5.1). It is hard to believe that Nature could somehow tune the self-enrichment process in planet hosts by not allowing them to have [Fe/H] > 0.5. Most probably the cutoff represents a rough upper limit to metallicities in the solar neighbourhood. If [Fe/H] ~ 0.4 represents the 'present day' state of galactic chemical evolution, then certain trends should appear for all other chemical species. How then can we disentangle the abundance anomalies produced by the presence of planets?

Perhaps the easiest way is to study those trends which are difficult to interpret in the framework of standard galactic chemical evolution models. Given the constant rate of Type II and Type Ia SN during the last 10 Gyr of galactic evolution, constant IMF and SFR, we would not expect any significant change in the slope of [α/Fe] versus [Fe/H], where the αs are O, Si, S, Mg, Ca and Ti. However, observations (Gonzalez et al. 2001; Sadakane et al. 2002; Bodaghee et al. 2003; Beirao et al. 2005; Gilli et al. 2006) show a sudden change at [Fe/H] = 0 in the slopes of [Si/Fe], [Ti/Fe] and [Mg/Fe] versus [Fe/H], while [Ca/Fe] slowly decreases with [Fe/H]. It is not clear why Si, Ti and Mg should drastically change their slopes at [Fe/H] = 0 and become flat. Moreover, high quality observations by Ecuvillon et al. (2006) demonstrate that [O/Fe] continues to decrease at [Fe/H] > 0 without showing the flattening out found in previous studies (Nissen & Edvardsson 1992). Galactic chemical evolution models predict similar trends for O, Si and all the other α-elements (Tsujimoto et al. 1995) down to [Fe/H] = 0. Why should C and N have a different behaviour at high metallicities, both being volatiles and having similar production sites in the Galaxy? There could be three reasons for these anomalies; (a) models of galactic chemical evolution are very uncertain at high metallicities, (b) abundance trends in metal-rich stars are affected by the presence of planets and (c) abundance analysis of metal-rich stars is not reliable.

5.4.2. Astrobiology

The abundance anomalies or the correlation between [Fe/H] and the presence of giant planets have certain implications for astrobiology and even SETI. As long ago as the 1960s Drake (1965) and Shklovski & Sagan (1966) independently proposed a method which extraterrestrial intelligent civilizations could employ in order to announce their existence. They could add some short-lived isotope(s) into the atmosphere of their sun with the hope that possible observers would detect the absorption spectral lines of that element and realize their artificial origin. The amount of matter required to produce detectable absorption lines of some rare elements/isotopes is not that large and any developed civilization should be able to handle this task (Drake 1965).

The possible correlation between [Fe/H] and the formation of giant planets may have an impact on the formation of terrestrial planets and their habitability via a shielding effect. Terrestrial planet formation should strongly depend on metallicity as well. A giant planet can relocate planetesimals and comets via scattering into the inner region of the planetary system. This process can protect inner habitable terrestrial planets against the impacts of the comets. Numerical simulations (Ida, Junko & Lin 2001) show that a planet with large semi-major axis and/or large mass may eject planetesimals and prevent pollution of metals onto the host star. These authors proposed that the habitability may be regulated by a giant planet(s) since the shielding effect does not only inhibit impacts onto the host star but also prevents inner terrestrial planets from being impacted by cometary bodies and asteroids.

The abundances of elements in the Sun and on the primitive Earth was suited to the creation and evolution of a biosphere. Even the rare elements, those especially suited to the extraction of energy using nuclear processes (like U and Th) are present on the Earth in suitable concentration for the development of life. It is well known that the rock masses that form the external part of the Earth have been in rapid motion for hundreds of millions of years. This motion is so rapid that 2/3 of the crust has been recycled into the Earth's mantle in the last 200 million years. Many geological processes (such as volcanic activity, drift of continents, iron catastrophe[3]) in the lithosphere most probably result from the heat released by the radioactive elements, mostly thorium and uranium. Active volcanoes are a source of many volatile compounds (water, methane, carbon oxides, etc.) which accumulated with time and formed the primitive gaseous atmosphere and, somewhat later, the liquid hydrosphere.

It is clear that there are certain chemical and physical preconditions for mankind's evolution and these preconditions depend on the chemical composition and evolution of the protoplanetary matter. The products of supernova explosions many billions of years ago have influenced the evolution of life on our planet. Abundance studies in stars with exoplanets may help us to understand the origin and chemical evolution of planetary systems and learn about the evolution of biosystems in the Universe.

Acknowledgments

I would like to thank my colleagues Nuno Santos, Michel Mayor, Rafael Rebolo, Alexandra Ecuvillon and Ramón García López for many useful discussions.

REFERENCES

BEICHMAN, C., BRYDEN, G., RIEKE, G. H., *et al.*, 2005, *ApJ* **622**, 1160.

BEIRAO, P., SANTOS, N. C., ISRAELIAN, G. & MAYOR, M., 2005, *A&A* **438**, 251.

BODAGHEE, A., SANTOS, N. C., ISRAELIAN, G. & MAYOR, M., 2003, *A&A* **404**, 715.

CHEN, Y. Q., NISSEN, P. E., BENONI, T. & ZHAO, G., 2001, *A&A* **371**, 943.

CODY, A.-M. & SASSELOV, D., 2005, *ApJ* **622**, 704.

DELIYANNIS, C., CUNHA, K., KING J. & BOESGAARD, A., 2000, *AJ* **119**, 2437.

DRAKE, F., 1965, in *Current Aspects of Exobiology*, ed. G. Mamikunian, Oxford: Pergamon Press, p. 323.

ECUVILLON, A., ISRAELIAN, G., SANTOS, N. C., MAYOR, M., GARCÍA LÓPEZ, R. J., REBOLO, R. & RANDICH, S., 2004a, *A&A* **418**, 703.

ECUVILLON, A., ISRAELIAN, G., SANTOS, N. C., MAYOR, M., VILLAR, V. & BIHAIN, G., 2004b, *A&A* **426**, 619.

---

[3] 'Iron catastrophe' denotes the infall of very heavy elements like iron into the planet's interior during its formation.

Ecuvillon, A., Israelian, G., Santos, N. C., Shchukina, N., Mayor, M. & Rebolo, R., 2006, *A&A* **445**, 633.
Forestini, M., 1994, *A&A* **285**, 473.
García López, R. J. & Pérez de Taoro, M. R., 1998, *A&A* **334**, 55.
Gilli, G., Israelian, G., Ecuvillon, A., Santos, N. & Mayor, M., 2006, *A&A* **449**, 723.
Gonzalez, G., 1997, *MNRAS* **285**, 403.
Gonzalez, G. & Laws, C., 2000, *AJ* **119**, 390.
Gonzalez, G., Laws, C., Sudhi, T. & Reddy, B. E., 2001, *AJ* **121**, 432.
Ida, S., Junko, K. & Lin, D. N. C., 2001, *A Symplectic Odyssey*. Proceedings of the 33rd Symposium on Celestial Mechanics, Gunma, Japan, eds. E. Kokubo, T. Ito and H. Arakida, Mitaka, Tokyo, Japan: NAO, p. 110.
Israelian, G., Santos, N. C., Mayor, M. & Rebolo, R., 2001, *Nature* **411**, 163.
Israelian, G., Santos, N. C., Mayor, M. & Rebolo, R., 2003, *A&A* **405**, 753.
Israelian, G., Santos, N. C., Mayor, M. & Rebolo, R., 2004, *A&A* **414**, 601.
Laws, C., Gonzalez, G., Walker, K. M., Tyagi, S., Dodsworth, J., Snider, K. & Suntzeff, N., 2003, *AJ* **125**, 2664.
Mandell, A., Ge, J. & Murray, N., 2004, *AJ* **127**, 1147.
Mayor, M. & Queloz, D., 1995, *Nature* **378**, 355.
Murray, N. & Chaboyer, B., 2002, *ApJ* **566**, 442.
Nissen, P. E. & Edvardsson, B., 1992, *A&A* **261**, 255.
Pollack, J. B., Hubickyj, O., Bodenheimer, P., et al., 1996, *Icarus* **124**, 62.
Reddy, B., Lambert, D., Laws, C., Gonzalez, G. & Covey, K., 2002, *MNRAS* **335**, 1005.
Ryan, S., 2000, *MNRAS* **316**, L35.
Sadakane K., Ohkubo, M., Takeda, Y., Sato, B., Kambe, E. & Aoki, W., 2002, *PASJ* **54**, 911.
Sandquist, E. L., Dokter, J. J., Lin, D. N. C. & Mardling, R., 2002, *ApJ* **572**, 1012.
Sadakane, K., Ohkubo, M., Takeda, Y., et al., 2002, *PASJ* **54**, 911.
Santos, N. C., Israelian, G. & Mayor, M., 2000, *A&A* **363**, 228.
Santos, N. C., Israelian, G. & Mayor, M., 2001, *A&A* **373**, 1019.
Santos, N. C., García López, R. J., Israelian, G., et al., 2002, *A&A* **386**, 1028.
Santos, N. C., Israelian, G., Mayor, M., Udry, S. & Rebolo, R., 2003, *A&A* **398**, 363.
Santos, N. C., Israelian, G. & Mayor, M., 2004a, *A&A* **415**, 1153.
Santos, N. C., Israelian, G., García López, R. J., et al., 2004b, *A&A* **427**, 1085.
Santos, N. C., Israelian, G., Mayor, M., Bento, J. P., Almeida, P. C., Sousa, S. G. & Ecuvillon, A., 2005, *A&A* **437**, 1127.
Shklovski, I., 1966, *Intelligent Life in the Universe*, New York: Deli, p. 406.
Smith, V. V., Cuhna, K. & Lazzaro, D., 2001, *AJ* **121**, 3207.
Takeda, Y. & Honda, S., 2005, *PASJ* **57**, 65.
Takeda, Y. & Kawanomoto, S., 2005, *PASJ* **57**, 45.
Takeda, Y., Sato, B., Kambe, E., et al., 2001, *PASJ* **53**, 1211.
Tsujimoto, T., Yoshii, Y., Nomoto, K. & Shigeyama, T., 1995, *A&A* **302**, 704.

6. Brown dwarfs: the bridge between stars and planets

RAFAEL REBOLO

Brown dwarfs are objects with masses, effective temperatures and luminosities intermediate between those of very low-mass stars and the most massive giant planets. In the last decade, numerous searches have revealed their ubiquitous presence in star forming regions and stellar clusters, orbiting stars and free-floating in the field. Hundreds of brown dwarfs have been identified via direct imaging techniques. Brown dwarfs appear to be as numerous as stars. Follow-up spectroscopic observations have been crucial for establishing their properties. New spectroscopic classes have been required for these objects. The L and T brown dwarfs form a unique laboratory in which to test substellar atmospheric and evolutionary models. We briefly review the photometric and spectroscopic properties, the multiplicity, mass function and possible formation scenarios of these substellar objects. Old low-mass brown dwarfs are expected to cool down to atmospheric temperatures similar to those of the planets in the Solar System. Their atmospheric properties will guide future planet searches.

6.1. Introduction

Brown dwarfs populate the mass domain between that of very low-mass stars and giant planets. They share some characteristics with stars and others with planets. Current models of stellar evolution predict a minimum mass of ~ 73 $M_{\rm Jup}$ for stable hydrogen burning to take place in the interior of a solar metallicity self-gravitating object (e.g. Baraffe et al. 1998). This is the mass generally adopted in defining the frontier between stars and brown dwarfs for solar metallicity. Because of the lack of hydrogen burning, brown dwarfs progressively cool and dim. Their atmospheric properties evolve drastically with age leading to unique, very distinctive features that can be used to distinguish them from the lowest mass stars.

The frontier between brown dwarfs and giant planets is much more subtle and cannot be established so obviously. In principle, we might expect that very low-mass brown dwarfs and massive giant planets, such as those detected by radial velocity surveys of solar-type stars, would have different formation mechanisms. In reality, our limited knowledge of these mechanisms and their implications on the properties of the resulting objects prevent a clear distinction on this basis. Brown dwarfs are likely to form as stars, i.e. as a result of the fragmentation of molecular clouds. Massive giant planets most probably form in protoplanetary discs via gravitational instability or the accretion of planetesimals. While objects originating through any of these mechanisms might possibly have different physical conditions in their interiors and even in their atmospheres, current observations are far from telling us their origin unambiguously. A particularly difficult case for classification are objects with masses in the range 5–15 $M_{\rm Jup}$ ($M_{\rm Jup} = M_\odot/1047$) found orbiting stars, but also free-floating, isolated from stars. Such low-mass objects may have formed from the direct fragmentation of molecular clouds or, alternatively, could originate in protoplanetary discs and, due to dynamic interactions, be ejected far from the gravitational

Extrasolar Planets, eds. Hans Deeg, Juan Antonio Belmonte and Antonio Aparicio.
Published by Cambridge University Press.
© Cambridge University Press 2007.

influence of the parent star. Since at present we cannot establish the formation process of these objects, a criterion based on history or origin – which undoubtedly has physical merit – is not applicable in practice. A classification criterion based on the existence or absence of burning of nuclei in the interiors has the advantage of a direct relation to mass, which is a primary parameter determining the evolution of substellar objects. Deuterium is the most fragile isotope in Nature; objects unable to burn this isotope will not burn any other nuclei, therefore it seems natural to consider the minimum mass for deuterium burning as a possible mass limit for giant planets. According to models of Chabrier et al. (2000), this minimum mass for solar metallicity is 13 $M_{\rm Jup}$. Here, by 'brown dwarf' we mean any object able to produce stable core fusion of deuterium but not hydrogen, irrespective of how it formed or whether it is free-floating in the field, or bound to a binary or multiple system.

The physical and chemical properties of brown dwarfs and extrasolar giant planets are generally described by the same theory (see, for example, Burrows et al. 2001), whatever the mass or origin of the object. These properties evolve drastically with age as brown dwarfs progressively contract and cool off. Most brown dwarfs will reach at sufficiently late times a radius similar to that of Jupiter and develop convective cores of metallic hydrogen and helium with maximum central temperatures below $\sim 3 \times 10^6$ K. Old brown dwarfs (age > 1 Gyr) span a luminosity range 10^{-4}–$10^{-7} L_\odot$ and atmospheric effective temperatures in the range 200–1000 K. The evolution of the temperature, radius and luminosity of brown dwarfs as a function of mass has been considered in great detail (Burrows et al. 1997; Chabrier et al. 2000 and references therein). Theoretical power laws describing the relations between luminosity, mass, age, effective temperature and radius at late-time cooling phases can be found, for instance, in Burrows et al. (2001).

6.2. Early searches and first brown dwarf discoveries

The existence of brown dwarfs was proposed in the early 1960s (Kumar 1963), and theoretical evolutionary tracks were discussed by Hayashi & Nakano (1963). For decades, many searches in the field, in star-forming regions and around stars, proved unsuccessful in their attempt to unveil these rather low-luminosity objects. Young stellar clusters and star-forming regions were the focus of many of these searches because at early ages brown dwarfs are expected to be still contracting and hence be much more luminous. Before 1995, a number of candidates were found in young star forming-regions like ρ Ophiuchi (Rieke & Rieke 1990), Taurus (Stauffer et al. 1991), stellar clusters like the Pleiades (Jameson & Skillen 1989; Stauffer et al. 1989, 1994) and α Persei (Rebolo et al. 1992). However, the effective temperatures and luminosities of these candidates were too high to ensure a substellar nature. This was also the case for searches in the field in the early 1990s. Observations around stars produced interesting brown dwarf candidates like GD 165b, a companion to a white dwarf (Zuckerman & Becklin 1992). The spectral energy distribution of this object was rather different from that of the coolest stars known at the time. Several years later, it was realized that GD 165b belongs to the L-type class (Kirkpatrick et al. 1999), which is not exclusive of brown dwarfs. Very low-mass stars can reach such late spectral types. The substellar nature of this object is thus still uncertain.

6.2.1. Spectroscopic signatures of substellarity

An accurate direct mass determination for the first brown dwarf candidates was not possible because they were either single objects or members of binaries with very long orbital periods. Mass estimates using evolutionary models were so uncertain that no brown dwarf could be successfully identified prior to 1995. Strictly speaking, the substellar nature of an object can be established only if it is proved that stable burning of light hydrogen cannot take place in the interior. In practice, an observable is required to infer this circumstance. Fortunately, several spectroscopic tests were developed in the 1990s that were potentially able to establish the nature of a brown dwarf candidate in the absence of a direct mass determination. We will summarize two here: the Li test (Rebolo et al. 1992; Magazzú et al. 1993) and the methane test (Tsuji 1995).

At solar metallicity, the effective temperatures of stars with the minimum mass for hydrogen burning are predicted in the range 1700–1750 K and the luminosity at $\sim 6 \times 10^{-5} L_\odot$ (Burrows et al. 2001). Below these temperatures methane bands can form and leave an imprint in the spectrum of a cool object. Observations of these bands in the 1–2.5 µm range in objects less luminous than the limit given above provides a test of substellarity. It should be recalled, however, that due to uncertainties in silicate grain physics and opacities it cannot be ruled out that the effective temperature of a star may be as low as 1600 K (Chabrier & Baraffe 2000).

The detection of spectroscopic signatures of lithium in the atmosphere of a sufficiently cool dwarf is an indication that lithium burning via the ^{7}Li(p,α)^{4}He reaction has not taken place in the interior. Lithium preservation in such fully convective low-mass objects implies that the maximum central temperature achieved during evolution is lower than $\sim 2.5 \times 10^6$ K, i.e. below the required temperature for stable hydrogen burning ($\sim 3.2 \times 10^6$ K). Solar-metallicity brown dwarfs with masses below 63 $M_{\rm Jup}$ would preserve most of their original lithium (see, for example, Magazzú et al. 1993; Burrows et al. 2001; Chabrier & Baraffe 2000), while very low-mass stars destroy this element very efficiently on very short timescales (less than ~ 100 Myr). The amount of Li depletion in more massive brown dwarfs is a function of age. Most of the lithium depletion in such brown dwarfs takes place between an age of 100 and ~ 500 Myr. Observations of brown dwarfs in stellar clusters spanning this mass range are potentially able to trace the lithium depletion mechanism, which in turn can tell us about the evolution of the physical conditions in substellar interiors and set constraints on the age of these stellar systems. The lithium test showed that many of the coolest dwarfs considered in the early 1990s as good brown dwarf candidates had destroyed a large fraction of their initial Li content and were therefore most likely stars (Martín et al. 1994).

6.2.2. Unveiling the substellar realm

The search for fainter members of the Pleiades cluster led to the discovery in 1995 of the first object that could be recognized as a brown dwarf. According to state of the art evolutionary models available at that time, the photometric (luminosity) and spectroscopic (M8 spectral type) properties of a new radial velocity/proper motion member of the Pleiades, Teide 1 (Rebolo et al. 1995), could only be explained if the mass was clearly below the minimum mass for hydrogen burning. The estimated mass for Teide 1 was in the range 20–50 $M_{\rm Jup}$. The discovery of Teide 1 was closely followed (less than two months later) by the discovery of Gl 229 B (Nakajima et al. 1995; Oppenheimer et al. 1995), a very cool low luminosity ($7 \times 10^{-6} L_\odot$) object orbiting an M-dwarf star in the solar neighbourhood. The large physical separation of the components in the Gl 229 system (~ 44 AU) prevented a dynamical measurement of the mass, but according to evolutionary models the low luminosity could only be explained if it were a brown dwarf with a mass in the range 30–40 $M_{\rm Jup}$. The spectrum of Gl 229 B exhibits methane bands (Oppenheimer et al. 1995). Observations of the lithium resonance doublet at 6708 Å soon confirmed the substellar nature of Teide 1 (and Calar 3, a twin object also discovered in the Pleiades, Rebolo et al. 1996). However, the low temperature of Gl 229 B makes it extremely difficult to detect the lithium feature in its spectrum.

In the last decade, hundreds of brown dwarfs have been discovered in star clusters, in the field and in much lesser numbers as companions to stars. Brown dwarfs appear to be very common. Stellar clusters, in particular, provide evidence that brown dwarfs can form in the whole mass range down to the deuterium-burning limit. Field counterparts of such low-mass brown dwarfs have not been detected yet, partly because their luminosities and effective temperatures are much lower than those of the coolest known brown dwarfs ($T_{\text{eff}} \sim 700$ K). Efficient surveys require a good knowledge of atmospheres in a tempeature domain previously unexplored. It is likely that such very low-mass brown dwarfs elude detection by current large-sky near-infrared surveys such as 2MASS and DENIS, but could possibly be detected by the highly sensitive mid-infrared instruments on board the *Spitzer* satellite providing invaluable information for future exoplanet searches. In the following sections we briefly review the basic properties of the brown dwarfs discovered so far, the status of searches around stars, in star clusters and in the field, and the various mechanisms proposed to explain their formation.

6.3. Brown dwarfs: photometric and spectroscopic properties

Brown dwarf searches in the field (Ruiz et al. 1997; Delfosse et al. 1997) and around stars (Rebolo et al. 1998) showed objects with new spectral characteristics intermediate between those of the young brown dwarfs detected in the Pleiades cluster and the very cool Gl 229 B. The discovery of many of these ultracool objects by the 2MASS survey (Kirkpatrick et al. 1999; Reid et al. 1999) led to the establishment of the new spectral class L (Martín et al. 1997, 1999; Kirkpatrick et al. 1999). In Figure 6.1 we can see absolute magnitude versus the $I - J$ colour for late M-, L- and T-dwarfs with available parallaxes. In L-dwarfs the characteristic red TiO and VO bands of M-dwarfs disappear from the optical spectrum due to condensation in dust grains (see Chabrier et al. 2000), and broad absorption resonance lines of Na I and K I become dominant. Lines of other alkali metals (Cs I and Rb I) and bands of FeH, CrH and H_2O are also present in the optical red. Strong H_2O absorption bands and CO overtone bands at 1–2.5 µm are also present in the spectrum. Remarkably, L-dwarfs present rather red near-infrared colours, $1.3 < J - K < 2.3$ and $3 < I - J < 4$ (see Figure 6.2), which increase from early to late spectral subclasses. These colours favour a relatively easy identification in the large-sky survey databases. Effective temperatures for L-dwarfs can be estimated from the following relationship with spectral type: $T_{\text{eff}} = (2380 \pm 40) - (138 \pm 8)$ SpT, where SpT ranges from 0 to 8 for spectral types L0 to L8 (Burgasser 2001).

About one-third of the L-dwarfs show detections of lithium. These are most probably either brown dwarfs with masses below $\sim 65 M_{\text{Jup}}$ or young more massive brown dwarfs that have not yet depleted a large amount of their initial lithium content. The lithium-depleted L-dwarfs are either stars with masses close to the minimum mass for hydrogen burning or sufficiently old massive brown dwarfs that have destroyed most of the initial lithium content.

Gl 229 B is the prototype of another new class of objects: the T-dwarfs. It took four years to find other objects with similar spectroscopic characteristics. Finally, several searches (Burgasser et al. 1999, 2000; Cuby et al. 1999; Leggett et al. 2000) proved the existence of such analogues with a space density similar to that of L-dwarfs. T-dwarfs are characterized (see, for example, Burgasser et al. 2002) by strong CH_4 absorptions in the H and K bands, stronger water bands than in L-dwarfs, and blue infrared colours $J - K \sim 0$ with large increase in flux from the red part of the optical spectrum to the

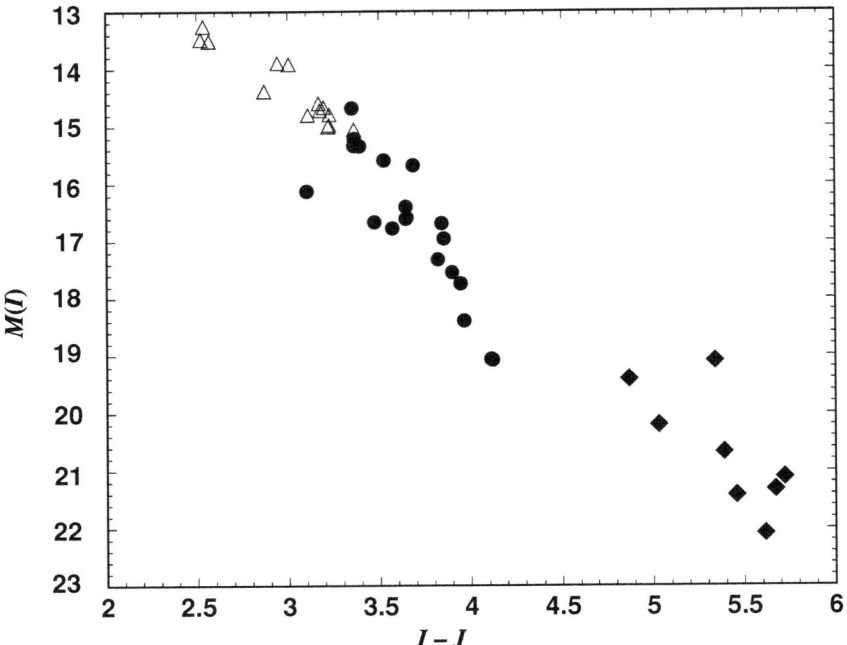

FIGURE 6.1. Absolute I-band magnitude versus $I - J$ colour for a sample of late M-dwarfs (triangles), L-dwarfs (dots) and T-dwarfs (diamonds) with known parallaxes.

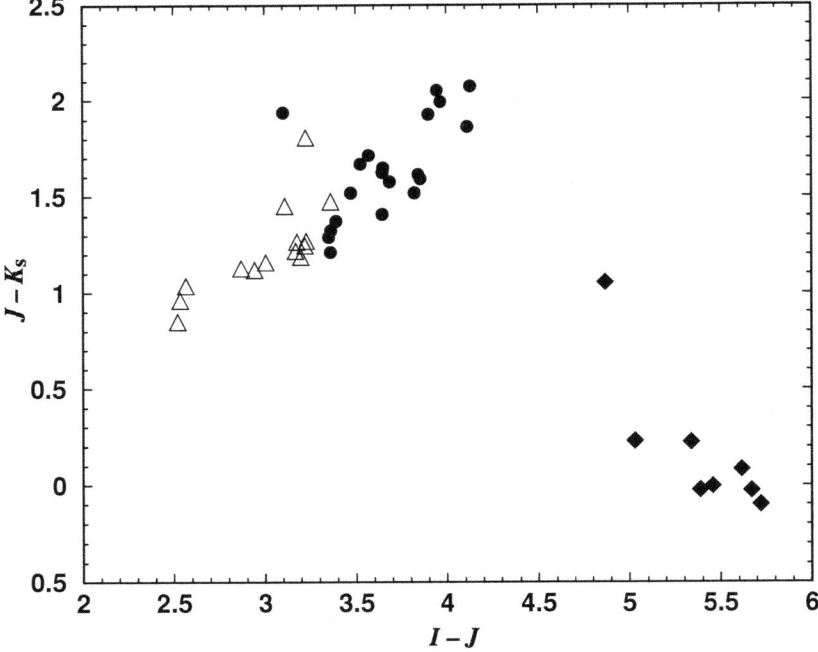

FIGURE 6.2. $J - K_s$ versus $I - J$ for a sample of late M-dwarfs (triangles), L-dwarfs (dots) and T-dwarfs (diamonds) with known parallaxes.

near infrared ($I - J > 4$). In Figure 6.2 we can see the different location of late M-, L- and T-dwarfs in the near-infrared colour–colour diagram; in particular, we can notice the bluing of the $J - K_s$ colour in T-dwarfs as compared with L-dwarfs. A relationship between effective temperature and spectral type derived from the K-band bolometric correction shows a monotonic behaviour throughout the L–T sequence (Nakajima et al. 2004). T-dwarfs are classified from T0 to T8. The effective temperatures of the latest T-dwarfs are in the range 800–700 K and the transition between L- and T-dwarfs takes place at $\sim 1400 - 1300$ K (Dahn et al. 2002). We can appreciate in Figure 6.1 the extremely low luminosity in the red optical of the coolest objects (the reddest in $I - J$). More than 40 T-dwarfs have been discovered so far, but the edge of the *stellar* mass function has not been reached yet, either in the field or in stellar clusters. Cooler and fainter objects are likely to exist in the brown dwarf realm with absolute magnitude $M_J \geq 18$ and temperatures similar to those of the planets in the Solar System.

Much effort has been devoted to interpreting L- and T-dwarfs in terms of atmospheric chemistry including formation of dust clouds (see Burrows et al. 2001). Tsuji & Nakajima (2003) propose a grid of self-consistent non-grey unified cloudy models (700 K $< T_{\text{eff}} <$ 2600 K) in which a thin dust cloud is always formed near the dust condensation temperature and will therefore be located relatively deep in the photosphere for the cooler T-dwarfs, while it will appear in the optically thin regime in the warmer L-dwarfs. In this scenario, the rapid bluing from the late L- to the early T-dwarfs is a direct result of the migration of the dust cloud from the optically thin ($\tau < 1$) to thick ($\tau > 1$) regimes, while luminosity and effective temperature lower only slightly. The nature of these clouds is not yet clear. The photometric variability detected in ultracool dwarfs (see, for example, Bailer-Jones & Mundt 2001) may be an indication that clouds follow the meteorological activity typical of the planets in the Solar System.

6.4. Brown dwarfs in multiple systems

Multiple systems offer clear advantages for a better determination of the physical properties of brown dwarfs and provide important information for studying their formation mechanisms. In such systems brown dwarfs have been found as companions to stars with a wide range of separations, but mostly at wide separations ($\rho \geq 40$ AU). Binary brown dwarfs appear to be common both in the field and as companions to stars and as separated binaries in systems with near-unity mass ratios.

6.4.1. Brown dwarf companions to stars

In spite of the great success of radial velocity searches to detect planets around solar-type stars (more than 150 planets are already known), these surveys have discovered only a small number of companion brown dwarfs ($m > 13$ M_{Jup}). At least 7% of solar-type stars appear to have planets at separations less than 5 AU with masses $M \sin i = 0.25 - 13$ M_{Jup}. The mass distribution of these planets decreases as the mass increases, such that brown dwarfs are detected around fewer than 0.5% of solar-type stars at separations smaller than 4 AU. Brown dwarf companions appear to be ~ 10 times less frequent than planetary-mass companions and ~ 20 times less frequent than stellar companions at such separations. The lack of brown dwarfs at these distances, termed by Marcy & Butler (2000) as the 'brown dwarf desert', could possibly indicate a different formation mechanism for brown dwarf and planetary companions to solar-type stars. There are, however, some *oases* in the 'desert'. For example, a brown dwarf ($M \sin i = 26$ M_{Jup}) has been found orbiting at 1.85 AU from the star HD 137510 (Endl et al. 2004).

At much larger separations, brown dwarfs appear to be slightly more frequent. Since the first detections of brown dwarfs around stars (Gl 229b orbiting at $\sim$44 AU, Nakajima et al. 1995; G 196–3b at $\sim$300 AU, Rebolo et al. 1998), about 20 wide brown dwarf companions to stars have been directly imaged. Interestingly, some are found at very large separations. For example, ε Indi B is a gravitationally bound companion to the high proper motion star ϵ Indi A at a separation of $\sim$1500 AU (Scholz et al. 2003). Indeed, ε Indi B has itself turned out to be a binary brown dwarf formed by a T1 and a T6 (McCaughrean et al. 2004).

About 1% of the stars seem to have brown dwarf companions at separations larger than 75 AU (McCarthy & Zuckerman 2004), and about 1% of early M-dwarfs appear to have them at separations between 10 and 50 AU (see, for example, Forveille et al. 2004). The abundance of brown dwarfs around very low-mass stars may be higher, as suggested by the discovery of a brown dwarf orbiting the M8-dwarf LHS 2397a at a separation of 2.34 ± 0.14 AU (Freed et al. 2003; Masciadri et al. 2003) and also by the astrometric discovery of a potential brown dwarf companion of the nearby M-dwarf star GJ 802 (Pravdo et al. 2005). For comparison, about 35% of M-dwarfs are binaries with most frequent separations in the range 3–30 AU. Efforts to uncover the mass and radial distribution of extrasolar planets and brown dwarfs within a few AU of M-stars are in progress (see, for example, Guenther & Wuchterl 2003; Kenyon et al. 2005). These searches are at too early a stage to establish any significant comparison with the frequency of brown dwarfs or planets in more massive stars. It is interesting to note that at much larger separations direct imaging of very young stars with available facilities can provide the detection of planetary-mass companions and that current searches may have succeeded with the discovery of a low-mass proper motion companion of the star GQ Lup (Neuhäuser et al. 2005). This very young K7eV-type classical T Tauri star (age $\sim$0.1–2 Myr) is located in the Lupus I cloud. The spectral type of the companion is estimated in the range M9 to L4 and its mass, according to evolutionary models, could be as low as a few times the mass of Jupiter. Another interesting object is the companion to AB Pic, a very young (age $\sim$30 Myr) star, member of the Tucana–Horologium association. Chauvin et al. (2005) have detected an object with a mass in the frontier between planets and brown dwarfs at a projected physical separation of $\sim$280 AU from the primary.

6.4.2. Binary brown dwarfs

The binary properties of brown dwarfs (binary fraction, separation distribution and mass ratio distribution) seem to be different from those of low-mass stars. The fraction of substellar field binaries for separations between 5 and 20 AU is $\sim$10–20% (Bouy et al. 2003; Burgasser et al. 2003; Close et al. 2003; Gizis et al. 2003), peaking around 2–4 AU. For comparison, the binary fraction of field low-mass stars is at least a factor of two higher, and the peak of the stellar companion distribution for both G- and M-dwarf primaries is at about 30 AU (Duquennoy & Mayor 1991; Fischer & Marcy 1992).

Brown dwarf binaries may exhibit very small separations. An extreme case (separation $\rho = 0.03$ AU) is PPl 15 in the Pleiades, a likely pair of brown dwarfs according to the radial velocity determinations of Basri & Martín (1999), which, however, appear to have too low a Li abundance for the suggested masses of the components (Basri et al. 1996; Rebolo et al. 1996). It is important to confirm whether both components of the system have preserved Li as expected for bona fide Pleiades brown dwarfs. New radial velocity searches have possibly found other close binary brown dwarfs in the Chamaeleon I star-forming region (Joergens et al. 2001) and in σ Orionis (Kenyon et al. 2005).

Using direct imaging techniques, a relatively large number (higher than 30) of binary brown dwarfs have been discovered in the field (see, for example, Close et al. 2003), in stellar clusters (e.g. Martín et al. 2003), or as companions to more massive primaries (e.g. Burgasser et al. 2005 and references therein). Binaries have been found among L- and T-brown dwarfs down to masses ~ 30 $M_{\rm Jup}$ with separations from a few to hundreds of AU. Future searches will benefit enormously from recent advances in adaptive optics systems with laser guide stars, as shown in the case of Kelu-1, one of the first free-floating brown dwarfs discovered (Ruiz et al. 1997). This brown dwarf was considered a rather peculiar case for its apparent overluminosity, frequently interpreted as evidence for youth. Very recently, using a laser guide system and adaptive optics at Keck, Liu & Leggett (2005) have resolved Kelu-1 into a 0.29" (5.4 AU) binary with near-infrared flux ratios of ~ 0.5 mag. The estimated spectral types of each component are L1.5–L3 and L3–L4.5 with masses in the range 50–70 $M_{\rm Jup}$ and 45–65 $M_{\rm Jup}$, respectively. The nearest brown dwarf binary to the Sun, ε Ind Ba and Bb (McCaughrean et al. 2004), is formed by two brown dwarfs of spectral types T1–6 with masses close to 50 and 30 $M_{\rm Jup}$, respectively. The physical separation is ~ 2.6 AU and the orbital period is ~ 15 yr. There is increasing evidence that the binary fraction of brown dwarfs in stellar systems is higher than that of field brown dwarfs. Burgasser et al. (2005) find values of $45^{+15}_{-13}\%$ and $18^{+7}_{-4}\%$ for substellar binaries in stellar systems and in the field, respectively.

There are also examples of wide binaries, such as 2MASS J 11011926–7732383AB, found by Luhman (2004a) towards the Chamaeleon I cloud. The angular separation of these two M7/8 brown dwarfs is 1.44 arcsec, which corresponds to ~ 240 AU at the distance of this star-forming region. About 15% of field brown dwarfs appear to be binaries with separations in the range 5–20 AU (see, for example, Bouy et al. 2003). It has been claimed that the overall binary fraction in young clusters such as the Pleiades could be as high as 50% (Pinfield et al. 2003). But this claim is based on the photometric properties of Pleiades brown dwarf candidates that are still rather uncertain. Direct imaging searches in the cluster give a binary fraction of 15% (Martín et al. 2003). The corresponding stellar binary fraction is in the range 50–60% in solar-type stars and decreases to $\sim 35\%$ in M types. Different surveys suggest that the distribution of orbital semi-major axes of brown dwarfs peaks at much smaller values than in stars.

Most remarkably, the study of substellar binaries has led to the first dynamical determination of the mass of a brown dwarf. Using Adaptive Optics techniques at the Keck telescope (see Figure 6.3), Zapatero Osorio et al. (2004) have measured the orbital motion of the two components of the binary Gl 569B. The total mass of the system is 0.125 ± 0.005 $M_\odot$ and the orbital period is 876.0 ± 9.2 days. Radial velocity measurements of each of the components have allowed a determination of the masses resulting in $m_1 = 0.068 \pm 0.011$ $M_\odot$ and $m_2 = 0.057 \pm 0.011$ $M_\odot$. The lightest component is clearly located in the substellar domain. Additional radial velocity measurements have the potential to reduce significantly the error bar in the mass determination, and to clarify whether the more massive component is also a brown dwarf.

6.4.3. Planets around brown dwarfs

Deep searches around young brown dwarfs have also led to the discovery of a giant planet candidate imaged at the VLT as a common proper motion companion of the young M8 brown dwarf 2MASS WJ 1207334–393254 (Chauvin et al. 2004). The primary is a proper motion member of the TW Hydrae association (age 8 ± 4 Myr) with a mass of ~ 25 $M_{\rm Jup}$. The faint candidate companion would have a mass of 5 ± 2 $M_{\rm Jup}$. A modest-quality near-infrared spectrum suggests an L spectral type. This object would then have similar characteristics (mass and effective temperature) to the isolated planetary-mass objects discovered by Zapatero Osorio et al. (2000) in the σ Orionis cluster. It is important to

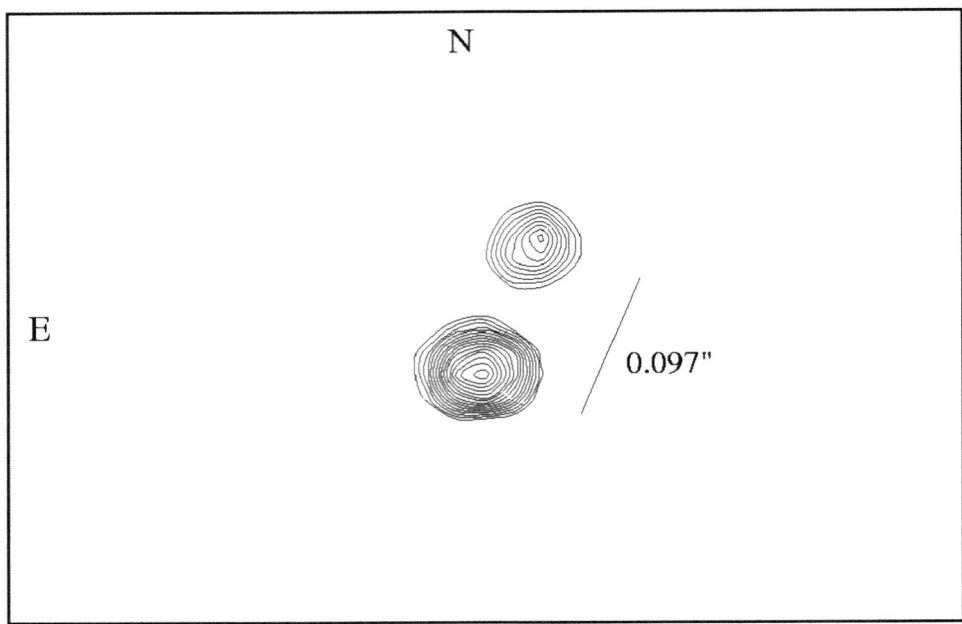

FIGURE 6.3. The binary brown dwarf Gl569 Bab resolved at Keck (Zapatero Osorio et al. 2004).

study the frequency and distribution of objects with these masses, be they free-floating or orbiting stars/brown dwarfs, in order to clarify their origin.

6.5. Searches in stellar clusters and young stellar associations: the brown dwarf mass function

At young ages, the higher luminosity of brown dwarfs allows the study of a low-mass domain still unexplored by large-scale field searches. Subsequent to the first discoveries, various surveys in the Pleiades cluster revealed a rather numerous population of brown dwarfs with masses down to $\sim 35\ M_{\rm Jup}$ (Zapatero Osorio et al. 1997; Bouvier et al. 1998; Martín et al. 1998). At low masses ($M \leq 0.3\ M_\odot$), the power-law index, α, of the mass function $N(M) \propto M^{-\alpha}$ is in the range 0.4–1. These results have also been confirmed by other more recent surveys (Nagashima et al. 2003; Jameson et al. 2002; Moraux et al. 2003). Very recently, a wide near-IR search covering $\sim 1.8\ {\rm deg}^2$ of the Pleiades cluster (Bihain et al. 2006) has led to the discovery of more than a dozen L-dwarf candidates ($J \geq 17.5$ and $I - J \geq 3.3$). Proper motion measurements have already verified the membership in the Pleiades of several of these candidates, plotted in the colour–magnitude diagram of Figure 6.4 for comparison with more massive Pleiades cluster members. For a cluster age of 120 Myr (estimated from the lithium depletion boundary), the estimated mass of the faintest proper motion member so far identified is $\sim 25\ M_{\rm Jup}$ according to evolutionary models of Chabrier et al. (2000).

In many other younger clusters and stellar associations, brown dwarfs have been detected down to masses in the range 15–30 $M_{\rm Jup}$. For instance, in ρ Ophiuchi (Williams et al. 1995; Luhman et al. 2000), Chamaeleon I (Comerón et al. 2000), Trapezium and IC 348 (Luhman et al. 2000), σ Orionis (Béjar et al. 1999, 2001) and Taurus (Luhman

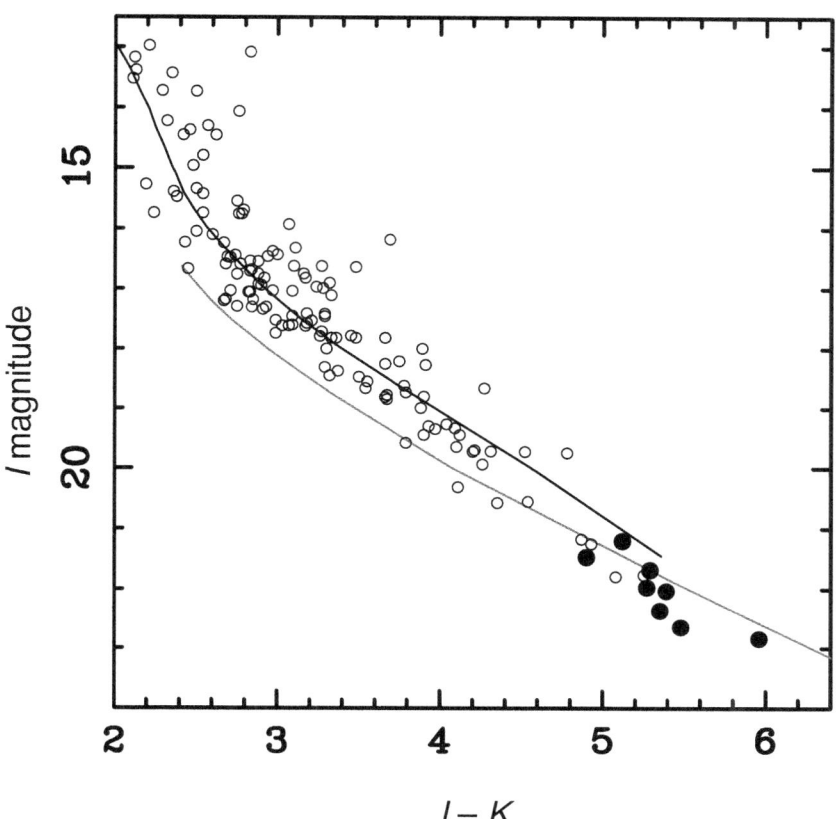

FIGURE 6.4. The Pleiades brown dwarf sequence in the I versus $I - K$ diagram. The new candidate L-type brown dwarfs from the survey by Bihain et al. (2006) are marked with filled circles. The upper solid line is the DUSTY model 100 My isochrone from Chabrier et al. (2000). The line below shows the location of field L-dwarfs shifted to the distance of the Pleiades.

2004b). Even though the stellar density of these clusters spans more than two orders of magnitude, the substellar mass functions are rather similar with values of α ranging between 0.5 and 1. Given the observational and theoretical difficulties associated with the determination of this exponent, such a small range of values suggests that the mass distribution of substellar objects is the result of a rather general – possibly universal – process. The number of brown dwarfs per star appears to fluctuate from cluster to cluster but only within a factor of $\sim$2–3. A power-law fit to the mass spectrum is just a convenient description of the distribution of masses; another possible representation based on a log-normal fit (see, for example, Moraux et al. 2003) also leads to a similar conclusion.

6.5.1. Isolated planetary-mass objects

Remarkably, the masses of the faintest unbound substellar objects discovered in stellar clusters appear to be below 10 $M_{\rm Jup}$. Zapatero Osorio et al. (2000) reported on very faint, extremely red objects in the σ Orionis cluster with L-type spectra, and masses down to $\sim$5 $M_{\rm Jup}$. Low-resolution optical and near-infrared spectroscopy has been obtained for most of these objects (Martín et al. 2001; Barrado y Navascués et al. 2003). Lucas

& Roche (2000) have also reported photometric candidates to planetary-mass objects in the Trapezium cluster. Zapatero Osorio et al. (2002) claimed the detection of a T-dwarf (S Ori 70) in the σ Orionis cluster. The near-infrared spectrum clearly shows the presence of methane bands. According to evolutionary tracks the mass is 3^{+5}_{-2} $M_{\rm Jup}$. Burgasser et al. (2004) have argued that it might be a foreground T-dwarf unrelated to the cluster, but this object displays a much more extreme $I - K$ colour than field T-dwarfs, which favours the cluster membership hypothesis. Further observations in the mid-infrared, feasible with the *Spitzer* satellite, would be very important in confirming and clarifying the origin of any possible infrared excess. More recently, Caballero et al. (2006) have extended the survey in σ Orionis, detecting candidates with $J = 20.0$–20.8 and colours typical of L6–L8 dwarfs. This suggests that the T-dwarfs in the cluster will be at $J \sim 21$–21.5 and that S Ori 70 would more probably be a binary system. Béjar et al. (2001) find an index $\alpha = 0.8 \pm 0.25$ for the substellar mass function in σ Orionis down to ~ 5 $M_{\rm Jup}$. At lower masses, the incompleteness of the current surveys prevents us from reaching any firm conclusion; the few additional objects in the Caballero et al. (2006) sample, however, are consistent with a smooth extrapolation of this mass function to lower masses, albeit with a lower α index, suggesting that the number of isolated planetary-mass objects in the cluster might be comparable to the number of brown dwarfs.

6.5.2. *Substellar discs*

Observational evidence for the presence of discs around young brown dwarfs in star clusters and associations has accumulated over the past few years. Observations in the near-IR (e.g. Muench et al. 2001) and mid-IR excess emission (Comerón et al. 1998, 2000; Natta et al. 2002; Jayawardhana et al. 2003; Mohanty et al. 2004) indicate that dusty discs are common around very young brown dwarfs (age of 1–2 Myr). For example, over 50% of the observed substellar sources exhibit infrared-excess emission in the Trapezium cluster. The observed spectral energy distributions (SEDs) and IR excess emission of some brown dwarfs have been modelled using flat and flared reprocessing discs (e.g. Natta & Testi 2001; Walker et al. 2004). Brown dwarf discs seem to be analogous to those in classical T Tauri stars and exhibit a similar range of disc geometries and dust properties. The disc lifetimes are also comparable to those of low-mass stellar counterparts in the T Tauri phase. Grain growth, dust settling, and evolution of disc geometry may appear on a timescale of 10 Myr.

The characteristic broad asymmetric Hα emission lines of classical T Tauri stars, usually interpreted as a spectroscopic signature of accretion, are also seen in brown dwarfs of any possible mass almost down to the deuterium-burning limit (e.g. Muzerolle et al. 2000; Jayawardhana et al. 2002, 2003; Barrado y Navascués et al. 2002; Natta et al. 2004). Similarly to T Tauri stars, brown dwarfs accrete material from the discs, albeit with a rate of mass accretion of a factor 10–100 times lower.

The irregular large-amplitude photometric variations observed in some young brown dwarfs might also be related to accretion (Scholz & Eislöffel 2004; Caballero et al. 2004). Emission-line variability studies of young brown dwarfs appear to give support to the magnetospheric accretion scenario (Scholz & Eislöffel 2005), and therefore the existence of large-scale magnetic fields in the substellar regime.

Submillimetre and millimetre emission are optically thin and measure the total dust mass in discs. Using the JCMT and IRAM, Klein et al. (2003) detected cold dust in the discs around the two young (age less than 10 Myr) brown dwarfs CFHT-BD-Tau 4 and IC 348 613 at flux levels of a few μJy. The masses of the discs were estimated to be in the range 0.4–6 $M_{\rm Jup}$. Such a large amount of circumstellar material suggests that planet formation might also take place around brown dwarfs. Dust masses were estimated to be a few times the Earth mass. Far-infrared observations with *Spitzer Space Telescope* and with ground-based radio telescopes may reveal the sizes and masses of brown dwarf discs and their structure, allowing us to determine whether most discs are truncated, information of paramount importance for constraining brown dwarf formation scenarios.

6.6. Formation mechanisms

We are far from a complete understanding of the origin of brown dwarfs. Do brown dwarfs form as small stars, i.e. from the turbulent fragmentation and collapse of molecular cloud cores? Alternatively, do brown dwarfs form like giant planets or in different ways? How can free-floating objects with inferred masses as low as 3–5 Jupiter masses form?

There is no generally accepted formation scenario and it is likely that there is more than one process able to originate brown dwarfs. One possibility is that they form in the same way as stars. Indeed, turbulent fragmentation can extend the conventional star formation scenario to much lower masses. Padoan & Nordlund (2004) find that, under the typical temperature, density and r.m.s. Mach number of star-forming regions, unstable protostellar cores of brown dwarf mass are formed with an abundance consistent with empirical determination of the substellar IMF. Analytical Jeans mass estimates by Low & Lynden-Bell (1976) and Rees (1976) predict that the minimum mass for a cloud fragment to overcome pressure and collapse under gravity could be as small as 3–7 Jupiter masses. This so called opacity limit for fragmentation is now challenged by the detection of free-floating objects with lower masses (e.g. Zapatero Osorio *et al.* 2000; Lucas & Roche 2000). However, if magnetic field tension effects are important in the cloud environment the initial masses of the smallest fragments could be as low as 1 Jupiter mass (Boss 2001).

Reipurth & Clarke (2001) proposed that early ejection from the gas reservoirs where star formation takes place could lead to the formation of brown dwarfs. Computer simulations to resolve the fragmentation process down to the opacity limit (Bate *et al.* 2003) show that very low-mass objects are often kicked out of new-born multiple systems before they can accrete enough mass to become stars. These authors predict that a large fraction of brown dwarfs are originally formed as members of multiple systems by the gravitational fragmentation of a common protostellar disc on rather short timescales. Subsequent dynamical encounters, gas accretion and/or interaction with discs would lead to the current mass distribution. The simulations predict that brown dwarfs and stars are produced in similar numbers. Few brown dwarfs are predicted to be binary and to have circumsubstellar discs. In fact, simulations by Bate & Bonnell (2005) obtain a binary fraction of only 8 per cent among very low-mass stars and brown dwarfs at separations around 10 AU. Only 20% of the low-mass binaries would be produced at separations between 1 and 4 AU. This value appears to be below the observed binary frequency at such small separations (Maxted & Jeffries 2005), posing a difficulty for the ejection models. Another important test for these simulations is the frequency and size of discs around brown dwarfs which in very young clusters (IC 348, Chamaeleon I, etc.) exhibit similar disc frequencies to stars. In the same ejection scenario, star disc encounters also truncate discs, which would typically have radii less than 20 AU. To date, the size of brown dwarf discs has not been measured. The advent of sensitive mid-infrared instruments operating at the diffraction limit of very large telescopes (0.1–0.2 arcsec) will shed light on this question.

An alternative mechanism for brown dwarf formation is the removal of accretion envelopes by collision or photoevaporation (Whitworth & Zinnecker 2004). Photoerosion and ionization by very massive stars may inhibit the accretion process. In the vicinity of an OB star, the outer layers of a low-mass prestellar core can be eroded by the ionizing radiation before they can accrete on to the protostar, leading to the formation of a free-floating brown dwarf. According to this scenario, brown dwarfs should be much more frequent in stellar systems containing large numbers of massive stars.

Finally, instabilities in circumstellar or cimcumbinary discs (Pickett et al. 2000; Jiang et al. 2004) have been considered as a mechanism for the formation of brown dwarfs. In principle, disc self-gravity may cause protostellar discs to fragment into bodies that can become stellar, substellar, or planetary-mass companions, depending partly on the site of formation in the disc. The possibility that giant gaseous planets can form from the fragmentation of a disc was already suggested by Cameron (1978) and has received much attention (e.g. Boss 1997; Mayer et al. 2002). However, it has been argued recently that the formation of giant planets by direct disc instability is impossible on thermodynamic grounds (Rafikov 2005), or that it is not a viable mechanism to produce substellar companions except at periods above 20 000 yr (Matzner & Levin 2005). Indeed, the much higher frequency of brown dwarfs at large separations of stars over those in very close orbits seems to argue against formation in circumstellar discs. The higher frequency of planets in small orbits ($\rho \leq 4$ AU) also supports the idea that planets form in a different way from brown dwarf companions.

The most direct way to address the question of the origin and nature of brown dwarfs is to investigate the properties of extremely young substellar objects in regions of active star formation. Brown dwarfs and planetary-mass candidates are found in regions less than 1 Myr old in large abundances, indicating that the process of formation is rather fast. Observations in various star-forming regions with different environmental conditions (different star densities, existence of O-type stars, etc.) are required to distinguish among the formation scenarios. Direct observations of the youngest phases of substellar objects, proto-brown dwarfs, will be crucial in evaluating these scenarios (Haisch et al. 2004; Duchene et al. 2004; Apai et al. 2005). The multiplicity fraction, separation and mass ratio distribution of brown dwarf binaries and brown dwarf companions, the properties of substellar discs and the determination of the substellar mass function in various environments will be the key to disentangling the origin of brown dwarfs.

Acknowledgments

I thank J. A. Caballero, V. J. S. Béjar, G. Bihain, S. Iglesias Groth and M. R. Zapatero Osorio for valuable discussions on this topic and help with the figures.

REFERENCES

Apai, D., Toth, L. V., Henning, T., Vavrek, R., Kovács, Z. & Lemke, D., 2005, A&A **433**, L33.
Bailer-Jones, C. A. L. & Mundt, R., 2001, A&A **367**, 218.
Baraffe, I., Chabrier, G., Allard, F. & Hauschildt, P. H., 1998, A&A **337**, 403.
Barrado y Navascués, D., Zapatero Osorio, M. R., Martín, E. L., Béjar, V. J. S., Rebolo, R. & Mundt, R., 2002, A&A **393**, L85.
Barrado y Navascués, D., Béjar, V. J. S., Mundt, R., Martín, E. L., Rebolo, R., Zapatero Osorio, M. R. & Bailer-Jones, C. A. L., 2003, A&A **404**, 171.
Basri, G. & Martín, E. L., 1999, AJ **118**, 2460.
Basri, G., Marcy, G. W. & Graham, J. R., 1996, ApJ **458**, 600.
Bate, M. R. & Bonnell, I. A., 2005, MNRAS **356**, 1201.

BATE, M. R., BONNELL, I. A. & BROMM, V., 2003, *MNRAS* **339**, 577.
BÉJAR, V. J. S., ZAPATERO OSORIO, M. R. & REBOLO, R., 1999, *ApJ* **521**, 671.
BÉJAR, V. J. S., MARTÍN, E. L., REBOLO, R., *et al.*, 2001, *ApJ* **556**, 830.
BIHAIN, G., REBOLO, R., BÉJAR, V. J. S., *et al.*, 2006, *A&A* **458**, 805.
BOSS, A. P., 1997, *Science* **276**, 1836.
BOSS, A. P., 2001, *ApJ* **551**, L167.
BOUVIER, J., STAUFFER, J. R., MARTÍN, E. L., BARRADO Y NAVASCUÉS, D., WALLACE, B. & BÉJAR, V. J. S., 1998, *A&A* **336**, 490.
BOUY, H., BRANDNER, W., MARTÍN, E. L., *et al.*, 2003, *AJ* **126**, 1526.
BURGASSER, A. J., 2001, Ph.D. Thesis, Caltech.
BURGASSER, A. J., KIRKPATRICK, J. D., & BROWN, M. E., 2002, *ApJ* **564**, 421.
BURGASSER, A. J., KIRKPATRICK, J. D., BROWN, M. E., *et al.*, 1999, *ApJ* **522**, 65.
BURGASSER, A. J., KIRKPATRICK, J. D., CUTRI, R. M., *et al.*, 2000, *ApJL* **531**, 57.
BURGASSER, A. J., KIRKPATRICK, J. D., REID, I. N., BROWN, M. E., MISKEY, C. L. & GIZIS, J. E., 2003, *ApJ* **586**, 512.
BURGASSER, A. J., KIRKPATRICK, J. D., McGOBERN, M. R, McLEAN, I. S., PRATO, L. & REID, I. N., 2004, *ApJ* **604**, 827.
BURGASSER, A. J., KIRKPATRICK, J. D. & LOWRANCE, P. J, 2005, *AJ* **129**, 2849.
BURROWS, A., MARLEY, M., HUBBARD, W. B., *et al.*, 1997, *ApJ* **491**, 856.
BURROWS, A., HUBBARD, W. B., LUNINE, J. I. & LIEBERT, J., 2001, *Rev. Mod. Phys.* **73**, 719.
CABALLERO, J. A., BÉJAR, V. J. S., REBOLO, R. & ZAPATERO OSORIO, M. R., 2004, *A&A* **424**, 857.
CABALLERO, J. A., MARTÍN, E. L., DOBBIE, D. D., *et al.*, 2006, *A&A* **460**, 635.
CAMERON, A. G. W., 1978, *Earth, Moon, Planets* **18**, 5.
CHABRIER, G. & BARAFFE, I., 2000, *ARAA* **38**, 337.
CHABRIER, G., BARAFFE, I., ALLARD, F. & HAUSCHILDT, P. H., 2000, *ApJ* **542**, 464.
CHAUVIN, G., LAGRANGE, A.-M., DUMAS, C., *et al.*, 2004, *A&A* **425**, L29.
CHAUVIN, G., LAGRANGE, A.-M., ZUCKERMAN, B., *et al.*, 2005, *A&A* **438**, L29.
CLOSE, L. M., SIEGLER, N., FREED, M. & BILLER, B., 2003, *ApJ* **587**, 407.
COMERÓN, F., RIEKE, G. H., CLAES, P., TORRA, J. & LAUREIJS, R. J., 1998, *A&A* **335**, 522.
COMERÓN, F., NEUHÄUSER, R. & KAAS, A. A., 2000, *A&A* **359**, 269.
CUBY, J. G., SARACCO, P., MOORWOOD, A. F., D'ODORICO, S., LIDMAN, C., COMERÓN, F. & SPYROMILIO, J., 1999, *A&AL* **349**, 41.
DAHN, C. C., HARRIS, H. C. & VRBA, F. J., 2002, *AJ* **124**, 1170.
DELFOSSE, X., TINNEY, C. G., FORVEILLE, T., *et al.*, 1997, *A&AL* **327**, 25.
DUCHENE, G., BOUVIER, J., BONTEMPS, S., ANDRÉ, P. & MOTTE, F., 2004, *A&A* **427**, 651.
DUQUENNOY, A. & MAYOR, M., 1991, *A&A* **248**, 485.
ENDL, M., HATZES, A. P., COCHRAN, W. D., *et al.*, 2004, *ApJL* **611**, 1121.
FISCHER, D. A. & MARCY, G. W., 1992, *ApJ* **396**, 178.
FORVEILLE, T., SÉGRANSAN, D., DELORME, P. *et al.*, 2004, *A&AL* **427**, 1.
FREED, M., CLOSE, L. M. & SIEGLER, N., 2003, *ApJ* **584**, 453.
GIZIS, J. E., REID, I., KNAPP, G. R., LIEBERT, J., KIRKPATRICK, J. D., KOERNER, D. W. & BURGASSER, A., 2003, *AJ* **125**, 3302.
GUENTHER, E. M. & WUCHTERL, G., 2003, *A&A* **401**, 677.
HAISCH, K. E., GREENE, T. P., BARSONY, M. & STAHLER, S. W., 2004, *AJ* **127**, 1747.
HAYASHI, C. & NAKANO, T., 1963, *Prog. Theor. Phys.* **30**, 460.
JAMESON, R. F. & SKILLEN, I., 1989, *MNRAS* **239**, 247.

Jameson, R. F., Dobbie, P. D., Hodgkin, S. T. & Pinfield, D. J., 2002, *MNRAS* **335**, 853.

Jayawardhana, R., Mohanty, S. & Basri, G., 2002, *ApJ* **578**, L141.

Jayawardhana, R., Ardila, D. R., Stelzer, B. & Haisch, K. E. Jr., 2003, *AJ* **126**, 1515.

Jiang, I.-G., Laughlin, G. & Lin, D. N. C., 2004, *Rev. Mex. A&A* **21**, 227.

Joergens, V., Guenther, E. Neuhäuser, R., et al., 2001, *Astronomische Gesellschaft Abstract Series* **18**, 402.

Kenyon, M. J., Jeffries, R. D., Naylor, T., Oliveira, J. M. & Maxted, P. F. L., 2005, *MNRAS* **356**, 89.

Kirkpatrick, J. D., Reid, I. N., Liebert, J., et al., 1999, *ApJ* **519**, 802.

Klein, R., Apai, D., Pascucci, I., Henning, Th. & Waters, L. B. F., 2003, *ApJ* **593**, L57.

Kumar, S. K., 1963, *ApJ* **137**, 1121.

Leggett, S. K., Geballe, T. R., Fan, X., et al., 2000, *ApJ* **519**, 802.

Liu, M. C. & Leggett, S. K., 2005, *ApJ* **634**, 616.

Low, C. & Lynden-Bell, D., 1976, *MNRAS* **176**, 367.

Lucas, P. W. & Roche, P. F., 2000, *MNRAS* **314**, 858.

Luhman, K. L., 2004a, *ApJL* **614**, 398.

Luhman, K. L., 2004b, *ApJ* **617**, 1216.

Luhman, K. L., Rieke, G. H. & Young, E. T., 2000, *ApJL* **540**, 1016.

Magazzú, A., Martín, E. L. & Rebolo, R., 1993, *ApJL* **404**, 17.

Marcy, G. & Butler, P., 2000, *PASP* **112**, 137.

Martín, E. L., Rebolo, R. & Magazzú, A., 1994, *ApJ* **436**, 262.

Martín, E. L., Basri, G., Delfosse, X. & Forveille, T., 1997, *A&AL* **327**, 29.

Martín, E. L., Basri, G., Zapatero Osorio, M. R., Rebolo, R. & García López, R. J., 1998, *ApJL* **507**, 41.

Martín, E. L., Delfosse, X., Basri, G., Goldman, B., Forveille, T. & Zapatero Osorio, M. R., 1999, *AJ* **118**, 2466.

Martín, E. L., Zapatero Osorio, M. R., Barrado y Navascués, D., Béjar, V. J. S. & Rebolo, R., 2001, *ApJ* **558**, L117.

Martín, E. L., Barrado y Navascués, D., Baraffe, I., Bouy, H. & Dahm, S., 2003, *ApJ* **594**, 525.

Masciadri, E., Brandner, W., Bouy, H., Lenzen, R., Lagrange, A. M. & Lacombe, F., 2003, *A&A* **411**, 157.

Matzner, C. D. & Levin, Y., 2005, *ApJ* **628**, 817.

Maxted, P. F. L. & Jeffreis, R. D., 2005, *MNRAS* **362**, L45.

Mayer, L., Quinn, T., Waldsley, J. & Stadel, J., 2002, *Science* **298**, 1756.

McCarthy, C. & Zuckerman, B., 2004, *AJ* **127**, 2871.

McCaughrean, M. J., Close, L. M., Scholz, R. D., et al., 2004, *A&A* **413**, 1029.

Mohanty, S., Jayawardhana, R., Natta, A., Fujiyoshi, T., Tamura, M. & Barrado y Navascués, D., 2004, *ApJ* **609**, L33.

Moraux, E., Bouvier, J., Stauffer, J. R. & Cuillandre, J.-C., 2003, *A&A* **400**, 891.

Muench, A. A., Alves, J., Lada, C. J. & Lada, E. A., 2001, *ApJ* **558**, L51.

Muzerolle, J., Briceño, C., Calvet, N., et al., 2000, *ApJ* **545**, L141.

Nagashima, C., Dobbie, P. D., Nagayama, T., et al., 2003, *MNRAS* **343**, 1263.

Nakajima, T., Oppenheimer, B. R., Kulkarni, S. R., Golimowski, D. A., Matthews, K. & Durrance, S. T., 1995, *Nature* **378**, 463.

Nakajima, T., Tsuji, T. & Yanagisawa, K., 2004, *ApJ* **607**, 499.

Natta A. & Testi L., 2001, *A&A* **376**, L22.

NATTA A., TERTI, L., COMERÓN, F., et al., 2002, A&A **393**, 597.

NATTA A., TESTI, L., MUZEROLLE, J., RANDICH, S., COMERÓN, F. & PERSI, P., 2004, A&A **424**, 603.

NEUHÄUSER, R., GUENTHER, E. W., WUCHTERL, G., MUGRAUER, M., BEDALOV, A. & HAUSCHILDT, P. H., 2005, A&A **435**, L13.

OPPENHEIMER, B. R., KULKARNI, S. R., MATTHEWS, K. & NAKAJIMA, T., 1995, Science **270**, 1478.

PADOAN, P. & NORDLUND, A., 2004, ApJ **617**, 559.

PICKETT, B. K., CASSEN, P., DURISEN, R. H. & LINK, R., 2000, ApJ **530**, 1160.

PINFIELD, D. J., DOBBIE, P. D., JAMESON, R. F., STEELE, I. A., JONES, H. R. A. & KATSIYANNIS, A. C., 2003, MNRAS **342**, 1241.

PRAVDO, S. H., SHAKLAN, S. B. & LLOYD, J., 2005, ApJ **630**, 528.

RAFIKOV, R. R., 2005, ApJL **621**, 69.

REBOLO, R., MARTÍN E. L. & MAGAZZÚ, A., 1992, ApJL **389**, 83.

REBOLO, R., ZAPATERO OSORIO, M. R. & MARTÍN, E. L., 1995, Nature **377**, 129.

REBOLO, R., MARTÍN, E. L, BASRI, G., MARCY, G. & ZAPATERO OSORIO, M. R., 1996, ApJL **469**, 53.

REBOLO, R., ZAPATERO OSORIO, M. R., MADRUGA, S., BÉJAR, V. J. S., ARRIBAS, S. & LICANDRO, J., 1998, Science **282**, 1309.

REES, M., 1976, MNRAS **176**, 483.

REID, I. N., KIRKPATRICK, J. D., LIEBERT, J., et al., 1999, ApJ **521**, 613.

REIPURTH, B. & CLARKE, C., 2001, AJ **122**, 432.

RIEKE, G. H. & RIEKE M. J., 1990, ApJL **362**, 21.

RUIZ, M. T., LEGGETT, S. K. & ALLARD, F., 1997, ApJL **491**, 107.

SCHOLZ, A. & EISLÖFFEL, J., 2004, A&A **419**, 249.

SCHOLZ, R. D. & EISLÖFEL, J., 2005, A&A **429**, 1007.

SCHOLZ, R. D., MCCAUGHREAN, M. J., LODIEU, N. & KUHLBRODT, B., 2003, A&AL **398**, 29.

STAUFFER, J., HAMILTON, D., PROBST, R., RIEKE, G. & MATEO, M., 1989, ApJL **344**, 21.

STAUFFER, J., HERTER, T., HAMILTON, D., RIEKE, G. H., RIEKE, M. J., PROBST, R. & FORREST, W., 1991, ApJL **367**, 23.

STAUFFER, J., LIEBERT, J., GIAMPAPA, M., MACINTOSH, B., REID, N. & HAMILTON, D., 1994, AJ **108**, 160.

TSUJI, T., 1995, In *The Bottom of the Main Sequence and Beyond*, ed. C. G. TINNEY, ESO Astrophysics Symposium, Garching, Heidelberg: Springer-Verlag, p. 45.

TSUJI, T. & NAKAJIMA, T., 2003, ApJ **585**, L151.

WALKER C., WOOD, K., LADA, C. J., et al., 2004, MNRAS **351**, 607.

WHITWORTH, A. P. & ZINNECKER, H., 2004, A&A **427**, 299.

WILLIAMS, D. M., COMERÓN, F., RIEKE, G. H. & RIEKE, M. J., 1995, ApJ **454**, 144.

ZAPATERO OSORIO, M. R., REBOLO, R., MARTÍN, E. L., et al., 1997, ApJL **491**, 81.

ZAPATERO OSORIO, M. R., BÉJAR, V. J. S.., MARTÍN, E. L., REBOLO, R., BARRADO Y NAVASCUÉS, D., BAILER-JONES, C. A. L. & MUNDT, R., 2000, Science **290**, 103.

ZAPATERO OSORIO, M. R., BÉJAR, V. J. S., MARTÍN, E. L., et al., 2002, ApJL **578**, 536.

ZAPATERO OSORIO, M. R., LANE, B. F., PAVLENKO, YA., MARTÍN, E. L., BRITTON, M. & KULKARNI, S. R., 2004, ApJ **615**, 958.

ZUCKERMAN, B. & BECKLIN, E. E., 1992, ApJ **386**, 260.

7. The perspective: a panorama of the Solar System

AGUSTÍN SÁNCHEZ-LAVEGA

We review the basic properties of the bodies constituting the Solar System as a reference for understanding the properties of the increasing number of extrasolar planets and planetary systems discovered.

7.1. Introduction

Space exploration has allowed us over the last 40 years to visit most of the different types of bodies that constitute the Solar System (which we define as comprising all those objects under the gravitational influence of the Sun). We have reached all the planets, except Pluto, and consequently most of their satellites, by means of fly-bys, orbital injection (Venus, Mars, Jupiter and Saturn), landing (the Moon, Venus, Mars and Titan) and probe sounding (Jupiter). We have sent vehicles to asteroids and comets (fly-bys), with impacts on the asteroid Eros and Comet Temple 1, and we have samples returned from the Moon, and from meteorites coming from Mars and the asteroid belt. All this has provided a large quantity of information, so the reader can find a large number of books dealing with the Solar System as a whole, or reviewing the properties of each individual constituent (Gehrels 1976, 1979; Burns 1977; Wilkening 1982; Morrison 1982; Hunten *et al.* 1983; Gehrels & Matthews 1984; Greenberg & Brahic 1984; Burns & Matthews 1986; Chamberlain & Hunten 1987; Kerridge & Matthews 1988; Vilas, Chapman & Matthews 1988; Atreya, Pollack & Matthews 1989; Binzel, Gehrels & Matthews 1989; Kieffer *et al.* 1992; Cruikshank 1995; Lewis 1997; Bougher, Hunten & Phillips 1997; Shirley & Fairbridge 1997; Beatty, Collins Petersen & Chaikin 1999; Weissman, McFadden & Johnson 1999; de Pater & Lissauer 2001; Cole & Woolfson 2002; Bertotti, Farinella & Vokrouhlicky 2003; McBride & Gilmour 2003; Encrenaz *et al.* 2004; Bagenal, Dowling & McKinnon 2004). These works will not be cited individually in the text that follows. A review discussing planetary diversity in our Solar System and in extrasolar planets has recently been presented by Sánchez-Lavega (2006).

The purpose of this chapter cannot be to provide an overview or resumé of all that is covered in these books, but more modestly to present the Solar System in a comparative way, i.e. grouping and comparing the properties of the various bodies, and so serve as a reference for the growing amount of information on other planetary systems and their planets. Since planets are the most massive and complex objects, and the first observational targets in extrasolar systems, we shall mainly focus on these. Some topics, such as the origin of planetary systems and planet formation, are the subject of other chapters of this book and so will not be addressed here. The outline of this chapter is as follows. In Section 7.2 I present the general features of the Solar System and its dynamical organization by gravity; in Section 7.3 I compare the physical properties of the various bodies; in Section 7.4 I present the current state of our knowledge of the atmospheres of planets and their satellites; in Section 7.5 I present a panorama of the electromagnetic

Extrasolar Planets, eds. Hans Deeg, Juan Antonio Belmonte and Antonio Aparicio.
Published by Cambridge University Press.
© Cambridge University Press 2007.

TABLE 7.1 Orbital characteristics of the planets

Planet	Distance to Sun ($\times 10^8$ km)	Orbital tilt (degrees)	Insolation (W/m^2)	Orbital period (years)	Number of satellites	Rings
Mercury	0.58	7	9040	0.24	0	–
Venus	1.08	177.4	2620	0.61	0	–
Earth	1.49	23.4	1370	1.00	1	–
Mars	2.28	24	590	1.88	2	–
Jupiter	7.78	3.08	50.6	11.86	60	a
Saturn	14.27	26.7	15.1	29.5	31	b
Uranus	28.69	97.9	3.72	84.01	22	c
Neptune	44.96	28.8	1.52	164.79	11	d
Pluto	59.0	17.15	0.88	248.54	1	–

Notes for rings: D (distance from planet's centre); τ (maximum optical depth); m (estimated mass in grammes)
$^a D = 1.4\text{–}3.2\ R_J;\ \tau = 3 \times 10^{-6};\ m = 10^{16}$ g
$^b D = 1.4\text{–}2.2\ R_S;\ \tau = 2.5;\ m = 2.8 \times 10^{25}$ g
$^c D = 1.5\text{–}1.95\ R_U;\ \tau = 2.3$
$^d D = 1.68\text{–}2.53\ R_N;\ \tau = 0.15$

environment of the planets and their interaction with the interplanetary medium; and finally I conclude with a brief overview of some planned exploration missions of the Solar System.

7.2. Global structure of Solar System

It is traditional with astronomical observations and discoveries to classify the objects forming the Solar System into groups that correspond to certain common properties. The nine known *planets* orbit around the Sun (from 0.4 AU to 30 AU) and, except for the two closest (Mercury and Venus), the rest are themselves orbited by smaller bodies, *satellites*, ranging in number from one (for the Earth and Pluto), to 60 or more in the case of Jupiter. In total about 125 satellites have been discovered so far. The first peculiarity is that two satellites (Ganymede and Titan) are larger than the planets Mercury and Pluto. In addition, Io, Europa, Ganymede and Titan are more massive than Pluto, and Titan possesses a dense atmosphere that is more to be expected on a planet than on a satellite.

Planets have in common only their low orbital eccentricity (~ 0, nearly circular orbits) and small orbital tilt (closer than 7° to the equatorial plane of the Sun (except for Pluto, tilted by 17°). The total mass of the planets is about 446.68 Earth masses (1 $M_\oplus = 5.976 \times 10^{24}$ kg) and those of the satellites 0.104 $M_\oplus$. The total mass of the planets and satellites represents merely ~ 0.001 that of the Sun. However, whereas the mass of the Solar System resides in the Sun, the bulk angular momentum resides in the giant planets owing to the slow rotation rate of the Sun around its axis (Table 7.1).

In addition to the planets and moons, there are three populations of smaller bodies (diameter ≤ 1000 km and ranging in number from 10^6 to 10^{11}) that orbit the Sun at different distances and are classified as: (1) *asteroids*, mainly located between 2 and 3.5 AU, with a total mass $\sim 5 \times 10^{-4}\ M_\oplus$; (2) *Kuiper Belt objects* (KBOs) or *trans-Neptunian objects* (TNOs), mainly located between 30 and 50 AU, with a total mass of $\sim 0.2\ M_\oplus$; (3) *comets*, located at distances ranging from 10^3 to 10^5 AU in the Oort cloud, with a total mass of $\sim 0.1\text{–}1\ M_\oplus$. The total estimated mass of these populations of objects

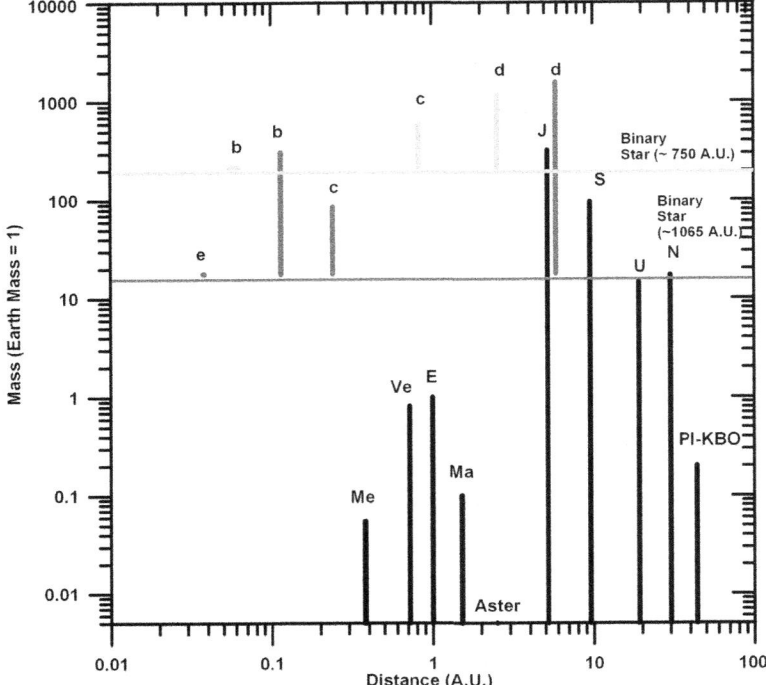

FIGURE 7.1. Mass (in Earth masses) versus distance (in AU) distribution in the Solar System (lower series) and in the two planetary systems: 55 Cancri (middle series: planets b, c, d, e) and Upsilon Andromedae (upper series: planets b, c, d).

is thus ~ 1 $M_\oplus$. Their orbits show a wide range of eccentricities and axial tilt as a result of their gravitational interaction with the massive objects, as described below.

The smallest bodies in the Solar System are the *meteoroids*, also called meteors when becoming incandescent by friction when entering the Earth's atmosphere, and meteorites when reaching the ground. They are typically less than metre-size. *Micrometeorites* and *dust* are particles with sizes in the range 10–100 µm and are distributed in the interplanetary medium, most densely in the inner Solar System, forming the *zodiacal light*. Finally there are local concentrations of 'particles' that surround the four major planets (Jupiter, Saturn, Uranus and Neptune) forming *rings* in their equatorial planes. The interplanetary medium is impregnated by electromagnetic radiation and magnetic fields from the Sun and to a lesser extent by some planets influencing the behaviour and motion of charged particles (plasma) in their environments. All these aspects are described in detail in the following sections.

There is at present evidence for the existence of other planetary systems (128 planetary systems, 147 planetary candidates and 15 multiple systems; Schneider 2005). In Figure 7.1 we compare the Solar System distribution of planetary mass versus orbital radius with two well-known exoplanetary systems (Schneider 2005). The distributions are quite different, reflecting most probably that these giants *migrated* from the outer to the inner parts of the planetary system owing to tidal interaction between the gaseous disc and the protoplanet (Thommes & Lissauer 2005). Some of the nearest and best studied young protoplanetary discs, such as those of β Pictoris and AU Microscopii, suggest the presence of planetesimal populations (asteroid- or KBO-like families; Okamoto *et al.* 2004; Liu 2004).

7.2.1. Mechanics: orbits

Since the population of objects in the Solar System larger than 1 km is most probably above 10^{12}, we are dealing with an N-body problem undergoing Newtonian gravitational interaction. However, $M_{\text{Solar System}} = 0.00134\ M_\odot$. This means that the centre of mass of the system is located inside the Sun (close to its surface) with all these bodies orbiting it. The same principle is valid for planets with satellites and rings (except for the Pluto–Charon system, whose mass ratio is ~ 8). The motions of planets and satellites reduce to a *two-body problem* (planets occupying the centre of mass of their systems), and the applicability of Kepler's laws is a direct consequence, with secondary bodies describing elliptical, parabolic or hyperbolic orbits. The object masses can be treated as single material points in a first approach to the orbital motion. A good limit to representing the gravitational influence by a secondary object (planet or satellite) relative to its primary (Sun or planet, respectively) is defined by the *Hill sphere*:

$$R_{\text{H}} = \left(\frac{m_2}{3(m_2 + m_1)}\right)^{1/3} a; \tag{7.1}$$

m_2 being the mass of the secondary, m_1 that of the primary, and a their mutual separation.

However, this is only a first approach to the motions. More detailed calculations of planetary motion must take into account the secondary *perturbations* between the bodies, and even *general relativity*, as in the case of Mercury's orbit because of the proximity and strong influence of the Sun. The following are some of the more important motions resulting from gravitational interaction.

(1) *Chaotic* trajectories result from very sensitive initial conditions and having unpredictable solutions. This is the case for the rotation axes of some bodies (e.g. the rotation state of Saturn's satellite Hyperion).

(2) The three-body problem gives rise to a variety of orbital solutions. Within the restricted three-body problem (one mass small relative to the other two), one solution gives rise to the existence of the *Lagrange stability points* (denoted L1, L2, L3, L4 and L5). An example is the Sun–Jupiter system with the Trojan asteroids located behind and ahead of the Jupiter orbit at $60°$ (points L4 and L5). Other examples are the planet–satellite systems Saturn–Tethys (satellites Calipso at L4 and Telesto at L5) and Saturn–Dioné satellite Helene). Other solutions are the *horseshoe orbits*, as observed between Saturn's satellites Janus and Epimetheus separated by 50 km in their orbits. They periodically (about every four years) interchange their orbits. The horseshoe is then the orbit described by one satellite as seen in the reference frame of the other.

(3) *Resonances* occur between two bodies if certain of their orbital parameters become commensurable. For example the *mean motion resonance* (the mean motion is defined as $n = 2\pi/P$, where P is the orbital period) relates two bodies' orbital periods by a given ratio, as for example the 3:2 resonance between Neptune and Pluto around the Sun, or the 4:2:1 between the satellites Io, Europa and Ganymede around Jupiter. This mechanism occurs between some extrasolar planets, as for example those of Gliese 876, HD 82943 and 55 Cancri (Schneider 2005 and references therein). The synchronous rotation of satellites (spin–orbit coupling 1:1, orbital period and spin having the same value) are other examples of resonance between planets and their satellites.

(4) Resonant perturbations in the asteroid belt produced by Jupiter force the asteroids into highly eccentric orbits, leading to collisions among them and finally producing

in the asteroid belt unstable regions with a low number of asteroids. These holes, known as *Kirkwood gaps*, correspond to the ratios 4:1, 3:1, 3:2 and so on, between the asteroid orbital period and that of Jupiter around the Sun.

(5) Resonant perturbations by a moon and a disc of particles also produce gaps (such as the Cassini and Encke divisions in Saturn's rings). These perturbations can also excite *spiral density waves* and *undulations*, such as produced by Mimas and Titan in Saturn's rings. Other satellites within a ring act as *shepherds* confining the ring material in narrow ringlets.

(6) Dissipative, non-gravitational forces resulting from the solar radiation acting on small bodies substantially modify their trajectories. These are: (a) the *radiation pressure* that moves micrometre-sized particles away from the Sun; (b) the *Poynting–Robertson* effect, which causes centimetre-sized particles to spiral inward towards the Sun by absorbing and reradiating solar radiation; (c) the *Yarkovski* effect, which changes the orbits of metre- to kilometre-sized objects owing to the temperature difference between the illuminated and non-illuminated surface.

(7) Comets suffer considerable *non-gravitational* forces from the momentum imparted to the nucleus as a reaction when sublimated gas and dust escape from it in narrow jets.

7.2.2. Mechanics of extended bodies

Observations of the Solar System show that bodies above a hundred kilometres in size have a nearly spherical shape, resulting from a balance between gravity, rotation and internal cohesion. Gravity tends to draw heavier elements towards the centre, producing *differentiation* and a layered, structured body. The centrifugal force resulting from the rapid rotation of the outer fluid planets makes them flattened at the poles with flattening $f = 1 - (R_p/R_e)$, with R_p and R_e being the polar and equatorial radii. The strength of this effect is measured by the rotational parameter,

$$q = \frac{\Omega^2 R_e^3}{GM_p}, \quad (7.2)$$

where Ω is the planetary angular velocity and M_p the planet's mass (Table 7.2). The orbits around such an oblate body tend to *precess*, as occurs with satellites orbiting the major planets (in particular Jupiter and Saturn). In the limit, the strong centrifugal forces could cause the body to be torn apart, the lowest rotation period for body cohesion being given by

$$\tau_{\rm rot}({\rm min}) = \left(\frac{3\pi}{G\langle\rho\rangle}\right)^{1/2}. \quad (7.3)$$

This can be taken as a lower limit for the rotation period expected in extrasolar planets and for expected values would be 1–3 hr.

Differential gravitational attraction between close objects (planets and satellites) raise tidal forces, forming a bulge along the line of interaction that deforms the body. Because of the different periods of rotation of the planet (spin) and orbit of the satellite around it, a tidal torque is exerted, modifying the orbits and spins of the interacting bodies. Tidal locking effects are involved in Mercury's 3:2 spin–orbit resonance (rotation period ∼58 days, orbital period ∼88 days). They are also the reason for Earth's long-term decrease in rotation rate (period ∼5 days at 4 Gyr ago) that is accompanied by the Moon's recession at a velocity of 3.74 cm per year. Equilibrium in the system will be reached in 50 000 million years when the rotation and revolution periods become equal to 47 days.

TABLE 7.2 Physical characteristics of planets and major satellites

Body	R_P [km]	M_P [g]	$\langle \rho \rangle$ [g cm^{-3}]	g [cm s^{-2}]	Rotation period	Rotation parameter (q)	Internal energy [Wm^{-2}]
Mercury	2439	3.30×10^{26}	5.43	370	58.65 days	9.5×10^{-7}	–
Venus	6050	4.86×10^{27}	5.20	884	243.01 days	6.1×10^{-8}	0
Earth	6378	5.98×10^{27}	5.51	981	23.93 hrs	0.0035	0.062[a]
Mars	3398	6.42×10^{26}	3.91	376	24.62 hrs	0.0046	0
Jupiter	71 300	1.90×10^{30}	1.33	2288	9.84 hrs	0.089	5.44
Io	1822	8.93×10^{25}	3.53	181	1.77 days (s)	0.0017	2.5
Europa	1565	4.80×10^{25}	2.99	131	3.55 days (s)	5×10^{-4}	0.02
Ganymede	2631	1.48×10^{26}	1.94	142	7.15 days (s)	1.8×10^{-4}	0.003
Callisto	2410	1.08×10^{26}	1.83	125	16.7 days (s)	3.6×10^{-5}	0.003
Saturn	60 100	5.68×10^{29}	0.69	950	10.23 hrs	0.16	2.01
Titan	2575	1.34×10^{26}	1.88	135	15.95 days	4×10^{5}	–
Uranus	25 500	8.68×10^{28}	1.32	869	17.9 hrs	0.027	0.042
Neptune	24 800	1.02×10^{29}	1.64	1100	19.2 hrs	0.018	0.433
Triton	1355	2.15×10^{25}	2.05	77	5.88 days (s)	2.6×10^{-4}	–
Pluto	1150	1.32×10^{25}	2.0	72	6.38 days	2.2×10^{-4}	–

Note: (s) indicates spin–orbit synchronization.
[a] Average. In geothermal areas ~ 1.7 Wm^{-2}.

A stable equilibrium has, however, been reached in the Pluto–Charon system because of their low mass difference.

Spin–orbit synchronism due to tidal interaction is expected to have occurred in 'hot jupiter' types of extrasolar planets owing to their proximity (<0.05 AU) to their parent stars (Guillot et al. 1996). This occurs when the synchronization time, $\tau_{\rm syn}$, is lower than the age of the system, τ_*,

$$\tau_{\rm syn} = Q\Omega_{\rm i}\left(\frac{M_{\rm p}}{GM_*^2}\right)\left(\frac{a^6}{R_{\rm p}^3}\right) \ll \tau_*, \quad (7.4)$$

$\Omega_{\rm i}$ being the initial angular rotation velocity of the planet, a the planet–star distance and Q a factor defined in Section 7.3 (see Eq. (7.9)). The 'hot jupiters' so far discovered seem to follow this synchronism (Sánchez-Lavega 2001).

On the other hand, solar gravitational tides in the dense atmosphere of Venus are probably the main reason for its low rotation period (243 days), lower than its orbital period of 224 days. For the other planets (from Mars to Neptune) the rotation rates are probably close to their primordial values.

The orbital inclination of the equator of a planet, or equivalently of its rotational axis, with respect to the orbital plane, lies in most cases within 30°. The exceptions are Venus (177°), Uranus (98°) and Pluto (119°) that rotate in a retrograde sense ('prograde' is west to east). The origin of these peculiar tilts is not well known, although for Venus tides raised in the massive atmosphere are the most probable reason. For Uranus, a close encounter with a massive object could have been the cause.

Gravitational tidal forces on a body of mass M and radius R are of the order of $\sim GMR/a^3$, a being the distance between the two bodies. They lead to internal heating of planetary bodies as, not being perfectly fluid, they change in shape, producing internal stresses and dissipating energy as heat (see Section 7.3.1). When the tidal force overcomes the self-gravity of a body, the *Roche limit* is reached and disruption of the body (for example a moon) can occur. This limit occurs for a simple fluid body at a distance

$$R_{\rm Roche} = 2.5 R_{\rm p}\left(\frac{\rho_{\rm p}}{\rho}\right)^{1/3}, \quad (7.5)$$

$\rho_{\rm p}$ and ρ being the density of the planet and moon respectively. The rings around the giant and icy planets lie within this distance, so at least some of them could have originated from a disintegrated satellite.

For smaller satellites, the internal cohesive forces are important and the critical satellite radius, R_0, is then determined by the *tensile strength*, ξ,

$$R_0 \approx \sqrt{\frac{\xi}{G\rho^2}}. \quad (7.6)$$

For example, for stones $\xi \sim 1000$ bar, for ices $\xi \sim 100$ bar and for porous bodies $\xi \sim 10$ bar, and thus for $\rho = 3$ g cm^{-3}, $R_0 \sim 30$–300 km.

7.3. Basic physical properties of Solar System bodies

7.3.1. *Planets and satellites*

Our Solar System is characterized by a large diversity in the physical properties of its major constituents (planets and satellites). The planets and major satellites differ in the basic properties of size, mass, chemical composition, internal energy source, rotation rate

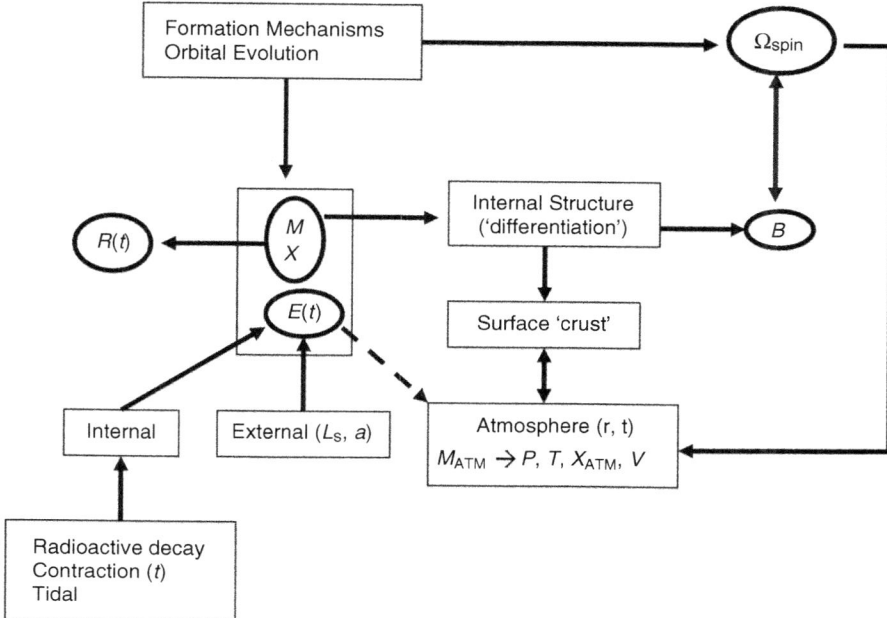

FIGURE 7.2. Flow diagram showing the basic mechanisms intervening in planetary structure. The most important are the mass (M), composition (X) and energy sources (E). These parameters determine the radius R (varying in time), the internal structure and atmosphere and crust morphology (if not a giant). The formation mechanisms play an important role in the rotational angular velocity (Ω) and in the magnetic field (B).

and magnetic environment. Their visual aspect (surfaces or atmospheres) reflect mainly the differences in the first four properties (Figure 7.2, Table 7.2).

Clearly there is a first dichotomy if we attend to the mean density between the terrestrial planets Mercury, Venus, Earth and Mars (Figure 7.3, mean densities $\sim$3.9–5.5 g cm^{-3}) and fluid giants Jupiter, Saturn, Uranus and Neptune (Figure 7.4, mean densities $\sim$0.7–1.6 g cm^{-3}). The reason is simply the proportions of refractory to volatile elements. Within the terrestrial group, the ratio of metals (basically Fe and Ni) to rocks (mainly SiO_2 and MgO) varies with distance, as shown in Table 7.3. On the other hand, the satellites have densities between $\sim$3.5 g cm^{-3} (in the case of Io and the Moon) and $\sim$1 g cm^{-3} (most satellites). With the exception of the Moon and Io, most satellites have a composition of $\sim$80% of volatile ices. This distribution can be simply explained by the distance to the Sun of these bodies and their subsequent condensation of refractories and volatiles as temperature decreases with distance. The 'ice-line', where ices can form, separates both regions and is presently at about 4 AU from the Sun. The same kind of density gradient can be found within the four Galilean satellites of Jupiter. Figure 7.5 shows the radius–mass relationship for the terrestrial planets and major satellites.

In contrast the giant planets are basically made of lighter hydrogen and helium gas. The He mass mixing ratio (i.e. the mass of helium atoms over total mass) is $\sim$0.24 on Jupiter, between 0.18 and 0.25 on Saturn, and close to the protosolar values ($\sim$0.27) on Uranus and Neptune. Most 'heavy' elements (i.e. those elements other than hydrogen and helium, such as C, N, S) are enriched on Jupiter and Saturn by a factor of two when compared to solar values, but increase up to 30–60 times for carbon on Uranus

FIGURE 7.3. Comparative view on the scale of the terrestrial planets and major satellites. Upper row: Earth, Mars, Mercury, the Moon. Middle row: the Galilean satellites Io, Europa, Ganymede and Calisto. Lower row: Venus (atmosphere and surface), Titan (atmosphere and surface), Triton (upper) and Pluto (lower). NASA and ESA images.

and Neptune. Because of the observational bias towards great masses, exoplanets so far discovered are massive (Burrows *et al.* 2001; Hubbard *et al.* 2002). For the cases with measured radius, their mean density ranges from 0.35 to 1.3 g cm^{-3} (Figure 7.6).

As these bodies formed they heated up, and their interiors melted and reorganized into layers of different densities. The energy accumulated during this process has been mostly released in the terrestrial planets and satellites but is still present on the giants Jupiter, Saturn and Neptune (Uranus has lost most of it). These three planets emit an intrinsic flux as important as that received from the Sun (see Table 7.2 and Section 7.4.2). For example, Jupiter's interior cools at a rate ∼1 K per million years while still contracting at a rate of ∼3 cm per year. In the case of Saturn some internal differentiation process is necessary to explain its heat source (helium precipitation within the metallic hydrogen region has been proposed).

The cooling rate (the variation with time of the intrinsic internal luminosity) in giant extrasolar planets depends on the mass and atmosphere opacity and has been

FIGURE 7.4. Comparative view on the scale of the giant and icy planets. Upper row: Uranus (with its ring system and some satellites), Earth (for comparison) and Jupiter. Lower row: Saturn and Neptune. Keck (for Uranus), NASA and ESA for the other images.

theoretically estimated for isolated planets (Burrows et al. 2001). For close-in 'hot jupiter' families, the intense stellar irradiation can directly or indirectly 'inflate' the planet and reduce the cooling rate, or produce a strong hydrodynamic escape of the atmosphere, drastically influencing its evolution, as has been observed in the exoplanet HD 209458b.

Radioactive decay of long-lived elements ^{235}U, ^{238}U, ^{232}Th and ^{40}K (see also Section 7.3.3) is an important source of heat in the Earth's interior, and perhaps in other terrestrial planets and moons. The radioactive heat production rate is given by

$$Q_{\rm rad} = -f \sum_i X_{io} E_i \lambda_i e^{-\lambda_i t}, \qquad (7.7)$$

f being a composition factor (rock mass to total mass ratio per unit volume), X_{io} the initial concentration of the ith type of radioactive nuclei measured in kg of nuclei to kg of rock, E_i the energy released per kg of the i-nuclei upon decay and λ_i the decay constant. For example, a maximum $X_o E \lambda \sim 0.001$ is obtained for the element ^{40}K.

However, on Io and Europa the major internal source of energy is tidal heating, a strong source that is the origin of Io's volcanoes and Europa's young fractured surface

TABLE 7.3 Main internal composition of the terrestrial planets

Planet	Mantle (%)	Core (%)
Mercury	35 (MgO, SiO$_2$)	65 (Fe)
Venus	68 (SiO$_2$, MgO)	32 (Fe)
Earth	70 (SiO$_2$, MgO)	30 (Fe)
Mars	88 (SiO$_2$, MgO)	12 (Fe, S)

Note: main composition in percentage mass.

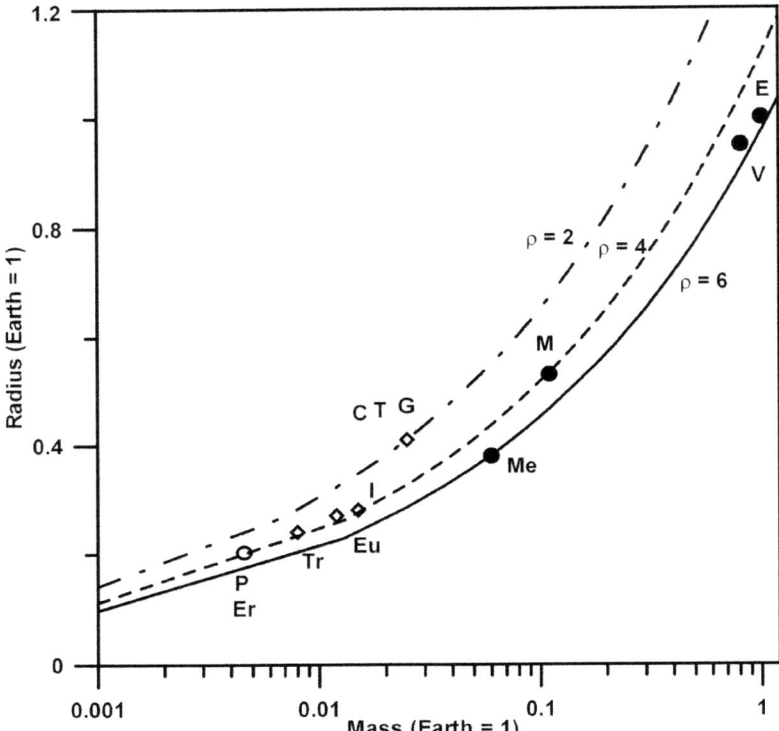

FIGURE 7.5. Radius versus mass distribution in the terrestrial planets and major satellites (in units of Earth values): filled circles for the inner planets according to their initials (Me, V, Ma, E), diamonds for the rest: Galilean satellites (I, Eu, G, C), Titan (T), Triton (Tr), Pluto (P) and Eris (Er). Lines of constant density 2, 4, and 6 g cm^{-3} are drawn.

that lacks large observable craters. The average tidal dissipation rate per unit volume of a body (satellite) is given by

$$Q_{\text{tide}} \approx 0.1 \rho n^5 R^4 e^2 \frac{1}{\mu \psi}, \tag{7.8}$$

where n is the mean orbital motion, e is the orbital eccentricity, μ is the shear modulus (or rigidity) and ψ is a dissipation function that depends on viscosity. It is customary to

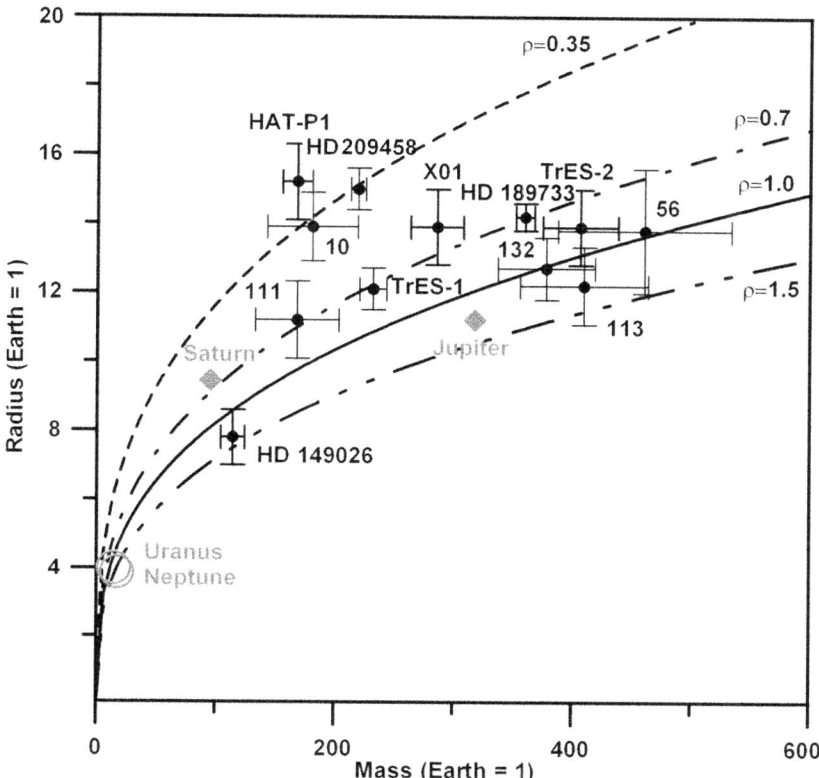

FIGURE 7.6. Radius versus mass distribution for the giant and icy Solar System planets and for the extrasolar planets with known radii from transit observations. Labels on planets that consist only of a number refer to the OGLE transit survey; e.g. 10 = OGLE-TR-10. Lines of constant density ranging from 0.35 to 1.5 g cm^{-3} are drawn.

give the amount of internal friction in terms of a quality factor

$$Q = \frac{2\pi E}{\Delta E}, \qquad (7.9)$$

where ΔE is the change of tidal energy, E, over one cycle of flexure (the lower the value of Q, the more energy is dissipated). For the Earth $Q \sim 13$, and for rocky and icy satellites $Q \sim 100$, but $Q \sim 10^4$–10^5 for gas and icy giants. This is the case for the innermost satellites of Jupiter (Io, Europa and Ganymede).

The internal structure of a planet (its mass distribution) is determined by the hydrostatic equilibrium law:

$$\frac{dP}{dr} = -\rho(r) \frac{GM(r)}{r^2}, \qquad (7.10)$$

P being the pressure and $\rho(r)$ and $M(r)$ the density and mass distribution with radial distance r. The pressure at the centre of a planet can be estimated by integrating the above equation for constant density, to get

$$P_c = \frac{3GM_p^2}{8\pi R_p^4}. \qquad (7.11)$$

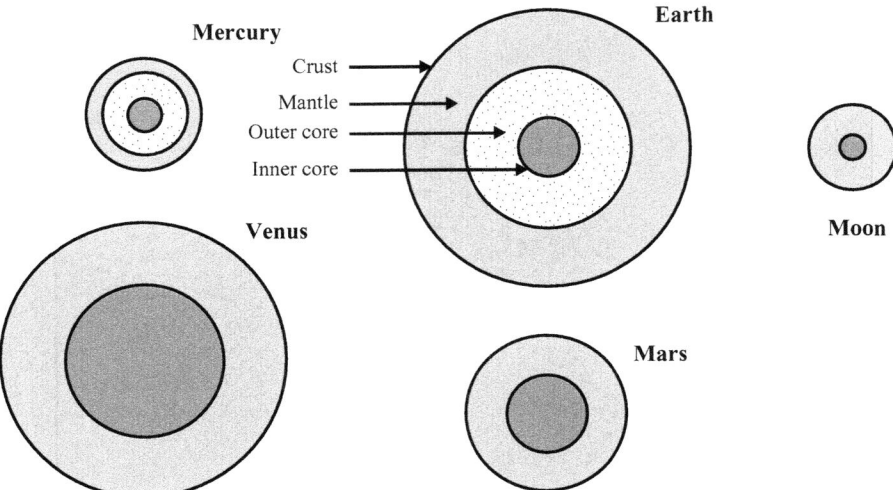

FIGURE 7.7. Layered internal structure in the terrestrial planets and the moon.

This gives $P_c \sim 3.6$ Mbar for the Earth and $P_c \sim 80$ Mbar for Jupiter, together with corresponding temperatures given at the centre of each planet by $T_c \sim P_c/\rho R_g$, leading to $\sim$6000 K (Earth) and 20 000 K (Jupiter).

Differentiation is the process of segregation of zones of different chemical or mineralogical properties of massive bodies. The metallic (e.g. Fe and Ni), densest elements sink towards the centre, giving rise to the existence of three main layers. For terrestrial planets and major satellites there is a *central core* (solid or liquid) where the internal heat resides. Surrounding this there is a *mantle* where heat is transported by conduction or by convection when fluid (as on Earth), and finally a *crust*, a thin, solid outer layer modelled by the internal processes (in the presence of an internal heat source) and by external (impact) activity (Figure 7.7). Geological structures form on the upper surface of the crust (see Table 7.4). *Tectonic* activity develops since the mantle is a moving fluid acting on the crust. But undoubtedly the most conspicuous feature of planetary surfaces are *impact craters* on bodies without a significant atmosphere or internal heat source. This crust is surrounded on most terrestrial planets and major satellites by a gaseous layer, the *atmosphere*, that will be the subject of the following section. Through winds and precipitation, it erodes the surface, a process known as *weathering*, modifying it on a long timescale.

The Galilean satellites differ among themselves. Io has a high density core and a lower density mantle. Europa and Ganymede also have a dense core and a mantle but both contain a subsurface layer of liquid or icy water some 75–150 km thick. In contrast Callisto shows surprisingly little differentiation, although it probably has a subsurface water layer. Titan also probably has a dense core made of iron and silicates surrounded by an icy (water and ammonia) mantle.

For the giant planets there is also internal differentiation, but of a different nature (Figure 7.8). First, there is yet no observational evidence for the existence of a central core in Jupiter, although its presence is suggested by theoretical models of the interior in the case of Saturn (a 'super-Earth' with a mass in the range $\sim$10 to 20 $M_\oplus$). Hydrogen being the main constituent, at least two layers form in both planets: an outer massive *atmosphere* formed basically by molecular hydrogen (H_2) and an internal mantle of highly

TABLE 7.4 Surfaces of terrestrial planets and major satellites

Planet/satellite	Surface composition	Main textures
Mercury	Silicates	Impact craters
Venus	Basalts, granites (?)	Basins & mountains
Earth	Basalts, granites, water	Oceans, polar icecaps, continents & mountains (plate tectonics)
Mars	Basalts, clays, ice	Impact craters, basins, ridges, mountains, volcanoes (inactive)
Io	Sulphur, SO_2 deposits, silicates	Volcanic activity
Europa	Water ice	Cracks, ridges
Ganymede	Dirty water ice	Impact craters
Callisto	Dirty water ice	Impact craters
Titan	Dirty water ice, methane	Methane icy deposits, hills
Triton	Water ice, methane	Impact craters

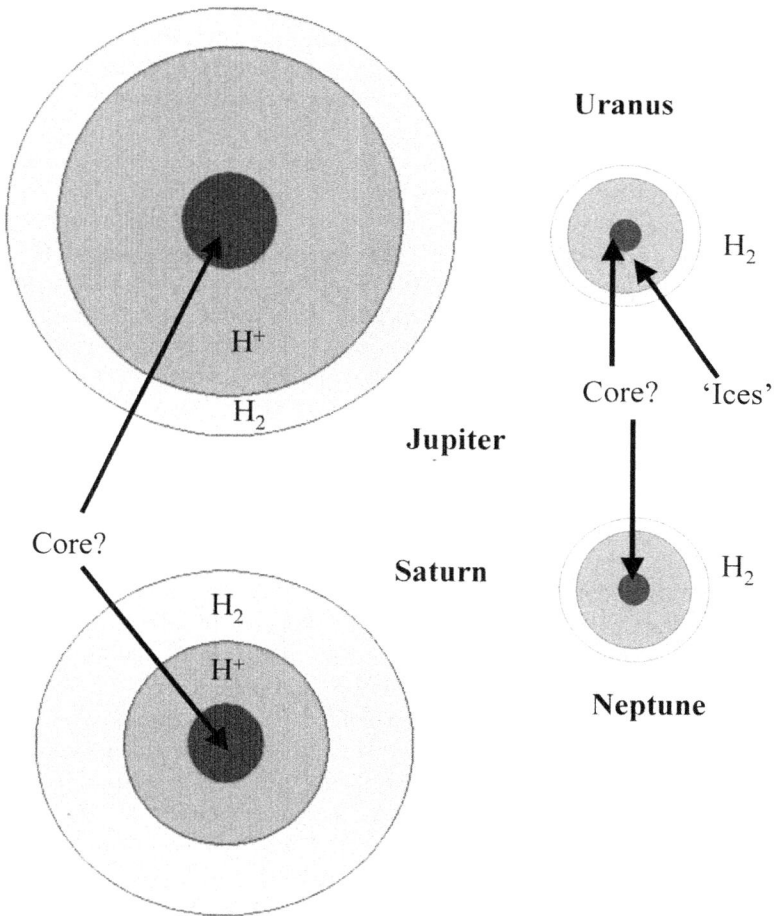

FIGURE 7.8. Layered internal structure in the giant and icy planets.

compressed metallic hydrogen (H$^+$), both in liquid state. Using a polytropic equation of state for the H$_2$ layer,

$$P \sim K\rho^n, \qquad (7.12)$$

(where $K = 2GR_\text{p}^2/\pi n \sim 2$), the internal density distribution within the planet can be calculated using the Lane–Emden equation,

$$\frac{d}{dr}\left(r^2 \rho^{n-2} \frac{d\rho}{dr}\right) = -\frac{4\pi G}{nK} r^2 \rho, \qquad (7.13)$$

to get

$$\rho(r) = \left(\frac{\pi M}{4R^3}\right) \frac{\sin(\pi\eta)}{\pi\eta}, \qquad (7.14)$$

where $\eta = D/R$, D being the thickness of the atmospheric layer. Laboratory experiments and theoretical quantum mechanical calculations show that transition from H$_2$ to H$^+$ occurs at $P \sim 1.5$–2 Mbar ($T \sim 6000$ K). Using this boundary condition, one can get the size of the atmospheric layer, η, for a given planetary mass M. This reasoning can also be used for the extrasolar giant planets so far discovered (Sánchez-Lavega et al. 2004).

The two icy planets (Uranus and Neptune) also have a similar structure, with an H$_2$ + He atmosphere, but the intermediate 'mantle' is now probably formed by compressed 'ices', an ionic 'sea' of methane, water and ammonia, perhaps mixed with rocks and some metallic hydrogen.

There is a general consensus on the origin of the terrestrial planets as a result of 'planetesimal accretion', but not for giants, where two theories are in dispute. The first assumes an early planetesimal accretion to form a super-Earth followed by runaway accretion of light gases from the protoplanetary nebula. The second hypothesis postulates the direct formation of a gaseous giant from a gravitational instability in the protoplanetary disc. This is the subject of other chapters of this book so I dwell no further on this. With respect to the origin of satellites, three basic scenarios have been at work: (1) *in situ* formation by accretion, as for the major satellites of the giant planet; (2) gravitationally captured bodies, such as Phobos and Deimos in the case of Mars and the external retrograde satellites of the giant planets (including the large satellite Triton of Neptune); and (3) the cataclysmic formation by impact on the planet of a protosatellite, followed by fragmentation and mixing, as was the case of the Moon and probably of the satellite Charon.

7.3.2. *Asteroids, Kuiper belt objects and comet populations*

These three populations are located at different distances from the Sun, as explained previously, and amount to a total mass of $\sim 1\ M_\oplus$. Because of their intermediate size and large numbers, their distribution and evolution are strongly subjected to the orbital mechanisms described in Section 7.2.1 and to collisions among themselves.

At present there are about 30 000 asteroids classified with determined orbits, mainly with distances between 2.1–3.2 AU from the Sun (resonances 4:1 to 2:1). An asteroid update catalogue can be obtained at the web address: cfa-www.harvard.edu/cfa/ps/mpc.html. Their sizes range from 950 km (for the largest, Ceres) to a few kilometres (Figure 7.9). They show a size distribution of the type $N(r) = N_0(R/R_0)^{-n}$ with $n = 3.5$ and R the average size. About 25 asteroids have sizes above 200 km. There is a large distribution in densities, ranging from carbonaceous objects of density $\rho = 1.2$ g cm^{-3} to metallic objects of density $\rho = 4$ g cm^{-3}, porosity playing a key role in the density distribution. The basic surface mineralogy comes from colours and spectra

FIGURE 7.9. Visual aspect on the scale of the asteroids Mathilde, Gaspra, Ida and Eros (the bar on the right corresponds to a length of 5 km). NASA images.

and divides the asteroids in two great groups: *carbonaceous C-type* (40% of all asteroids, low albedos ~0.05, composed of hydrated silicates with carbon and organics), and *stony S-type* (30–35%, albedos ~0.15, composed of olivine and pyroxene mixed with Fe and Mg). The rest are of various classes: D and P (low albedo, carbon and organic rich) and M (moderate albedo, metallic). The asteroids presumably grew from planetesimals but their mass was not great enough (although it was probably much greater in earlier times) to accrete and form a planet, presumably because of orbital perturbations by Jupiter.

Between the orbits of Saturn and Neptune (10–30 AU) there is a group of intermediate objects collectively known as *centaurs* located in chaotic planet-crossing orbits with short lifetimes ($\sim 10^6$–10^8 years). The best representative is Chiron, an object that shows comet-like activity.

There are about 700 trans-Neptunian or Kuiper belt objects catalogued, most of them lying within 50 AU. Dynamical calculations show that this population was much greater after the formation of the Solar System (reaching perhaps up to 10 $M_\oplus$) but because of interaction with the giants, particularly to the migration of Uranus and Neptune, about 99% of them have been expelled from the Solar System. They are classified according to their orbital characteristics into different groups: '*plutinos*' (in the orbit of Pluto at resonance 3:2 with Neptune and having a wide range of eccentricities ~0.1–0.35), and *classical, scattered* (high eccentricities ~0.4–0.8) and *resonant* populations. The largest of these objects known to date are Sedna (diameter ~1600 km) and Quaoar (~1300 km) although their diameters are uncertain because of their unknown albedos.

The most distant population of Solar System objects is that of the comets. Because of their large distances from the Sun (up to ~10^5 AU) in the Oort cloud, and their very low density, they could not have formed at such distances but probably evolved as icy planetesimals in the region of the giants, particularly Jupiter, which later ejected them. Their orbits are highly elliptical and are classified as *long-period* ($P > 200$ years, active with high eccentricities and orbits perturbed by the planets) and *short-period* ($P < 200$ years). At present, about 900 comets have well-determined orbits. Cometary *nuclei* have sizes ~1–20 km and very low albedos ~0.03 (Figure 7.10). They have densities in the range 0.2–1 g cm^{-3} and are composed mainly of water ice (~80%) with a mixture of CO and a large family of molecules with parent elements C, H, N and O. They are fragile and can be broken into fragments by the tidal forces of larger bodies, as occurred with Comet Shoemaker–Levy 9 before its impact with Jupiter in 1994. When the comet

FIGURE 7.10. Visual aspect of the nuclei of Comets Halley, Wild 2 and Borrelly, and of Comet Hale-Bopp showing the two types of tails. ESA and NASA images.

nucleus approaches the the Sun its surface temperature rises according to the $r^{-1/2}$ law, and at distances 2–3 AU volatile icy material sublimates at a rate proportional to T^{-4} according to the Stefan–Boltzmann law. The molecules escape as soon as their thermal velocity exceeds the escape velocity:

$$\left(\frac{8kT}{\pi m_0}\right)^{1/2} > \left(\frac{8\pi G}{3}\rho\right)^{1/2} R, \qquad (7.15)$$

k being the Boltzmann constant, m_0 the mass of the molecule and R the radius. They form a *coma* as the gases expand, and the parent molecules suffer photodissociation and photoionization forming a large number of radicals and ions. The nucleus and coma form the *head* of the comet. The ejection of gases and dust typically form two long tails that reach in some cases lengths $\sim 10^7$–10^8 km. (1) One tail is composed of neutral particles of dust (the *dust tail*) driven away from the nucleus by radiation pressure and solar gravitation forces, the ratio of the two forces being

$$\frac{F_{\rm rad}}{F_{\rm grav}} \propto \frac{1}{\rho a}, \qquad (7.16)$$

ρ and a being the density and size of dust particles (typically $a \sim 0.1$–1 μm). (2) The second tail is the *plasma or ion tail* formed by ions which are bound to interplanetary magnetic field lines and driven in the anti-solar direction by the solar wind. This tail has a bluish-yellowish colour due to fluorescent emission by CO^+ ions.

Asteroids and comets are the origin of most impact craters found on the surfaces of Solar System bodies. They have also a great influence on the evolution of life on Earth. The diameter D of a large impact crater can be used to estimate the associated impact energy through the empirical relationship

$$D \approx D_0 \left(\frac{E}{E_0}\right)^{0.294}, \qquad (7.17)$$

where $D_0 = 15$ km and $E_0 = 10^{20}$ J.

7.3.3. Meteorites and planetary rings

Meteorites are metre-size or smaller objects, similar in type to asteroids, that fall to Earth at speeds ~ 15–30 km s^{-1}. About 26 000 of them hit our planet each year.

Meteorites differ according to their composition: *irons* or metallic, *stones* similar to terrestrial rocks, and mixtures of these, the *stony-irons*. These come from differentiated parent bodies. The other great group (the most abundant) are the *chondrites*, so-called because they contain small quasi-spherical igneous inclusions known as chondrules. These are undifferentiated *primitive meteorites* that formed during the early life of the Solar System with a composition of refractory elements similar to that of the Sun. *Carbonaceous chondrites* are a particular type that contain several per cent of carbon by mass. Some meteorites are of lunar and martian origin, according to their composition, but most of them come from the asteroid belt. The radiometric dating of meteorites, based on the decay of radioactive elements, allows the determination of the age of the Solar System currently fixed in 4570 Myr. For example, the decay of ^{87}Rb to stable ^{87}Sr over a time t follows the relationship

$$(^{87}{\rm Sr}/^{86}{\rm Sr}) = (^{87}{\rm Sr}/^{86}{\rm Sr})_0 + (e^{\lambda t} - 1)(^{87}{\rm Rb}/^{86}{\rm Sr}), \qquad (7.18)$$

where the subscript 0 refers to the initial ratio at time t in the past and λ is the decay rate of ^{87}Rb. Measurements of these ratios allow us to extract the age t since λ is well known. The most accurate absolute age is obtained from the analysis of the ^{207}Pb/^{206}Pb ratio in a class of carbonaceous chondrites called *CAI* owing to the presence of 'calcium–aluminium inclusions'.

Micrometeorites are small dust particles of micron size distributed in the interplanetary medium that are observed as *zodiacal light* caused by the scattering of the sunlight. They are concentrated in the plane of the ecliptic in the inner part of the Solar System. In addition there are orbital concentrations of small particles called *meteor swarms* that follow the orbits of old or extinct comets. When colliding with the Earth they form *meteor showers*, apparently emerging from a single convergent point in the sky called the *radiant*, leaving a trail when becoming incandescent during the frictional entry into the Earth's atmosphere.

Finally, populations of small particles are found on the four giant planets surrounding their equatorial planes. They form *rings*, extending up to 2–3 planetary radii (Table 7.1), and are composed of a myriad of particles (micron and mm in size) and small blocks of material with a radial distribution in major rings and *ringlets* due to the gravitational action of the satellites. The most spectacular and massive are those of Saturn with a total mass of $\sim 10^{-8}$ Saturn masses. The rings of Saturn and Uranus contain the largest particles; for Saturn they are formed by water ice and can reach sizes ~ 10 m in the B and A rings, and 1 m for Uranus in its ε ring. There are two possible origins for the ring systems: (a) a large body that was broken apart after penetration of the Roche limit and the ensuing tidal disruption; (b) the fragments of a non-accreted satellite. It is tempting to suggest that rings should also be present in extrasolar giants, although their future detection will be possible only in the densest and most massive cases.

7.4. Planetary atmospheres

Atmospheres are the gaseous outer layers surrounding astronomical bodies, most often their directly observable parts. Basically most of the planets and great satellites with a mass large enough to retain volatiles against the action of escape processes (thermal or Jeans, see Eq. (7.13), impact or catastrophic, solar wind and EUV fluxes) and have an atmosphere, although its mass (manifested by its pressure at the base) varies by orders of magnitude from one body to another. Accordingly, we may make a first classification of atmospheres based on mass: (1) *thin atmospheres* ('exospheres') with low boundary surface pressure $P_s < 10^{-5}$ bar (e.g. Mercury, the Galilean satellites Io, Europa, Ganymede and Callisto, Triton and Pluto); (2) *intermediate atmospheres* with $P_s = 10^{-3}$–90 bar (e.g. Venus, Earth, Mars and Titan); (3) *massive and deep atmospheres* with $P_s >$ kbar–Mbar (e.g. Jupiter, Saturn, Uranus and Neptune). In Tables 7.5 and 7.6 we present a summary of the main properties of the atmospheres of the planets and satellites.

7.4.1. Origin and chemical composition

The chemical composition of an atmosphere is directly related to the origin of the planet. Thin atmospheres have their origin from the strong surface interaction with solar radiation and particles from the solar wind and magnetosphere (sputtering processes). Volcanic activity on Io is responsible for its sulphurous atmosphere. For Pluto and Triton the main gaseous supplier is the condensation–sublimation cycle of methane following insolation. Current terrestrial planetary atmospheres are of a secondary nature, i.e. they have evolved from original atmospheres formed following the accretion of planetesimals and the outgassing of volatiles during this high temperature phase. Direct surface interaction and tectonic activity, photochemistry and, in the case of the Earth, biological activity have been the major processes leading to present-day atmospheres. In contrast, the giant

TABLE 7.5 Atmospheric properties of terrestrial planets and major satellites

Planet/satellite	Composition (% mass)	P_s [bar]	T_s (K)	τ_{rad} [years]	Clouds composition (coverage)	Cloud optical depth	Adiabatic gradient [K/km]
Mercury	He (42), Na (42), O (15), H, K	10^{-12}	700 to 93[a]	—	Exosphere	—	—
Venus	CO_2 (96), N_2 (3)	90	730	10.8	SO_4H_2 (100%)	5–12	10.5
Earth	N_2 (77), O_2 (21) H_2O (100 ppm)	1	⟨288⟩[b]	0.08	H_2O (50%)	5–10	9.8
Mars	CO_2 (95), N_2(3), Ar (2)	0.007	⟨218⟩[b]	0.002	H_2O, CO_2, dust (10–100%)	0.3–6	4.5
Io	SO_2 (O_2, SO), (Na, K, S+, ...)	10^{-7}	120	—	Volcanic dust	0.1	—
Europa	O_2, (Na)	10^{-11}	95	—	Exospheres	—	—
Ganymede	O_2						
Callisto	O_2, CO_2						
Titan	N_2 (95), CH_4 (5)	1.5	92	7.9	CH_4 (clouds) C_nH_n (hazes)	1–10	1.3
Triton	N_2 (99)	10^{-5}	38	—	$CH_4 - N_2$ (?)	0.1–0.3	—
Pluto	CH_4, N_2	10^{-4}	40	5×10^{-4}	CH_4	>0.15	—

Notes: [a] Insolated–night hemispheres; [b] Average values.

TABLE 7.6 Atmospheric properties of the giant planets

Planet	Composition	P [bar]	T(K)	τ_{rad} [years]	Clouds	Optical depth	Adiabatic gradient [K/km]
Jupiter	H_2 (90), He (10) CH_4 (2000 ppm) NH_3 (300 ppm)	0.42	125	4.8	Hazes NH_3 NH_4SH H_2O	0.1–1 1–7	1.96
Saturn	H_2 (97), He (3) CH_4 (2000 ppm) NH_3 (300 ppm)	1.1	95	24.7	Hazes NH_3 NH_4SH H_2O	10–15	0.7
Uranus	H_2 (83), He (15) CH_4 (2)	1.2	59	142.9	Hazes CH_4	0.1–1 0.5–2	0.67
Neptune	H_2 (83), He (15) CH_4 (2)	1.5	59	113.2	Hazes CH_4	0.1 0.3–1	0.85

Note: Pressure, temperature and radiative time constant are given at the level of the upper main cloud formation (ammonia for Jupiter and Saturn, and methane for Uranus and Neptune).

planets have primary, nearly primordial, atmospheres that were acquired directly from the protoplanetary nebula.

The composition by mass (molecular weight) depends primarily on the temperature to planetary mass ratio, T/M. Lighter elements are expected to be present on massive and colder giant planets. Under thermodynamic equilibrium and from a solar element composition (dominated by H, He, C, N, O), the following reactions indicate the primary dominant molecules in an atmosphere:

$$CH_4 + H_2O \leftrightarrow CO + 3H_2, \quad (7.19a)$$

$$CO + H_2O \leftrightarrow CO_2 + H_2, \quad (7.19b)$$

$$2NH_3 \leftrightarrow N_2 + 3H_2, \quad (7.19c)$$

$$(T \text{ low}, P \text{ high}) \leftrightarrow (T \text{ high}, P \text{ low}).$$

The final relation indicates how temperature and pressure shift the reactions to the left or right.

7.4.2. Energy sources

Since planets and satellites rotate, there is first a mechanical (kinetic) energy in the atmosphere which varies as $\sim \Omega^2 M_{\text{atm}} D^2$, Ω being the rotation angular velocity, M_{atm} the atmospheric mass and D the atmospheric vertical extent. There is a second, thermal, energy source that can be of external or internal origin. For the external case (radiation), the flux and corresponding equilibrium temperature are given by

$$F_{\text{abs}} = \frac{(1 - A_b)}{4} \frac{L_\odot}{4\pi a^2}, \quad (7.20a)$$

$$T_{\text{eq}} = \left(\frac{F_{\text{abs}}}{\sigma} \right)^{1/4}. \quad (7.20b)$$

Here A_b is the planet's albedo, $L_\odot$ the solar luminosity, a is the distance and σ the Stefan–Boltzmann constant. The internal energy flux is given by

$$F_{\text{int}} = \frac{L_{\text{int}}(t)}{4\pi R_p^2}, \qquad (7.21)$$

where $L_{\text{int}}(t)$ is the internal luminosity, which is a function of time, as discussed in Section 7.3.1. When both sources are present it is convenient to define the planet's effective temperature:

$$T_{\text{eff}} = \left(\frac{F_{\text{abs}} + F_{\text{int}}}{\sigma}\right)^{1/4}. \qquad (7.22)$$

These energy sources heat the atmosphere and are the main drivers for their motions. A part of the energy budget is reflected (visible channel) and reradiated (infrared channel) into space (Figure 7.11a and 7.11b, central panel) and another important part is dissipated by frictional processes (internally in the atmosphere or with a surface).

Synthetic spectra for the visual and near infrared wavelengths have been presented by a number of authors for different types of giant extrasolar planets (e.g. Burrows *et al.* 2001; Burrows 2005). The spectra depend primarily on the mass and age of the planet, and on the temperature and composition of its atmosphere.

7.4.3. *Vertical thermal structure*

Since the bulk atmospheric composition is mainly molecular, these gases are radiatively active in the infrared, controlling the atmospheric temperature (e.g. CO_2, CH_4) and giving rise in some cases to a greenhouse effect. Other compounds are not so active but contribute significantly to atmospheric mass and pressure (e.g. N_2, He). Finally, some components become condensable at atmospheric temperature–pressure values, forming clouds and releasing latent heat (e.g. in order of decreasing temperatures for usual planetary conditions: H_2O, NH_3 and CH_4). According to the *Clausius–Clapeyron* equation, clouds will form whenever the partial pressure P_p of the vapour exceeds the saturation vapour pressure $P_v(T)$:

$$P_p = X_c P(T) \geq P_v(T), \qquad (7.23)$$

where X_c is the condensable mixing ratio and $P(T)$ the actual temperature profile in the atmosphere (see Figure 7.12). This simple relationship has been used to infer to a first order the basic composition of the expected clouds in giant extrasolar planets, and their reflective properties under certain assumptions (Burrows 2005 and references therein). According to the expected cloud structure, Sudarsky *et al.* (2003) have classified the giant exoplanets in five classes (from I –'Jupiter-like' to V –'hot jupiter'). At the high effective temperatures of hot jupiters (>1000 K), the clouds are expected to be formed by solid grains of iron (Fe) and silicates such as enstatite ($MgSiO_3$) and forsterite (Mg_2SiO_4). The presence of clouds affects planetary albedos and tends to smooth the spectral features in their reflection spectra.

Vertically, the atmosphere can be assumed on average to be in hydrostatic equilibrium,

$$\frac{dP}{dz} = -\rho g. \qquad (7.24)$$

Making use of the perfect gas law equation (R_g^* is the specific gas constant),

$$P = \rho R_g^* T, \qquad (7.25)$$

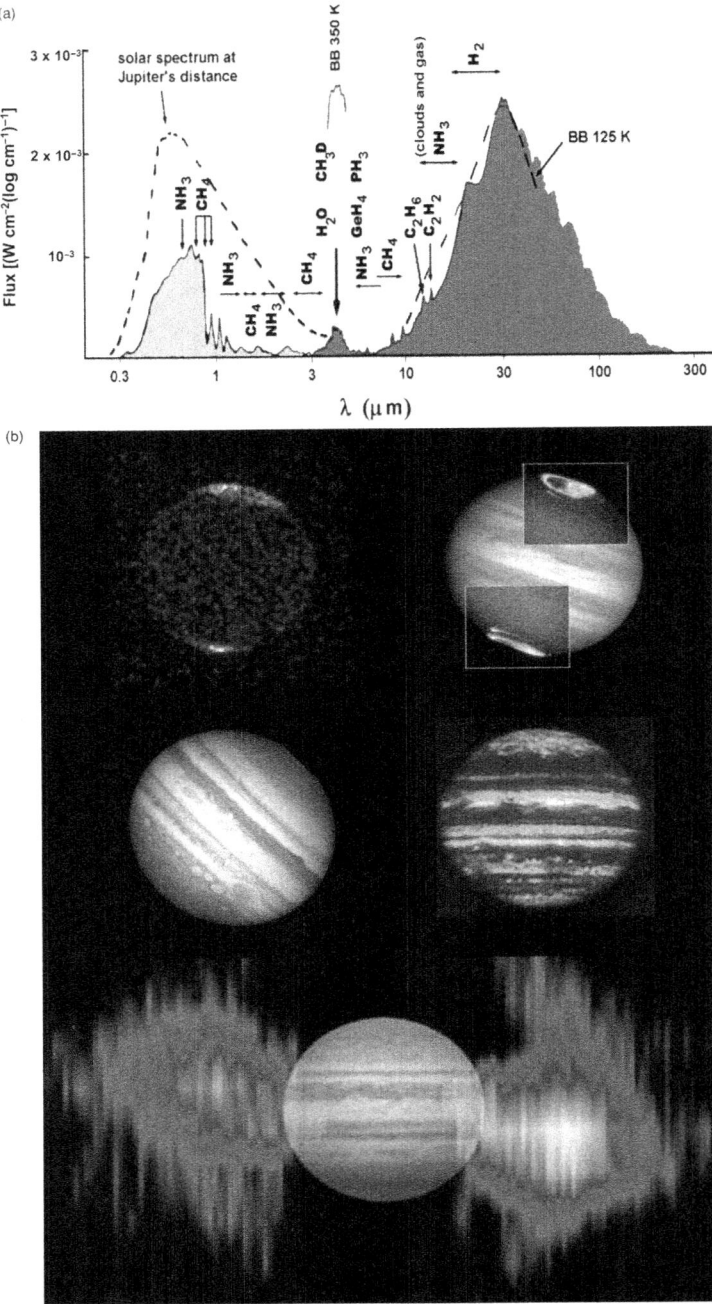

FIGURE 7.11. (a) Jupiter's spectrum in the visual and infrared with the absorbing molecules indicated. The theoretical reflected spectrum (dashed curve) is on the left with the lightly shaded area representing the fraction of the sunlight absorbed by the planet. The emission infrared spectrum (darkly shaded area) is fitted by a black body emission law for a temperature of 125 K. The emission peak close to 5 microns comes from lower (hotter) jovian regions basically concentrated in the equatorial region (called 'hotspots'). Adapted from Hanel et al. (1981). (b) Composite images of Jupiter as seen at different wavelengths. Upper row in X-ray and ultraviolet (superposed on a visible Jupiter image) showing the aurorae. Middle row in the visual (left) and infrared 5-microns (right) showing the clouds in reflected and transmitted light, respectively. Lower row at radio wavelengths, showing the radiation belts.

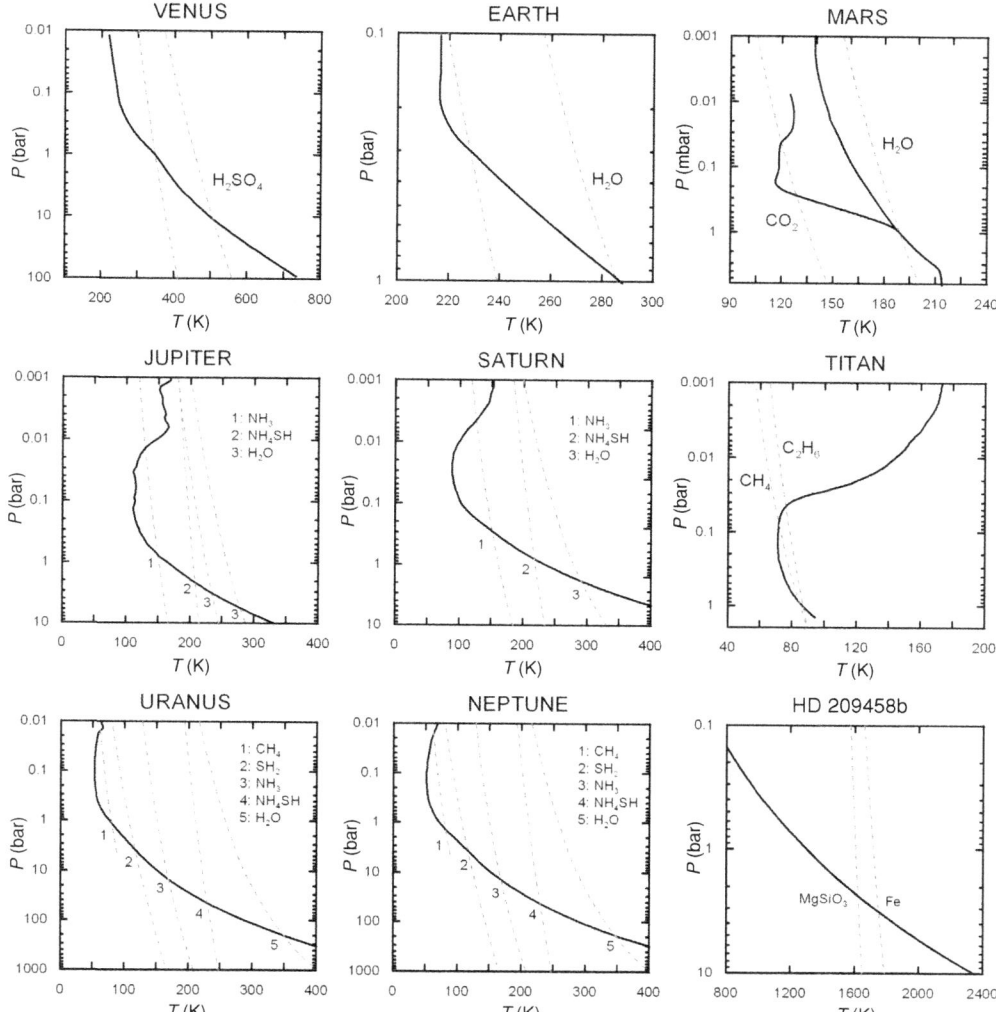

FIGURE 7.12. The continuous black lines trace the temperature versus altitude (pressure) in the atmospheres of planets, the satellite Titan and the extrasolar planet HD 209458b. The dashed lines show the vapour pressure curves for the abundances of the condensable species measured for the planets. The crossing point marks the altitude where the clouds of such compounds form. From Sánchez-Lavega et al. (2004).

we see that the pressure varies with altitude for isothermal conditions as

$$P(z) = P_0 \exp\left(-\frac{z}{H}\right), \tag{7.26}$$

where $H = R_g^* \, T/g$ is the density scale height. This parameter must be determined as a function of altitude, z, since real atmospheres are not isothermal (Figure 7.12).

The vertical temperature distribution within the atmosphere can be determined by first assuming radiative–convective equilibrium,

$$T(z) = T_0 + \Gamma(z - z_0), \tag{7.27}$$

where T_0 is a reference temperature (for terrestrial planets it corresponds to the surface temperature) and Γ is the vertical temperature gradient. We can obtain the temperature gradient in a convective atmosphere using the adiabatic relationship $T^\gamma P^{1-\gamma} =$ constant, where $\gamma = C_p/C_v$ is the ratio between the specific heats at constant pressure and constant volume. Differentiation with altitude gives

$$\frac{1}{T}\frac{dT}{dz} = \left(1 - \frac{1}{\gamma}\right)\frac{1}{P}\frac{dP}{dz}. \tag{7.28}$$

Using the hydrostatic relationship and the relationship $C_p - C_v = nR_g$, where n is the number of moles and R_g the gas constant, we get the dry adiabatic gradient, $\Gamma = dT/dz = -g/C_p$. Hence it is expected that the temperature decreases with altitude. However, a temperature inversion (where temperature increases with altitude) occurs in planetary atmospheres. The inversion takes place due to the absorption of radiation by different constituents (e.g. ozone on the Earth, hydrocarbons and aerosols in the giant planets). The inflexion point is known as the *tropopause* (most atmospheres have this point close to ~ 100 mbar) and divides the atmosphere into two main layers (Figure 7.12): below this point we find the *troposphere* ($dT/dz < 0$) and above it the *stratosphere* ($dT/dz > 0$). In the upper reaches of the atmosphere the density is very low and other layers are present (called, in ascending order, the mesosphere, the thermosphere and the ionosphere). The *ionosphere* is the layer of the atmosphere in which an ionized plasma forms as a consequence of the photoionization produced by solar extreme ultraviolet radiation. For example, on Earth the ionosphere is located at a mean altitude of $z \sim 300$ km, where the electron density $N(e^-) \sim 10^6$ cm^{-3}.

This behaviour is also expected from radiative–convective models, in the atmospheres of extrasolar giant planets (see Figure 7.12, last panel, and Burrows *et al.* 2001; Burrows 2005).

7.4.4. Dynamics: general circulation

Atmospheric dynamics can be described by three basic equations.
(1) The Navier–Stokes equation in a rotating frame

$$\rho\frac{\partial \boldsymbol{u}}{\partial t} + \rho(\boldsymbol{u}\nabla)\boldsymbol{u} + 2\rho\boldsymbol{\Omega} \times \boldsymbol{u} = -\nabla P + \rho \boldsymbol{g}_\text{eff} + \nu\nabla^2\boldsymbol{u}, \tag{7.29}$$

where $\boldsymbol{u}$ is the wind velocity, ν is the kinematic viscosity and the effective gravity $\boldsymbol{g}_\text{eff}$ includes the centrifugal term (planetary rotation).
(2) The thermodynamic equation (second law)

$$\frac{\partial T}{\partial t} + (\boldsymbol{u}\nabla)T = K\nabla^2 T + H_i, \tag{7.30}$$

where H_i includes the heating and cooling terms (see Section 4.2) and K is the thermal conductivity.
(3) The equation of continuity:

$$\frac{\partial \rho}{\partial t} = -\nabla(\rho\boldsymbol{u}). \tag{7.31}$$

This set of non-linear equations subjected to the appropriate boundary conditions can be numerically integrated using different approaches to retrieve the three-dimensional temporary dependent wind field $\boldsymbol{u}(\boldsymbol{r}, t)$.

Our first consideration is that the terrestrial planets and Titan have *shallow atmospheres* since $D \sim H \ll R$ and the global flow can be considered two-dimensional (horizontal) with zonal (west–east) and meridional (north–south) components. The basic nature of the flow can be addressed by quantifying the Rossby number, the ratio of the inertial acceleration to the Coriolis force in Eq. (7.29),

$$\text{Ro} = \frac{u}{fL}, \qquad (7.32)$$

where $f = 2\Omega \sin\varphi$ is the Coriolis parameter (φ is the latitude and L a typical scale of motion). In rapidly rotating bodies (Earth, Mars), Ro < 1 and the Coriolis force equilibrates the pressure gradient force. The system is then under *geostrophic* balance, the wind flows along isobars, and its speed can be estimated from the *thermal–wind equation* (the meridional temperature gradient produces a vertical wind shear) obtained from Eqs. (7.29) and (7.30),

$$\frac{\partial u}{\partial z} = -\frac{g}{fR_\text{p}T} \frac{\partial T}{\partial \varphi}. \qquad (7.33)$$

Integration of this equation vertically allows us to estimate the geostrophic wind speed,

$$u_\text{g} \approx -\frac{gH}{fL} \frac{\Delta T}{T}, \qquad (7.34)$$

where ΔT is the meridional temperature difference over a distance L. In slowly rotating bodies (Venus, Titan), Ro > 1 and the centrifugal force balances the pressure gradient force. The system in then in *cyclostrophic* balance, and the zonal velocity can be estimated from (7.29) and (7.30) as

$$u_\text{cl} = \pm\sqrt{\frac{gH\Delta T}{T}}. \qquad (7.35)$$

The measurements in the terrestrial planets of the dependence of the zonal wind velocity u with latitude are shown in Figure 7.13 and in Figure 7.14 we show their dependence with altitude.

On Mars there are in addition two other large-scale peculiar global motions: (a) a CO_2 *condensation–sublimation flow* that follows the seasonal insolation cycle and transports the condensates from pole to pole (this makes Mars' atmosphere change its surface pressure seasonally by 20%); (b) *thermal tide winds* produced by the strong temperature differences between the day and night hemispheres. Thermal tides are important in the tenuous atmosphere of Mars because of its short radiative time constant, τ_rad (roughly the time needed for a gas parcel to heat up or cool down by absorbing or emitting infrared radiation),

$$\tau_\text{rad} = \frac{C_\text{p}P}{\sigma g T^3}. \qquad (7.36)$$

Venus and Titan are in cyclostrophic balance and show a phenomenon called *superrotation* because the rotation period of the atmosphere is much lower than that of the planet (on Venus the upper atmosphere rotates in 4 days and the solid surface in 245 days). The wind velocity reaches 100 ms^{-1} at cloud level ($z = 70$ km) on Venus and blows from east to west, and 120 ms^{-1} at the same altitude on Titan but now flowing from east to west, in both cases in the sense of the planet's rotation.

In addition, there are meridional motions resulting from the heat imbalance between latitudes. Hot air close to the surface rises and flows along meridians towards regions

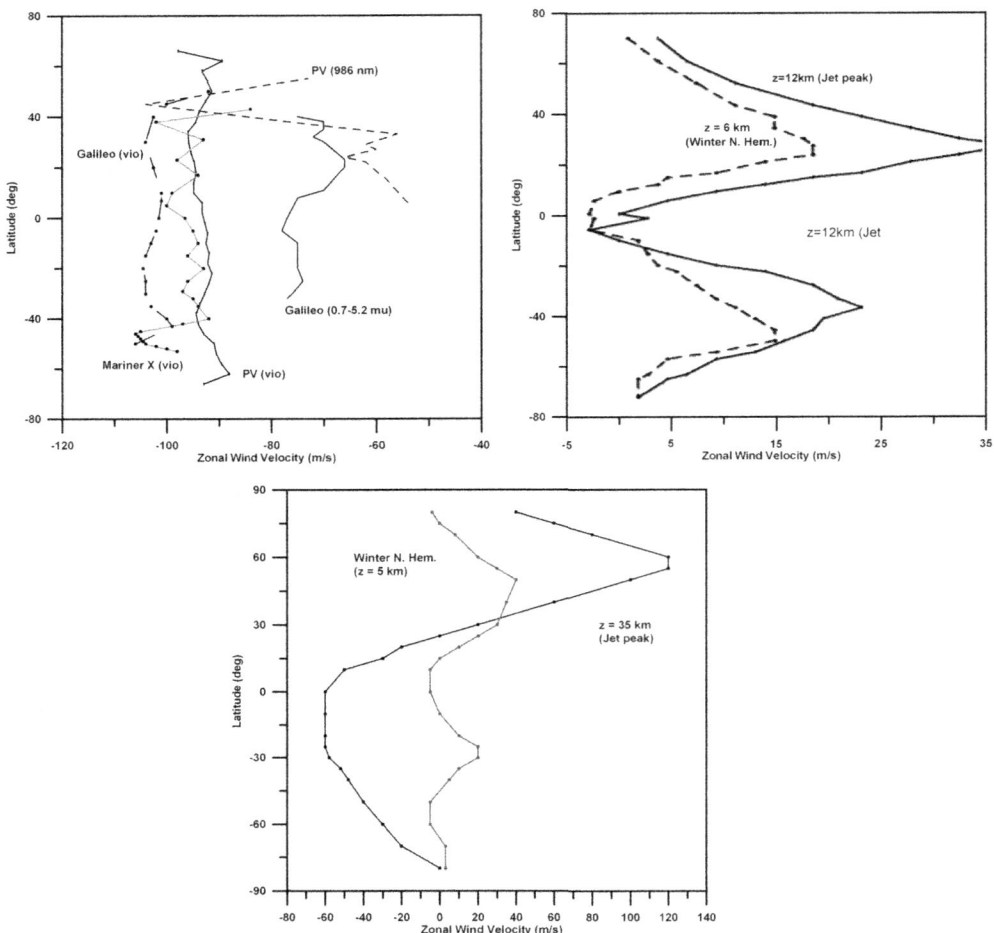

FIGURE 7.13. Zonal wind velocity profiles in the atmospheres of Venus (top left), Earth (top right) and Mars (top bottom). For Venus the data correspond to the two altitude levels, one at the maximum of the westward superrotation at $z = 70$ km corresponding to different dates and spacecraft (violet filters: *Mariner X*, *Pioneer-Venus* and *Galileo*). The second series corresponds to altitudes $z \sim 50$ km (*Pioneer-Venus*, 986 nm filter; *Galileo* in the range 0.7–5.2 microns). For the Earth the average zonal profiles are shown for altitudes $z = 6$ km and $z = 12$ km (where the jet peak forms) corresponding to the northern winter hemisphere epoch. For Mars the profiles correspond to altitudes $z = 5$ and 35 km (jet peak) in the northern winter hemisphere season.

of lower pressure; it then cools, subsiding and returning to lower altitudes. This motion is referred to as a *Hadley cell* circulation. The velocity of the meridional winds, v, can be roughly estimated from a balance between the radiative heating and the adiabatic cooling using Eq. (7.30):

$$v \approx \frac{\left(\dfrac{g}{P_0}\right) \sigma T_{\text{eff}}^4}{\left(\dfrac{C_{\text{p}}}{g}\right) T N^2}. \tag{7.37}$$

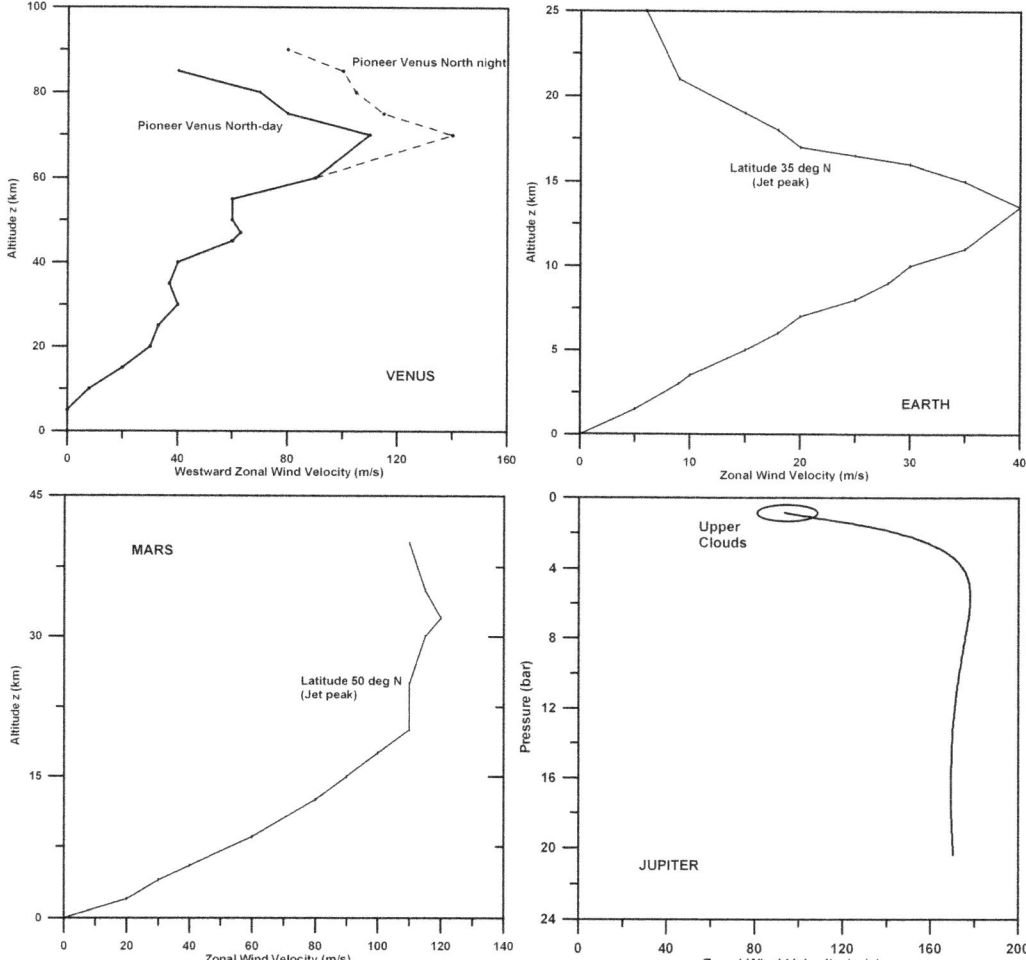

FIGURE 7.14. Plots of the vertical zonal wind velocity profiles in the atmospheres of Venus, Earth, Mars and Jupiter. For Venus the upper part of the profile differs in the northern day and night hemispheres as measured by *Pioneer-Venus* probes. For Earth and Mars the profiles correspond to the latitudes of the jet peaks (see Figure 7.13). For Jupiter the profile corresponds to a hotspot region at 7° North latitude of Jupiter as measured by the *Galileo* probe.

P_0 is the pressure level at which most of the solar radiation is absorbed and N is the Brunt–Väisälä frequency,

$$N^2 = \frac{g}{T}\left(\frac{\mathrm{d}T}{\mathrm{d}z} + \frac{g}{C_\mathrm{p}}\right), \qquad (7.38)$$

a parameter that indicates the degree of stability to vertical motions in the atmosphere. Typically, meridional velocities related to this mechanism are of the order $\sim$1–10 ms^{-1} in the terrestrial planets. The vertical velocity, w, scales accordingly:

$$w \sim v(H/L), \qquad (7.39)$$

where L is the meridional distance of the cell which depends on the rotation rate of the planet (which increases as Ω decreases). For example, on Venus $L \sim \pi R_p$ (from northern to southern regions) but on Earth and Mars $L \sim \pi R_p/3$ (centred on the equator) so the cell is restricted to the tropics.

The geostrophic condition applies to terrestrial and martian mid-temperate latitudes, with an eastward zonal jet per hemisphere. In tropical and equatorial latitudes the Hadley cell mechanism dominates, with the weak Coriolis force at these latitudes tilting the surface winds in the westward direction (trade winds). The Hadley cell mechanism, according to models, is directly involved in the superrotations of Venus and Titan.

The circulation in giant and icy planets is dominated at cloud levels ($P \sim 0.3$–2 bar) by an east–west (zonal) jet pattern alternating in direction with latitude. On Jupiter there are eight jets per hemisphere, four on Saturn, and one on Uranus and Neptune. However, a distinctive feature (Figure 7.15) is that the intense equatorial jet flows eastward on Jupiter ($u \sim 100$ ms^{-1}) and Saturn ($u \sim 450$ ms^{-1}), but strong westward equatorial jets are present on Uranus ($u \sim -100$ ms^{-1}) and Neptune ($u \sim -400$ ms^{-1}). The nature of the general circulation on the giant planets is unknown. The lack of a surface raises the basic preliminary question of how deep the winds extend below the clouds. In addition, because the giants, except Uranus, have an internal energy source as important in strength as the solar radiation (but both a thousand times weaker than on Earth), the second question is what energy source drives the motions and how this low energy generates wind speeds ten times higher than those on the Earth. Finally we do not know how the intense eastward equatorial jets on Jupiter and Saturn form or how they are maintained.

There are at present two basic hypotheses to explain these motions. One relies on the dominance of solar radiation acting on a thin upper shallow layer (depth $D \ll R$). In essence these models are a parametric version of Earth General Circulation Models (GCM). Under the model conditions, the potential vorticity, q, defined as

$$q = \frac{\zeta + f}{h}, \tag{7.40a}$$

$$\zeta = \hat{k} \nabla \times \vec{v} \tag{7.40b}$$

is a conserved quantity (h is the characteristic altitude in the fluid). Numerical simulations of the flow evolution on a rapidly rotating shallow spherical layer initially produce a two-dimensional turbulence pattern. This turbulence organizes itself into a zonally dominated flow (with the wind distributed in bands parallel to the equator). Vortices immersed within the bands formed from small-scale eddies merge and inject zonal momentum into the jets. They scale with the *Rossby deformation radius*,

$$L_D = \left(\frac{NH}{f}\right). \tag{7.41}$$

The zonal jets have width scales of the order of the so-called *Rhines scale*:

$$L_\beta = \sqrt{\frac{u}{\beta}}, \tag{7.42a}$$

$$\beta = \frac{2\Omega \cos \varphi}{R_p}. \tag{7.42b}$$

The number of jets per hemisphere is given approximately by

$$n_{\text{jets}} \approx \frac{R}{2L_\beta}. \tag{7.43}$$

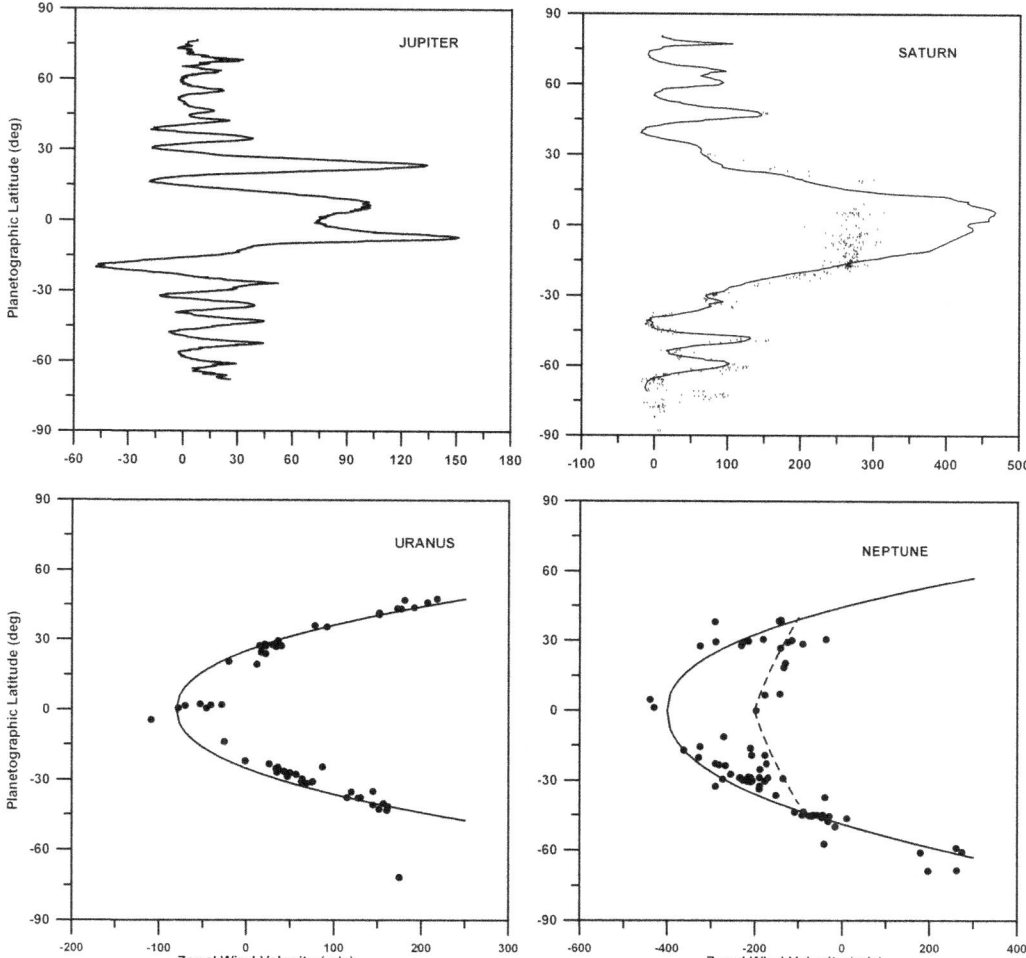

FIGURE 7.15. Zonal wind velocity profiles at cloud levels in the atmospheres of the giant and icy planets. For Jupiter the data come from *Hubble Space Telescope* (*HST*) observations which are the same as for *Cassini* (covering in total the period 1994–2000). For Saturn the continuous line is from *Voyager* data (1980–81) and the dots for *HST* (1994–2004). For Uranus they come from *Voyager* (1986, solid line), *HST* and ground-based data up to 2004 (dots). For Neptune the data come from *Voyager* (1986, dashed line), and *HST* and ground-based observations up to 1998 (dots); the continuous line is a polynomial fit.

These kinds of model are unable to produce the strong eastward equatorial jets observed on Jupiter and Saturn. The model also predicts variability in the circulation pattern with seasonal insolation changes, which are particularly important for Saturn. However, no seasonal changes have been observed in the jet patterns, although Saturn and Neptune undergo apparently significant variability in their equatorial jets (see Figure 7.15).

The second type of model assumes that the circulation is deep, extending through the entire H_2 molecular layer to pressures $P_{\text{base}} \sim 1$ Mbar. For Jupiter this represents $0.2\ R_J$ or $D\ (\gg H) \sim 12\,000$ km and for Saturn $0.5\ R_S$ or $D \sim 30\,000$ km. The motions are driven by the internal energy source transported convectively from the interior. Adopting the mixing-length theory for convection, the expected vertical velocities will be

$$w = \left[\frac{R}{\mu} \frac{F_{\text{int}}}{\rho C_{\text{p}}}\right]^{1/3}, \tag{7.44}$$

giving $w \sim 1$ ms^{-1} for both Jupiter and Saturn. However, the rapid rotation of the giants influences the radial direction of the convective motions. For an incompressible fluid Eq. (7.31) gives $\nabla \vec{v} = 0$, and then the rotation imposes the following constraint to the convective motions (Taylor–Proudman theorem):

$$(2\vec{\Omega} \bullet \nabla)\vec{u} = 0. \tag{7.45}$$

Accordingly, the motions will organize themselves in columns parallel to the rotational axis, giving rise to a secondary circulation in the form of counter-rotating cylinders coaxial with the rotation axis. The cylinders do not penetrate the metallic hydrogen layer and, extending along the H$_2$ layer, they intercept the planet's outward edge, giving rise to the alternating jet pattern. Numerical models and laboratory experiments confirm the formation of the columns and cylinders during developed convection in the rotating spherical shell. To characterize these motions, it is appropriate to introduce the following non-dimensional numbers:

$$\textit{Prandtl:} \quad \text{Pr} = \frac{\nu}{k}, \tag{7.46a}$$

$$\textit{Ekman:} \quad \text{Ek} = \frac{\nu}{2\Omega D^2}, \tag{7.46b}$$

$$\textit{Rayleigh:} \quad \text{Ra} = \frac{g\alpha(\mathrm{d}T/\mathrm{d}r)D^4}{\nu k}. \tag{7.46c}$$

Convection develops when the Rayleigh number exceeds a critical value (Ra$_{\text{crit}}$). This value depends on the rotational angular velocity of the planet according to the relationship

$$\text{Ra}_{\text{crit}} \approx \text{Ek}^{-4/3}. \tag{7.47}$$

Imposing different boundary conditions (rigid, free), these models are able to generate an eastward equatorial jet, but fail to reproduce the multi-jet pattern observed at other latitudes. They predict the existence of deep stable winds extending up to the H$^+$ layer. The only deep wind measurement by the *Galileo* probe on Jupiter, up to 24 bar (well below the sunlight penetration level) agrees with this prediction (Figure 7.13). One problem is that the values used for the above numbers, allowing computer calculations, differ significantly from the true value estimates on real planets: Ek $\sim 10^{-6}$ (Ek$_{\text{J}} \sim 10^{-10}$–$10^{-15}$), Ra $\sim 10^6$ (Ra$_{\text{J}} \sim 10^{12}$–10^{24}), Pr ~ 0.01–1 (Pr$_{\text{J}}$ is unknown). Other problems in these models relate to the effect of the compressibility of the atmosphere and to the possibility of a magnetic field interaction at the H$_2$–H$^+$ interface.

How these models apply to extrasolar planets is difficult to address since we do not know their rotation rates. Assuming rapid rotation, the dominance of one model or another will depend on the strength of the internal heat source (i.e. the age of the planet) against stellar radiation (Sánchez-Lavega *et al.* 2003; Sánchez-Lavega 2003). For 'hot jupiter' planets, spin–orbit synchronicity may be assumed, and the strong external energy should dominate the motions. Several models based on the shallow layer hypothesis have been published (Showman & Guillot 2002; Cho *et al.* 2003; Burkert *et al.* 2005). They coincide in the existence of a strong equatorial jet from the heated (day) to the colder (night) hemisphere with velocity $u \sim 1000$ ms^{-1} according to Eq. (7.35). This value is of the same order of magnitude as the sound speed, $c_{\rm s} = (\gamma g H)^{1/2}$, for the conditions obtaining on these planets.

The high temperatures of hot jupiters and the strong XUV flux leads to exospheric temperatures $\sim 10\,000$ K, higher than the blow-off temperature for hydrogen. The evaporation rate can be estimated by

$$\frac{{\rm d}M}{{\rm d}t} = \frac{3\beta^3 F_{\rm s}}{G\rho}, \qquad (7.48)$$

where β is the ratio of the radius at which most XUV radiation is absorbed to the planetary radius, ρ is the planet density and $F_{\rm s}$ is the XUV and Ly α flux at the planet's position. Vidal-Madjar et al. (2003) have detected a Ly α absorption in the atmosphere of the planet HD 209458b, with a large line width implying a high velocity of the hydrogen atoms escaping the planet. Vidal-Madjar et al. (2004) have also detected O and C in evaporation. These species must be carried out to the Roche lobe and beyond, most probably in a state of hydrodynamic escape, strongly influencing the planet evolution according to Eq. (7.48). Aerodynamic models predict escape velocities $V \sim 10$–1000 ms^{-1} (Yelle 2004). The detection of a thin sodium absorption line in the spectrum of this planet (Charbonneau et al. 2002), can give information on its abundance and on the presence of clouds at high levels in the atmosphere.

7.4.5. Meteorology

Meteorological phenomena develop on planets with intermediate and dense atmospheres, and manifest themselves basically in the densest tropospheric layers where the flow is subject to different types of dynamical instabilities. Here cloud formation serves to make meteorological phenomenology visible. In addition stratospheric dynamical systems are basically dominated by wavy motions and are important in transporting energy and momentum in most planets. The tropospheric meteorological systems are embedded within the dominant zonal circulation in all the Solar System planets.

In bodies with a solid surface the major meteorological phenomena can be classified as follows.

(1) *Baroclinic instability* occurs in rapidly rotating planets with large equator-to-pole temperature gradients and depends on the vertical wind shear (see Eq. (7.33)). The mean zonal flow becomes unstable in temperate latitudes on Earth and Mars forming alternating high and low pressure anticyclones and cyclones (with associated frontal systems) of scale $\sim L_{\rm D}$ (Eq. (7.41)). Geostrophic conditions apply and the air moves along the isobars circulating around the high and low pressure centres.

(2) The tropical and equatorial latitudes are dominated by the Hadley cell mechanism. On Earth water–moist air convergence in the lower part of the cell produces the *Intertropical Convergence Zone* (ITCZ), a band of dense clouds close to the equator and correspondingly a dry air band near the tropical latitudes. It must be thought that the Earth weather system is extremely complex due to the oceanic–continental mass distribution and vegetation, which leads to additional temperature contrasts and to spatial and temporal variability of water vapour distribution. For example the *monsoon* in the Indic region responds to these contrasts. The *El Niño Southern Oscillation* forms as a result of the atmosphere–ocean interaction. *Hurricanes* are meteorological phenomena apparently unique to Earth and are fed by the latent heat released by water moisture convection when $T_{\rm sea} > 27$ °C, and so they are a seasonal phenomenon. They form at neighbouring equatorial latitudes and show a prominent cyclonic flow at the surface with tangential velocities up to ~ 100 ms^{-1} for the strongest events.

(3) Surface features play an important role in the regional meteorology of Earth and Mars, with cloud and wave formation triggered when the flow encounters mountain obstacles. Local weather and seasonal effects are coupled on Mars to produce the most peculiar phenomena in its atmosphere, *dust storms*. When the wind blows at speeds $u \sim 50\text{--}100$ ms^{-1} at the edge of the polar cap, the soil dust is injected into the atmosphere up to altitudes $z \sim 30$ km, remaining suspended there for a few months. Major dust storms, which occur regularly, spread the dust over a planet-wide scale, covering the whole martian surface.

(4) Waves of different types abound on the terrestrial planets. On rapidly rotating planets, the most common is probably the *Rossby wave* formed as a consequence of the changing Coriolis force with latitude. On Venus, equatorial waves manifest themselves at cloud level, forming by contrast a characteristic 'Y pattern'.

(5) Strong vortices form in the polar areas of the three terrestrial planets, in particular on Earth and Venus (where it has a dipolar structure). Less is known about Titan's meteorology, but ground-based observations and the first *Cassini* images show the existence of methane cloud activity (storms?) in the southern polar area, and a well known detached haze layer formed by hydrocarbons and extending to high altitudes ($z = 30\text{--}300$ km).

The giants planets are surface-free, so ground meteorological phenomena are excluded. The main features that form at cloud level are as follows.

(1) Eddies, cyclones and anticyclones with closed circulation and sizes ranging from few km to a thousand km. The major vortices are longlived anticyclones, such as the jovian *Great Red Spot* (GRS), an elliptical structure with an east–west size $\sim 25\,000$ km, where the winds blow along its periphery at 100 ms^{-1}. The physical nature of these vortices is still unknown. Vortices also occur on Saturn and Neptune (e.g. the *Great Dark Spot*, GDS), but their longevity is apparently much less.

(2) Convective storms, probably fuelled by 'moist' ammonia and water vapour latent heat release, occur on Jupiter and Saturn. Major events are the *Great White Spots* (GWS), that occur sporadically on Saturn (mainly at equatorial latitudes), attaining a size of $20\,000$ km before they spread zonally. Related to convective stormy activity is the presence of *lightning*, detected as flashes of visible light in Jupiter's active areas. Convective activity due to methane condensation is probably the origin of some of the bright spot activity on Uranus and Neptune.

(3) Waves of different types have been detected at cloud level and in the temperature maps of Jupiter and Saturn. Most significant are those of Saturn, for example the mid-latitude northern *ribbon* that moves with a speed of 145 ms^{-1}, and the *hexagon* that surrounds the northern pole at $78°$ north, remaining stationary with respect to the planet's rotation.

7.5. Magnetic environment

7.5.1. Planetary magnetic fields

Several planets have magnetic fields as measured directly with magnetometers on board spacecraft, or indirectly by the radio-emission of moving charged particles in the field, or by the presence of auroral phenomena. Table 7.7 lists the data for objects with known magnetic fields. These fields have essentially a dipolar nature so their spatial structure can be represented by an equation of the type

$$B(r,\theta) = \frac{M_{\rm d}}{r^3}(1+3\sin\theta) \approx \Omega R^2, \qquad (7.49)$$

TABLE 7.7 Planetary magnetic environment

Planet	M_D[a]	B [gauss]	Angle (Ω, B)	R_{mag} in R_p (source[b])	Particles in ions/second (composition)
Mercury	0.0007	0.003	14°	1.5 (W)	(H$^+$)
Venus	<0.0004	<0.00003	–	–	–
Earth	1	0.305	10.8°	10 (W, A)	2×10^{26} (O$^+$, H$^+$)
Mars	<0.0002	<0.0003	–	–	–
Jupiter	20 000	4.28	9.6°	80 (W, A, Io)	>10^{28} (O$^+$, S$^+$)
Saturn	600	0.22	<1°	20 (W, A, S)	2×10^{26} (H$^+$, O$^+$, OH$^+$)
Uranus	50	0.23	58.6°	20 (W, A)	10^{25} (H$^+$)
Neptune	25	0.14	47°	25 (Triton)	10^{25} (H$^+$, N$^+$)

[a] Dipolar moment M_D relative to that of the Earth ($M_{Earth} = 7.91 \times 10^{25}$ gauss cm^3).
[b] The magnetopause size is given in planetary radius. The plasma sources are: W (solar wind), A (atmosphere–ionosphere), S (satellites and rings).

where r is the distance to the planet centre, θ is the co-latitude and M_d is the dipolar magnetic moment. The magnetic dipole and the rotational axis are not aligned, as seen in Table 7.7. The nature of the field can be investigated from the magnetic diffusion equation,

$$\frac{\partial \vec{B}}{\partial t} = \nabla \times (\vec{V} \times \vec{B}) + \lambda \nabla^2 \vec{B}, \qquad (7.50)$$

where λ is the magnetic diffusivity (for materials under compressed planetary conditions $\lambda \sim 5$–50 m^2 s^{-1}). Ignoring the convective term, we can derive a characteristic magnetic decay time,

$$\tau_{mag} \sim \pi^2 r_0^2 / \lambda, \qquad (7.51)$$

where r_0 is the size of the magnetic area (the core of a terrestrial planet or the H$^+$ region of a giant), leading to $\tau_{mag} \sim 50$ kyr–1 Myr, which are below the Solar System age (4.5 Gyr). Thus, we may conclude that the current existence of a magnetic field needs a generating source, a self-generated dynamo mechanism being the most plausible hypothesis.

For a dynamo to work we need: (a) an electrically conductive fluid medium; (b) kinetic energy provided by planetary rotation; and (c) an internal energy source to drive convective (turbulent) motions within the fluid. Condition (c) is fulfilled when the Rayleigh number is above a critical value (see Eqs. (7.46c) and (7.47)). For planetary interiors this number can be rewritten as

$$Ra = \frac{g_0}{\rho_0 T_0 C_p k^2 \nu} F_{int} d^4, \qquad (7.52)$$

where d is the depth of the convecting layer. According to the mixing length theory, the convective velocities will be of the order (see Eq. (7.44))

$$V(\text{conv}) \sim \left(\frac{F_{\text{int}}}{\rho H}\right)^{1/3}. \tag{7.53}$$

Numerical calculations and observations suggest that the dynamo is self-excited when the magnetic Reynolds number

$$R_{\text{m}} = \frac{V\ell}{\lambda} \tag{7.54}$$

is in the range ~ 1–10 (ℓ is a characteristic magnetic length). The Elsasser number,

$$\Lambda = \frac{B^2}{2\Omega\mu_0\rho\lambda}, \tag{7.55}$$

which measures the strength of the magnetic field against rotation and diffusion, is for the dynamo conditions of the order of 1. This result can be used to make a crude estimate of the order of magnitude for the dynamo magnetic field:

$$B = \sqrt{2\Omega\mu_0\rho_0\lambda}. \tag{7.56}$$

The magnetic moment, $M_{\text{d}} = Br_0^3$, on a planetary body depends on the precise relationships $B(\Omega, \lambda)$ and $r_0(M_{\text{p}}, F_{\text{int}})$; that is a complex problem. However, it has been observed that Solar System bodies with dynamo fields obey an empirical relationship between the magnetic moment and angular momentum, L_0, given by

$$M_{\text{d}} = 4 \times 10^{-9} L_0^{0.83}, \tag{7.57}$$

with L_0 given in kg m^2 s^{-1} and M_{d} in gauss m^3 (Vallée 2003).

7.5.2. Interaction with the solar wind

The Sun emits a flux of $\sim 2 \times 10^8$ cm^{-2} s^{-1} of ionizing particles equivalent to 5 ions cm^{-3} (95% protons and 5% α particles) called the *solar wind*. At 1 AU its average velocity is $V = 400$ km s^{-1}. During periods of solar activity these numbers are a factor of 10 greater. The flow carries with it an interplanetary magnetic field of strength $B_{\text{IP}} \sim 1$–10 nT. When this flow and the interplanetary magnetic field encounter a planet, three main types of interaction can occur depending on the planet's nature.

(1) For magnetized planets the *Lorentz force*, $\boldsymbol{F} = q\boldsymbol{V} \times \boldsymbol{B}$, acting on solar wind particles deflects the flow, forming a *bow shock*. This gives rise to the *magnetosphere*, a cavity where the planet's magnetic field dominates the motion of charged particles and where a large number of electromagnetic processes occur (electric currents, plasma waves, etc.). Its edge is called *the magnetopause* and its size can be estimated by assuming a balance between the radiation pressure and the magnetic pressure:

$$\frac{1}{2}\rho_{\text{sw}}v^2 = \frac{B^2}{8\pi}, \tag{7.58}$$

and since $B^2 = M_d^2/R^6$, the magnetopause size, R_{mag}, can be estimated by

$$R_{\text{mag}} = \left(\frac{M_d^2}{4\pi\rho_{\text{sw}}v^2}\right)^{1/6}. \tag{7.59}$$

(2) In non-magnetized objects possessing an atmosphere (such as Venus, Mars and Titan), the solar wind interacts directly with the atmosphere's outer ionized part (*the ionosphere*). This layer can be characterized by its conductivity (σ), associated electric field $\boldsymbol{E} = -\boldsymbol{V} \times \boldsymbol{B}_{\text{ip}}$, and current density, $\boldsymbol{J}$. The solar wind is then deflected away from the ionosphere by a force $\boldsymbol{F} = \boldsymbol{J} \times \boldsymbol{B}_{\text{ip}}$.

(3) In non-magnetized objects without atmospheres the interaction depends on the body's electrical nature: (a) conductive bodies (e.g. Europa, Ganymede and Callisto) deflect the solar wind at the surface; (b) insulating obstacles (e.g. the Moon) absorb the ions of the solar wind forming a plasma cavity.

On planets with atmospheres and strong magnetic fields, *aurorae* form around the polar areas (Figure 7.11b, upper images). Aurorae are the 'optical' manifestation of the interaction of energetic (accelerated) particles (electrons, ions or neutrals) that follow the magnetic field lines, with the upper neutral atoms and molecules of the atmosphere. This interaction generates emission in an ample wavelength range (X-rays, UV, visible, IR and radio). On the Earth, Jupiter and Saturn they have a ring-like (oval) morphology centred on the magnetic poles when observed outside the planet. The most energetic events occur on Jupiter, where particles reach typical kinetic energies $E_k \sim 1$–10 keV with maximum values E_k (max) ~ 100 keV and power $\sim 10^{12}$ W.

Particles from the solar wind and other sources (mainly high energy protons) become trapped in the magnetic field lines close to the planet, and form a radiating plasma flow called *radiation belts* (Figure 7.11b, lower image). On Earth they are located at a distance $R \sim 2$–3 $R_\oplus$ and on Jupiter at $R < 5$ R_J. For example, electrons moving along the magnetic field lines emit radio waves at the cyclotron frequency $f_c = eB/2\pi m_e$ with an energy power

$$P(\text{rad}) \sim B^2 v_{\text{sw}} R_{\text{mag}}. \tag{7.60}$$

The magnetic environment of the planets also interacts with the particles of the rings and with the satellites with orbits within the magnetosphere. The most intense interaction occurs between Jupiter and its satellite Io due to the large outflow of ions escaping from the satellite. Io has a plasma torus extending along its orbit, formed by S^+ and O^+ ions and a neutral cloud of Na (with S, O) supplied by its volcanic activity. Jupiter and Io are connected by a plasma current of intensity $I \sim 1$–3×10^6 A occupying a cross-sectional area of 5×10^4 km^2.

The existence of magnetic fields in giant exoplanets has been addressed following the preceding scale arguments (Sánchez-Lavega 2004). In the case of 'hot jupiters' the proximity to the parent star could cause the planet's magnetic field to influence the star's chromosphere, manifested as variability in the spectral lines synchronized to the planet's orbital period, as has been recently claimed for HD 179949 (Shkolnik *et al.* 2003). For extrasolar planets with intense magnetic fields, radio-emission from accelerated particles could be detected in the near future because of the power dependence of the emission on the square of the magnetic field intensity, as shown in Eq. (7.60) (Farrell *et al.* 1999; Bastian *et al.* 2000).

7.6. Conclusions

The future of Solar System exploration is promising, with many space missions planned (up to 2025) by NASA, ESA and national space agencies (up to 2025): Mercury orbiters (*Messenger, Beppi–Colombo*), Venus (*Venus Express*), the Moon (*Smart 1, Chandrayaan-1, Lunar Reconnaissance Orbiter*), Mars (rovers and landers, *Phoenix, Mars Science Laboratory, Exomars*), the asteroids (*Dawn* to orbit Vesta and Ceres), Jupiter (*Juno, JIMO* (?), *ESA mission*), Saturn (*Cassini* up to 2008), Pluto and KBOs (*Pluto Express*, and visit of the heliopause), comets (*Deep Impact, Stardust, Rosetta*). Many other missions will simultaneously explore the magnetic interplanetary environment and its interaction with the Sun's activity.

In addition, ground-based and space-borne telescopes will play a fundamental role in the study and classification of the population of minor bodies (asteroids, NEOs, NEAs, KBO-TNOs and comets). Large telescopes with adaptive optics will also be important in the study of the evolution of the dynamics of the atmospheres of the giant planets and probably the only source of information for the atmospheres of Uranus and Neptune.

Acknowledgements

My special thanks to Juan Antonio Belmonte and Hans Deeg for inviting me to participate in the excellent Winter School. This work was supported in part by research projects MCYT AYA2003-03216, FEDER funds and Group UPV 2004.

Notes

The definition of a "planet" adopted in August 2006 in Prague during the International Astronomical Union (IAU) General Assembly states that bodies in our Solar System fall into one of the following categories:

(1) A **planet** is a celestial body that: (a) is in orbit around the Sun, (b) has sufficient mass for its self-gravity to overcome rigid body forces so that it assumes a hydrostatic equilibrium (nearly round) shape, and (c) has cleared the neighbourhood around its orbit. The eight planets are: Mercury, Venus, Earth, Mars, Jupiter, Saturn, Uranus, and Neptune.

(2) A "**dwarf planet**" is a celestial body that (a) is in orbit around the Sun, (b) has sufficient mass for its self-gravity to overcome rigid body forces so that it assumes a hydrostatic equilibrium (nearly round) shape, (c) has not cleared the neighbourhood around its orbit, and (d) is not a satellite. At present the there are three dwarf planets: Ceres, Pluto, and Eris.

(3) All other objects, except satellites, orbiting the Sun shall be referred to collectively as "**Small Solar System Bodies**". These currently include most of the Solar System asteroids, most trans-Neptunian Objects (TNOs), comets, and other small bodies.

REFERENCES

ATREYA, S. K., POLLACK, J. B. & MATTHEWS, M. S. (eds.), 1989, *Origin and Evolution of Planetary and Satellite Atmospheres*, Tucson: The Arizona University Press.

BAGENAL, F., DOWLING, T. & MCKINNON, W. (eds.), 2004, *Jupiter: the Planet, Satellites and Magnetosphere*, Cambridge: Cambridge University Press.

BASTIAN, T. S., DULK G. A. & LEBLANC, Y., 2000, *ApJ* **545**, 1058.

BEATTY, J. K., COLLINS PETERSEN, C. & CHAIKIN, A., 1999, *The New Solar System*, Cambridge: Cambridge University Press.

BERTOTTI, B., FARINELLA, P. & VOKROUHLICKY, D., 2003, *Physics of the Solar System*, Astrophysics and Space Science Library, Dordrecht: Kluwer, Vol. 293.

BINZEL, R. P., GEHRELS, T. & MATTHEWS, M. S. (eds.), 1989, *Asteroids II*, Tucson: The Arizona University Press.

BOUGHER, S. W., HUNTEN, D. M. & PHILLIPS, R. J. (eds.), 1997, *Venus II*, Tucson: The Arizona University Press.

BURKERT, A., LIN, D. N. C., BODENHEIMER, P. H., JONES, C. A. & YORKE, H. W., 2005, *ApJ* **618**, 512.

BURNS, J. A. (ed.), 1977, *Planetary Satellites*, Tucson: The Arizona University Press.

BURNS, J. A. & MATTHEWS, M. S. (eds.), 1986, *Satellites*, Tucson: The Arizona University Press.

BURROWS, A., 2005, *Nature* **433**, 261.

BURROWS, A., HUBBARD, W. B., LUNINE, J. I. & LIEBERT, J., 2001, *Rev. Mod. Phys.* **73**, 719.

CHAMBERLAIN, J. W. & HUNTEN, D. M., 1987, *Theory of Planetary Atmospheres*, Orlando: Academic Press.

CHARBONNEAU, D., BROWN, T. M., NOYES, R. W. & GILLILAND, R. L., 2002, *ApJ* **568**, 377.

CHO, J. Y.-K., MENOU, K., HANSEN, B. M. S. & SEAGER, S. et al., 2003, *ApJ* **587**, L117.

COLE, G. H. A. & WOOLFSON, M. M., 2002, *Planetary Science*, Bristol: Institute of Physics.

CRUIKSHANK, D. P. (ed.), 1995, *Neptune*, Tucson: The Arizona University Press.

ENCRENAZ, T., BIBRING, J. P., BLANC, M., BARUCCI, M. A., ROQUES, F. & ZARKA, PH., 2004, *The Solar System*, third edition, Berlin: Springer-Verlag.

FARRELL, W. M., DESCH, M. D. & ZARKA, P., 1999, *J. Geophys. Res.* **104**, 14025–14032.

GEHRELS, T. (ed.), 1976, *Jupiter*, Tucson: The Arizona University Press.

GEHRELS, T. (ed.), 1979, *Asteroids*, Tucson: The Arizona University Press.

GEHRELS, T. & MATTHEWS, M. S. (eds.), 1984, *Saturn*, Tucson: The Arizona University Press.

GREENBERG, R. & BRAHIC, A. (eds.), 1984, *Planetary Rings*, Tucson: The Arizona University Press.

GUILLOT, T., BURROWS, A., HUBBARD, W. B., LUNINE, J. I. & SAUMON, D., 1996, *ApJ* **459**, L35.

HANEL, R., CONRATH, B., HERATH, L., KUNDE, V. & PIRRAGLIA, J., 1981, *J. Geophys. Res.* **86**, 8705.

HUBBARD, W. B., BURROWS, A. & LUNINE, J. I., 2002, *Ann. Rev. Astron. Astrophys.* **40**, 103.

HUNTEN, D. M., COLIN, L., DONAHUE, T. M. & MOROZ, V. I. (eds.), 1983, *Venus*, Tucson: The Arizona University Press.

KERRIDGE, J. F. & MATTHEWS, M. S. (eds.), 1988, *Meteorites and the Early Solar System*, Tucson: The Arizona University Press.

KIEFFER, H. H., JAKOSKY, B. M., SNYDER, C. W. & MATTHEWS, M. S. (eds.), 1992, *Mars*, Tucson: The Arizona University Press.

LEWIS, J. S., 1997, *Physics and Chemistry of the Solar System*, San Diego: Academic Press.

LIU, M. C., 2004, *Science* **305**, 1442.

MCBRIDE, N. & GILMOUR, I., 2003, *An Introduction to the Solar System*, The Open University, Cambridge: Cambridge University Press.

MORRISON, D. (ed.), 1982, *Satellites of Jupiter*, Tucson: The Arizona University Press.

OKAMOTO, Y. K., KATUZA, H., HONDA, M., et al., 2004, *Nature* **431**, 660.

PATER, I. DE & LISSAUER, J. J., 2001, *Planetary Sciences*, Cambridge: Cambridge University Press.

SÁNCHEZ-LAVEGA, A., 2001, *A&A* **377**, 354.

SÁNCHEZ-LAVEGA, A., 2003, in *Highlights of Spanish Astrophysics III*, eds. J. Gallego, J. Zamorano & N. Cardiel, Dordrecht: Kluwer, p. 385.

SÁNCHEZ-LAVEGA, A., 2004, *ApJ* **609**, L8.

SÁNCHEZ-LAVEGA, A., 2006, *Contemporary Physics*, **47**, 157.

SÁNCHEZ-LAVEGA, A., HUESO, R. & BAEZA, S., 2003, in *Towards Other Earths: DARWIN/TPF and the Search for Extrasolar Terrestrial Planets* ESA-SP 359, Noordwijk: ESA Publishing Division, 569.

SÁNCHEZ-LAVEGA, A., PÉREZ-HOYOS, S. & HUESO, R., 2004, *Am. J. Phys.* **72**, 767.

SCHNEIDER, J., 2005, *Encyclopedia of Extrasolar Planets*, exoplanet.eu.

SHIRLEY, J. H. & FAIRBRIDGE, R. W., 1997, *Encyclopedia of Planetary Sciences*, Heidelberg: Chapman and Hall.

SHKOLNIK, E., WALKER, G. A. H. & BOHLEUDER, D. A., 2003, *ApJ* **597**, 1092.

SHOWMAN, A. P. & GUILLOT, T., 2002, *A&A* **385**, 166.

SUDARSKY, D., BURROWS, A. & HUBENY, I., 2003, *ApJ* **588**, 1121.

THOMMES, E. W. & LISSAUER, J. J., 2005, in *Astrophysics of Life*, eds. M. Livio, I. N. Reid & W. B. Sparks, STScI Symp. No. 16, Chapter 4, Cambridge: Cambridge University Press.

VALLÉE, J. P., 2003, *New Astron. Rev.* **47**, 85.

VIDAL-MADJAR, A., LECAVELIER DES ETANGS, A., DÉSERT, J.-M., et al., 2003, *Nature* **422**, 143.

VIDAL-MADJAR, A., DÉSERT, J.-M., LECAVELIER DES ETANGS, A., et al., 2004, *Astrophys. J.* **604**, L69–L72.

VILAS, F., CHAPMAN, C. R. & MATTHEWS, M. S. (eds.), 1988, *Mercury*, Tucson: The Arizona University Press.

WEISSMAN, P. R., MCFADDEN, L. A. & JOHNSON, T. V., 1999, *Encyclopedia of the Solar System*, San Diego: Academic Press.

WILKENING, L. L. (ed.), 1982, *Comets*, Tucson: The Arizona University Press.

YELLE, R. V., 2004, *Icarus* **170**, 167.

8. Habitable planets around the Sun and other stars

JAMES F. KASTING

This chapter gives an overview about planetary habitability, which is based on the assumption that a habitable planet is one that supports liquid water on at least part of its surface. The factors that have kept Earth habitable throughout most of its life are reviewed, as well as those that made present-day Mars and Venus uninhabitable. These serve also as indicators for the expected width of the habitable zone around solar-like stars. The last two sections cover the causes for low abiotic O_2 abundances expected in Earth's early atmosphere, and the (biological) origin of the current high concentrations of O_2 and O_3. Implications for the detectability of biological activity on extrasolar planets are discussed.

8.1. Introduction

In this chapter, I have tried to present an overview of the topic of planetary habitability. This topic can be broken down into three related questions: (1) what are the factors that have kept the Earth habitable throughout most of its lifetime? (2) what has caused our neighbouring planets, Mars and Venus, to be uninhabitable? and (3) what are the chances that habitable planets exist around other main sequence stars, and how might we tell if they are inhabited? I will briefly address each question, recognizing that it will be impossible to do justice to any of them in the space of one short chapter. References to the relevant literature are provided, and this should allow the interested reader to pursue these topics further.

I begin this discussion by defining what I mean by the term 'habitable'. In the context of this chapter, a habitable planet is one that supports liquid water on at least part of its surface. This definition, of course, presumes that all life is basically like us, i.e. it is carbon-based and requires liquid water during some phase of an organism's life cycle. This assumption may or may not be true, but I will make no further apologies for it here, as this is the only definition that makes practical sense until we learn otherwise. Subsurface liquid water is also possible – indeed, both Mars and Jupiter's moon Europa may be habitable in this sense – but this is less interesting for astronomers because it is difficult or impossible to learn anything about this from remote observations. There are, of course, additional requirements for habitability by some organisms, especially those which we consider to be 'advanced'. In particular, nearly all multicellular organisms (except for certain filamentous bacteria) require molecular oxygen, O_2, in order to respire. We humans are among such organisms, and so the question of atmospheric O_2 concentrations is of particular interest to us.

The five talks that I presented at this workshop covered a broad variety of topics that relate to the questions listed above. This review will touch on each of these. However, special emphasis will be given to the question of abiotic O_2 concentrations. This is partly because the other climate-related topics have been reviewed elsewhere (Kasting 1987, 2002; Kasting & Catling 2003; Catling & Kasting 2007) and partly because the O_2

Extrasolar Planets, eds. Hans Deeg, Juan Antonio Belmonte and Antonio Aparicio.
Published by Cambridge University Press.
© Cambridge University Press 2007.

question is directly relevant to interpreting the data from future space-based telescopes such as NASA's *Terrestrial Planet Finder (TPF)* and ESA's *Darwin* mission. O_2, along with its photochemical product O_3, is one potential bioindicator that may tell us whether life exists on planets outside our Solar System. Here, I will attempt to provide some insight into how to interpret such a signal if we someday manage to observe it.

8.1.1. *Long-term climate evolution on Earth*

It is convenient to begin with the subject of long-term climate evolution. Both the Earth and the Solar System itself have been in existence for some 4.55 billion years. (I shall henceforth adopt geologists' notation and use the notation 'Ga', or 'giga-annum', to mean billions of years before present. Thus, the age of the Solar System is 4.55 Ga). Liquid water has been present on or near Earth's surface since about 4.4 Ga, based on the oxygen isotope composition of zircon (zirconium silicate) minerals (Valley *et al.* 2002). This should not be viewed as surprising, as models of Earth's formation predict that the planet's surface should have cooled quickly once the main accretion period was over (Zahnle *et al.* 1988). This does, however, help to confirm the prevailing view that most of Earth's volatiles, including water, were delivered early in the planet's history. According to (Morbidelli *et al.* 2000), most of Earth's water came from volatile-rich asteroids from the main asteroid belt region (2–3.5 AU). Comets are now thought to have delivered less than 10 per cent of Earth's water, based on the high D/H ratios of the three comets that have been studied (Morbidelli *et al.* 2000).

The climate record

The first 700 million years of Earth history are called the Hadean Eon (Figure 8.1). Little is known about this time period because few rocks have been preserved, other than the zircon inclusions just mentioned. Consequently, little is known about the climate during this time, either. Some authors (e.g. Walker 1985; Kasting & Ackerman 1986) have suggested that the Earth was quite warm ($\sim$80–90 °C) as a consequence of the greenhouse effect of a dense CO_2-rich atmosphere. Others (Sleep & Zahnle 2001) have suggested that most of Earth's carbon was sequestered in the mantle, and that the Earth was therefore cold because of the faintness of the young Sun (more on this below). My own opinion is that the early Earth was warm, but this is at best an educated guess. Perhaps future observations of other young, Earthlike planets will eventually provide insight into this earliest period of Earth history.

The rock record begins to come into existence around 3.8 Ga, and the period from 3.8–2.5 Ga is called the Archaean. Even here we are not sure whether the climate was warm or cold. Diamictites (rocks containing unconsolidated fragments resembling glacial till[1]) have been found in the Witwatersrand Supergroup in South Africa (Crowell 1999) in strata dated at 2.8 Ga. If these are glacial in origin – something that is not assured because there is no corroborating evidence – then the climate was already relatively cool by this time. In more recent history, polar ice appears to have been absent prior to $\sim$35 million years ago (Ma), when surface temperatures were about 5 degrees warmer than at present. The present mean surface temperature is 288 K, or 15 °C.

The next main eon of Earth history, 2.5–0.6 Ga, is called the Proterozoic Eon. By this time, the rock record is much more complete, so we have a much better idea of what conditions were like. Most of this period was ice-free, indicating that the climate was warm. However, both the beginning ($\sim$2.3 Ga) and end (0.75–0.6 Ga) of this time

---

[1] Glacial till is rocks that are dug up and transported by glaciers and deposited in piles termed 'moraines'.

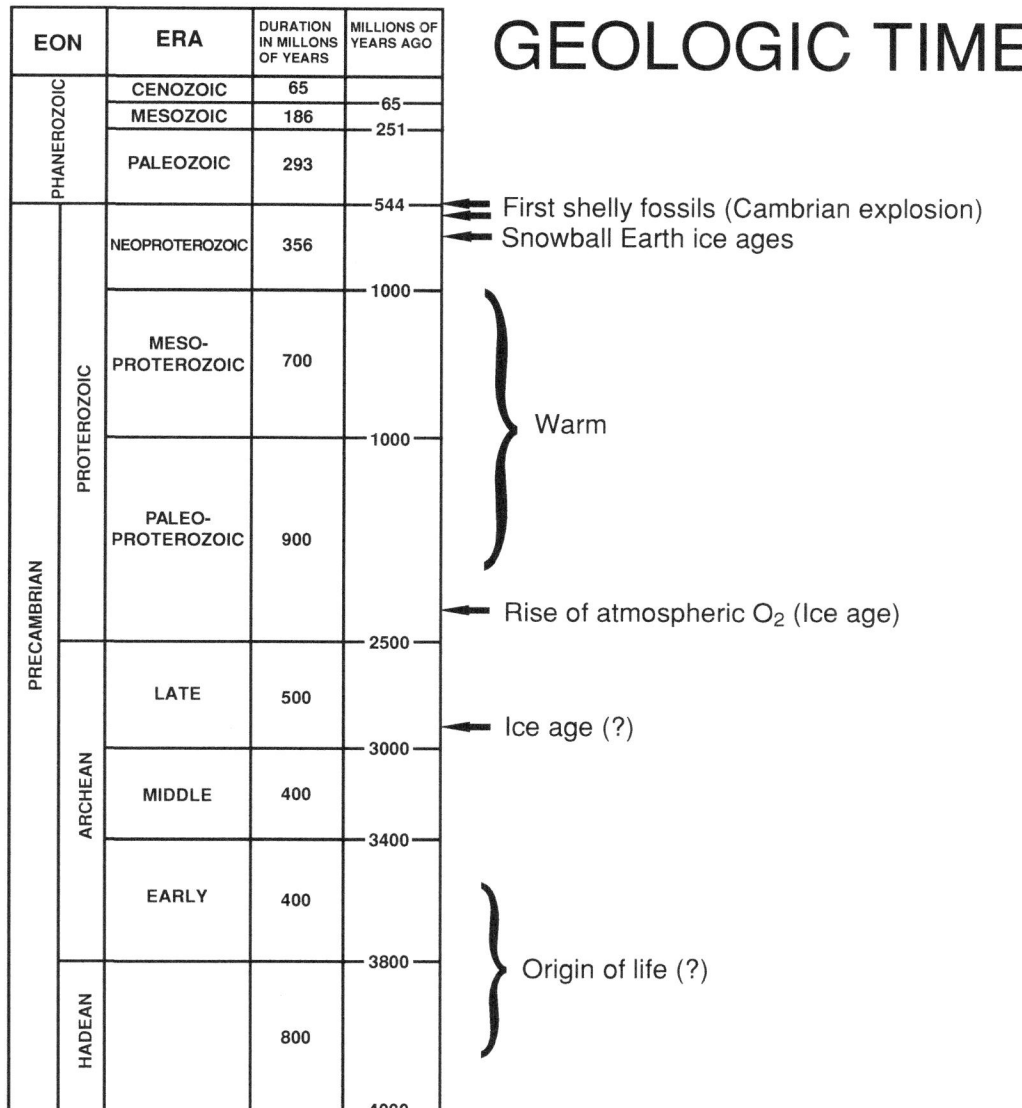

FIGURE 8.1. Geologic timescale showing warm and cold intervals.

period were marked by severe glaciations. In both cases, palaeomagnetic data imply that continental-scale ice sheets existed within the tropics (Kirschvink 1992; Hoffman et al. 1998; Evans et al. 1977). This has led to the idea that these represent so-called 'Snowball Earth' episodes, in which Earth's entire surface was frozen. Although this idea remains controversial, I personally think that it is correct, as climate models predict that this type of behaviour is actually to be expected (Budyko 1969; Sellers 1969; Caldeira & Kasting 1992; Pollard & Kasting 2005).

Finally, the last 540 million years of Earth history is termed the Phanerozoic Eon. During this time, Earth's climate has alternated between warm and cold periods, with two extended and one brief glaciation interspersed between longer periods of warmth.

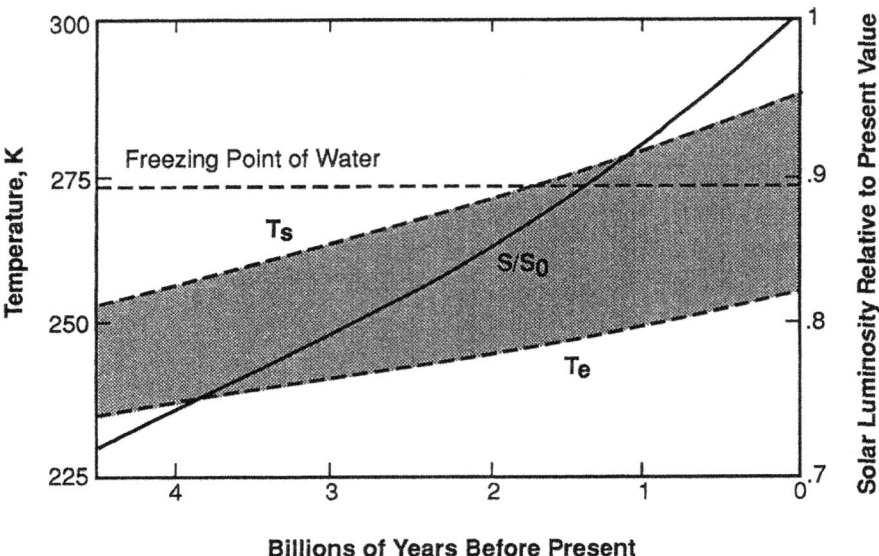

FIGURE 8.2. Diagram illustrating the faint young Sun problem. The solid curve represents solar luminosity relative to today's value. The two dashed curves represent Earth's effective radiating temperature, T_e, and its mean surface temperature, T_s. The shaded area in between shows the magnitude of the greenhouse effect. Constant CO_2 (330 ppmv) and a fixed relative humidity distribution were assumed in the (one-dimensional) climate model. From Kasting et al. (1988), based on Gough (1981).

We are currently still in a glacial period, as evidenced by the presence of continental ice sheets on Greenland and Antarctica. This use of the term 'glacial period' should not be confused with the more pronounced glacial periods of the Pleistocene, when the polar ice sheets covered much of Europe and North America.

The faint young Sun problem

Climate history is intriguing from a theoretical standpoint because the Sun has evidently been getting brighter throughout this entire time. Figure 8.2 shows a solar luminosity curve based on Gough (1981). According to this calculation, the Sun was about 30 per cent less luminous when it formed. It has been gaining in luminosity as a consequence of the gradual conversion of hydrogen into helium within its core.

It has occasionally been suggested that the young Sun may have been brighter than the standard model because it was initially more massive, then lost significant amounts of mass by way of an intense solar wind (e.g. Graedel et al. 1991; Boothroyd et al. 1991). The luminosity of a star like the Sun is proportional to roughly the fourth power of its mass ($M_\odot^4$). Planetary orbits migrate outward as the star loses mass in such a way as to conserve angular momentum; hence, the semi-major axis of Earth's orbit, a, should vary as $M_\odot^{-1}$. Putting these dependencies together with the inverse square law suggests that the radiation flux hitting Earth's surface should vary as $M_\odot^6$. Fortunately, data have recently become available that help to constrain this possibility. Wood et al. (2002) used the *Hubble Space Telescope* to measure excess 'astrospheric' absorption in Ly α radiation coming from a set of nearby young solar-type stars. The upper limit on integrated mass loss over time is about 3 per cent of a solar mass. If this same limit applied to the young

Sun, then solar flux at the Earth may originally have been higher by a factor of $1.03^6 \cong 1.2$, largely cancelling the luminosity decrease predicted by the standard solar model. Nearly all of the observed mass loss appears to occur within the first 200 million years of a star's history, however, corresponding to a date of 4.35 Ga for Earth. This is too early in Earth's history to affect anything in the rock record other than the zircons mentioned earlier. The faintness of the young Sun is therefore a real problem that must be addressed in any model of Earth's climate history.

Also shown in Figure 8.2 is Earth's effective radiating temperature, T_e, and its surface temperature, T_s. T_e is determined by balancing incoming solar energy at visible/near-IR wavelengths with outgoing energy at thermal-IR wavelengths, treating the planet as a black body:

$$\sigma T_e^4 = \frac{S}{4}(1-A). \qquad (8.1)$$

Here, σ is the Stefan–Boltzmann constant (5.67×10^{-8} W m^{-2} K^{-4}), S is the solar flux at Earth's orbit (currently 1370 W m^{-2}), and A is Earth's albedo (~ 0.3). Plugging in these values yields $T_e = 255$ K. The observed mean surface temperature is $T_s = 288$ K. The difference between these values, 33 K, is caused by the greenhouse effect of Earth's atmosphere. To generate Figure 8.2, two different greenhouse gases were assumed, CO_2 and H_2O, and a one-dimensional (globally averaged) climate model was used to calculate surface temperature. The CO_2 concentration was held fixed, while H_2O was made variable by assuming a fixed distribution of relative humidity. This causes the H_2O abundance to increase (decrease) as the climate warms (cools), creating a positive feedback loop that amplifies the effect of solar luminosity changes on Earth's surface temperature.

The CO_2–climate feedback

Taken literally, Figure 8.2 predicts that Earth's surface would have been frozen prior to ~ 2.0 Ga. We have already seen, though, that this was *not* the case. If anything, the first half of Earth's history was warmer than the second. It is not difficult to think of mechanisms for warming the early climate. Rather than listing all the possibilities (see Kasting & Catling 2003 for these), I shall focus on the two most likely mechanisms. The first involves significantly higher concentrations of atmospheric CO_2. There is a good reason why CO_2 levels may have been higher in the distant past, as first explained by Walker *et al.* (1981). It is illustrated in Figure 8.3, which shows the processes involved in the inorganic carbon cycle, sometimes called the carbonate–silicate cycle. This cycle is thought to exert most of the control over atmospheric CO_2 concentrations on timescales exceeding ~ 1 million years. (The faster organic carbon cycle is important on shorter timescales, including human ones; however, on long timescales it controls atmospheric O_2, not CO_2.) The carbonate–silicate cycle begins when CO_2 dissolves in rainwater to make carbonic acid, H_2CO_3. This weak acid dissolves silicate rocks on the continents, releasing Ca^{++}, Mg^{++} and HCO_3^- (bicarbonate) ions into groundwater. This water flows down to the oceans in streams and rivers. There, various organisms, notably the planktonic foraminifera, use these substances to make shells out of calcium carbonate, $CaCO_3$. When the organisms die, they fall into the deep ocean. There, most of them dissolve; however, a small percentage are preserved in sediments on the ocean floor. The sea floor spreads away from the midocean ridges as part of the plate tectonic cycle that renews our surficial geology. At certain plate boundaries, the sea floor is subducted down into the mantle. When this occurs, the carbonate sediments are heated up and undergo metamorphism, reforming as silicate minerals and releasing gaseous CO_2, which goes back into the atmosphere from volcanoes.

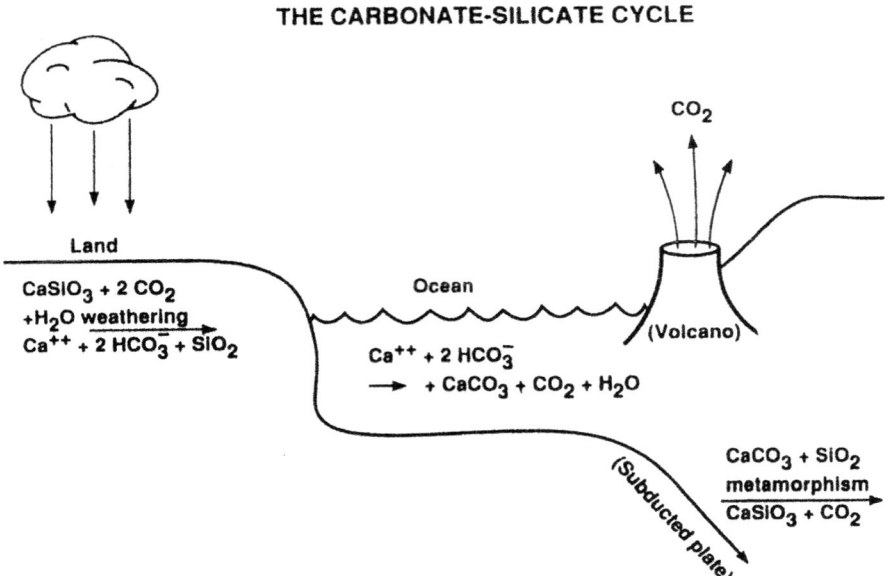

Figure 8.3. Diagram illustrating the carbonate–silicate cycle.

The implications of this cycle for the faint young Sun problem are reasonably clear. When the climate gets colder, the hydrologic cycle slows down, and the rate of silicate weathering slows down with it. Volcanism continues, however, partly because the mean lifetime of sea floor is much longer than 1 million years (more like 60 million) and partly because there is additional volcanic release of CO_2 from hotspots like Hawaii and from the midocean ridges themselves. Hence, when the climate cools, the CO_2 loss process slows down while the production of CO_2 continues unabated. The atmospheric CO_2 concentration then increases until the surface becomes warm again and equilibrium can be re-established in the cycle. During most of Earth's history, this cycle has prevented us from getting too cold. Occasionally, however, the controls break down for reasons that are still debated, and Earth slips into a global Ice Age. These are the Snowball Earth episodes, mentioned earlier, that occurred near the beginning and end of the Proterozoic. When this happens, silicate weathering ceases almost entirely, and volcanic CO_2 accumulates in the atmosphere until the greenhouse effect becomes large enough to melt the ice. This requires about 0.1 bar of CO_2 and 30 million years, according to Caldeira & Kasting (1992). (But see Pierrehumbert 2004; Pollard & Kasting 2005, for variations on these numbers.) Evidence that such a process actually did occur is provided by 'cap carbonate' deposits overlying the glacial diamictites of the Late Proterozoic (Hoffman et al. 1998).

The negative (stabilizing) feedback on climate provided by the carbonate–silicate cycle is important for another reason: it has implications for the outer edge of the liquid water 'habitable zone' around the Sun and other stars. Imagine what would happen if one could slowly push Earth outward from its current orbit, leaving everything else about the planet unchanged. The climate would cool, and CO_2 would build up in the atmosphere until equilibrium was re-established in the carbonate–silicate cycle. Eventually, this stabilizing mechanism would break down because the CO_2 would start to condense. *However, current estimates of the habitable zone width* (Kasting et al. 1993) *are rather generous as a consequence of this feedback mechanism, which was neglected in earlier estimates* (Hart 1978, 1979).

The importance of CH_4

For Earth, CO_2 is probably not the only greenhouse gas that was present at higher concentrations in the distant past. Methane (CH_4) is another greenhouse gas that is of minor importance today but which could have been much more abundant earlier in Earth's history. Unlike CO_2, CH_4 has predominately biological sources. The largest of these today are cows and rice paddies. What these two sources have in common is anaerobic (O_2-free) microenvironments in which *methanogens* can live. Methanogens, or methanogenic bacteria, are micro-organisms that make their metabolic living by producing CH_4. The starting materials for methanogenesis vary; however, nearly all methanogens can utilize the reaction: $CO_2 + 4H_2 \to CH_4 + 2H_2O$. Both CO_2 and H_2 are produced by volcanism; hence, the starting materials for methane production were probably present on the early Earth.

Methanogens are also thought to be evolutionarily ancient, based on analysis of their ribosomal RNA (rRNA) (Woese & Fox 1977). Ribosomes are organelles within cells in which proteins are made. Ribosomal RNA evolves slowly over time, as most changes to the protein-manufacturing apparatus within a cell are fatal. Hence, differences in rRNA sequences reflect evolutionary changes in the distant past. Methanogens are all found on a single, low-lying branch of the rRNA tree, consistent with the idea that they evolved early in Earth's history.

The other reason why CH_4 is likely to have been more abundant on the early Earth is that atmospheric O_2 was scarce until about 2.3 Ga. (More on this below.) The absence of O_2 changes the photochemical lifetime of CH_4. Today, most CH_4 is destroyed in the troposphere by reaction with the hydroxyl radical, OH. The reaction sequence is:

$$O_3 + h\nu(\lambda < 310\,\text{nm}) \to O_2 + O(^1D),$$
$$O(^1D) + H_2O \to 2OH,$$
$$CH_4 + OH \to CH_3 + H_2O,$$
$$CH_3 + O_2 + M \to CH_3O_2 + M \to \cdots \to CO\,(\text{or}\,CO_2) + H_2O, \qquad (R1)$$

where M represents a third molecule, necessary to carry off the excess energy to make two other molecules stick in a collision. On the early Earth both O_2 and O_3 were scarce; hence, this reaction sequence did not operate effectively. OH would still have been produced by direct photolysis of water vapour:

$$H_2O + h\nu \to H + OH. \qquad (R2)$$

However, most of this OH would have reacted with H_2 rather than CH_4. The main loss process for CH_4 in such an atmosphere is photolysis at wavelengths shorter than 145 nm. Most of the solar flux below this cut-off, about 5×10^{11} ph cm^{-2} s^{-1}, is in the Ly α line at 121.6 nm.

The consequence of all this is that the predicted lifetime of CH_4 in the early atmosphere is $\sim$10 000 years, as compared to 10–12 years today (Pavlov *et al.* 2001). Hence, the same biological production rate that leads to 1.7 ppmv (parts per million by volume) of CH_4 in the modern atmosphere would have produced over 1000 ppmv, equivalent to a volume mixing ratio of 10^{-3}, in the Archaean atmosphere. This is enough CH_4 to have had a substantial effect on climate, as shown in Figure 8.4. This figure, from Pavlov *et al.* (2000), shows one-dimensional climate model calculations of mean surface temperature as

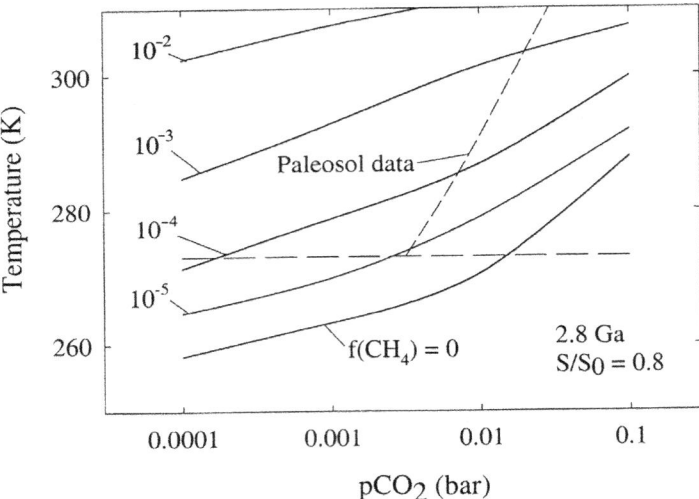

FIGURE 8.4. Solid curves represent mean surface temperature calculated by a one-dimensional climate model for various amounts of atmospheric CO_2 and CH_4. The time is 2.8 Ga, and the solar luminosity is 80% of the present value. The two dashed curves represent the freezing point of water and the upper limit on atmospheric pCO_2 derived from paleosols (see text). From Pavlov et al. 2000. New unpublished calculations show that the greenhouse effect due to CH_4 is overestimated here. However, some of it can be recovered by including other hydrocarbon gases, especially C_2H_6.

a function of CO_2 partial pressure and CH_4 mixing ratio. The calculation was performed for a time around 2.8 Ga when solar luminosity was ~80 per cent of the present value. The horizontal dashed line represents the freezing point of water. The other dashed curve is an upper limit on atmospheric CO_2 derived from palaeosols (ancient soils) by Rye et al. (1995). This limit should be regarded with caution, as it is based on the absence of a particular mineral (siderite, $FeCO_3$) and there may be other explanations for this; however, it provides some support for the idea that methane levels were high at this time. In Figure 8.4, one should note that the zero-CH_4 curve (bottommost solid curve) crosses the freezing point of water at a CO_2 partial pressure well above the palaeosol upper limit. By contrast, the $fCH_4 = 10^{-3}$ curve achieves surface temperatures of 280–290 K even for CO_2 levels no higher than today (~370 ppmv, or 3.7×10^{-4} bar). This, again, supports the hypothesis that atmospheric CH_4 was abundant at this time.

The actual Archaean climate story may have involved yet another complexity. When CH_4 becomes more abundant than CO_2 in the atmosphere, its photochemistry changes radically (Zahnle 1986; Pavlov et al. 2001). Instead of oxidizing to CO_2 or CO, it polymerizes to form higher hydrocarbons. These hydrocarbons can condense to form organic haze, similar to the haze observed today on Saturn's moon, Titan. The haze absorbs sunlight high up in the stratosphere and re-radiates it back to space, thereby producing an anti-greenhouse effect that cools the surface (McKay et al. 1991; Pavlov et al. 2001). The climate may thus have been stabilized by a negative feedback loop involving methane production, haze formation and surface temperature.

At some time near 2.3 Ga, atmospheric O_2 levels increased dramatically and the whole situation changed. Atmospheric CH_4 levels dropped, the greenhouse effect was diminished, and the climate became cold. This conjecture is supported by direct evidence from rocks in the Huronian Supergroup in southern Canada (Figure 8.5), first

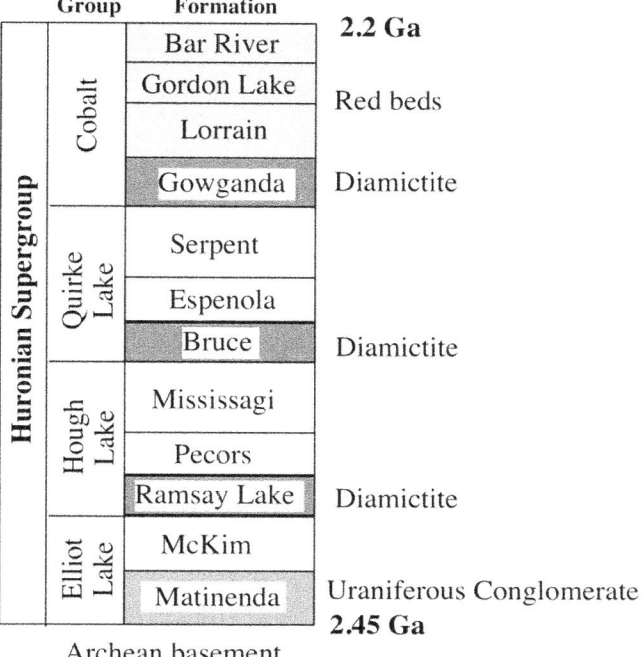

FIGURE 8.5. Stratigraphic sequence from the Huronian Supergroup in southern Canada. The entire sequence was deposited between 2.45 Ga and 2.2 Ga. The three glacial diamictite layers are underlain by sediments containing detrital uraninite and pyrite (indicating low atmospheric O_2) and are overlain by red beds (indicating high O_2). After Young 1991.

studied by Roscoe (1969, 1973). These rocks, which were laid down between 2.45 and 2.2 Ga, contain three glacial layers (diamictites). Below the lowermost glacial layer, in the Matinenda Formation, one finds detrital pyrite and uraninite. As discussed further below, these minerals are considered as evidence for low atmospheric O_2 levels. Above the uppermost glacial layer, one finds the Lorraine red bed formation, which is evidence for high atmospheric O_2. Hence, this stratigraphic sequence supports the story outlined above, in which CH_4 played a major role in climate during the Archaean and earliest Proterozoic. Indeed, CH_4 may have continued to play a significant role in climate during the mid-Proterozoic as well because of enhanced production from marine sediments (Pavlov et al. 2003); however, this goes beyond the scope of this chapter.

8.2. Venus, Mars and the habitable zones around the Sun and other stars

The discussion of the previous section explains why Earth has remained habitable throughout most or all of its history. Abiotic climate stabilization by CO_2 and possible biotic stabilization by CH_4 both seem to have been involved. There are clearly limits, however, to how robust these climate stabilization mechanisms may be, as evidenced by the observation that they do not appear to have worked on our neighbouring planets, Mars and Venus. Venus is a dry burning hell with no liquid water and a surface temperature of 730 K. Mars is a frozen desert with a mean surface temperature of 218 K.

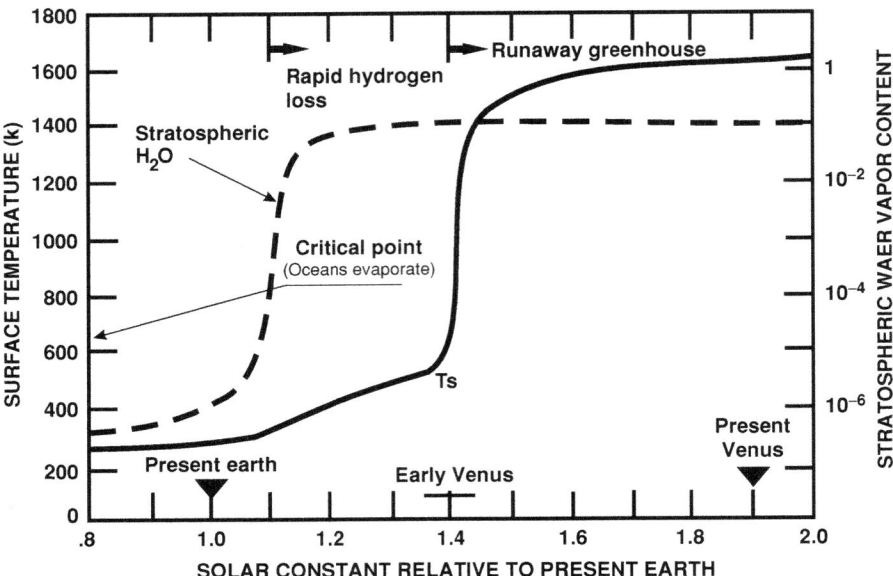

FIGURE 8.6. Diagram illustrating the 'runaway' and 'moist' greenhouse effects on Venus. The horizontal axis is the solar flux relative to the present flux at Earth. The solid curve represents surface temperature; the dashed curve represents stratospheric H_2O mixing ratio. After Kasting 1988.

8.2.1. *Venus: runaway and 'moist' greenhouses and loss of water*

The story of how Venus came to be the way it is has been told and retold several times. Important early work was by Ingersoll (1969) and Rasool & DeBergh (1970). In both papers, the authors assumed that Venus accreted wet and lost its water by way of a runaway greenhouse effect. This idea is consistent with the concept, mentioned earlier, that the water came originally from the asteroid belt region. If the H_2O-bearing planetesimals came from beyond the orbit of Mars, then it is likely that all four terrestrial planets would have accreted substantial amounts of water. Some planets (e.g. Earth) may have received more than others, however, if the impactors included large planetesimals, so that the statistics of small numbers were involved (Morbidelli *et al.* 2000). Evidence that Venus has lost at least a modest amount of water is provided by the extremely high D/H ratio in Venus' atmosphere, about 150 times that of Earth (Donahue *et al.* 1982). If both planets started out with the same D/H ratio in their initial water inventories, then Venus must have had at least 150 times more H_2O than it has at present (~30 ppmv in the lower atmosphere). Loss of some D, along with H, would allow Venus' initial water inventory to be much larger, perhaps approaching that of Earth (Kasting & Pollack 1983). (Earth has ~10^5 times as much water as Venus does today.)

A more recent attempt to study Venus' atmospheric evolution was made by Kasting (1988). Figure 8.6 summarizes the results of that study. In these calculations, which were begun while I was working with James Pollack at NASA Ames, we performed a hypothetical numerical experiment in which we pushed the present Earth closer to the Sun, keeping track of its surface temperature and planetary radiation balance. (In practice, this was done as an inverse calculation in which the planet's surface temperature was varied, and a one-dimensional climate model was used to calculate the solar flux required to sustain it.) The solar flux, of course, increases as the inverse square of the orbital distance as the planet moves inwards. The solid curve in Figure 8.6 shows surface temperature as a function of the effective solar flux, S/S_0 (the solar flux relative to that of present Earth). The surface temperature increases slowly at first, then jumps to extremely high values ($\sim$1500 K) when S/S_0 reaches 1.4. The last of the water disappears when the surface temperature rises above the critical temperature for water, 647 K. The surface pressure of this steam atmosphere (which assumes a full terrestrial ocean of water, 1.4×10^{21} l) is 220 bar. This is a true 'runaway greenhouse' atmosphere, in my preferred terminology. Venus' orbital radius is 0.723 AU, so $S/S_0 = 1.91$, placing the planet well within the runaway greenhouse regime.

The limit of planetary habitability is reached much sooner than this, however. The dashed curve in Figure 8.6, which goes with the scale on the right, represents the calculated stratospheric H_2O mixing ratio. This mixing ratio increases rapidly near $S/S_0 = 1.1$ and approaches unity for $S/S_0 > 1.2$. The key physics here, which was first elucidated by Ingersoll (1969), is that the tropopause 'cold trap' for water vapour becomes ineffective at high surface temperatures. The technical reason for this behaviour is that the large latent heat release from cloud formation lowers the tropospheric lapse rate (dT/dz), so that the convective troposphere extends up to very high altitudes. Once water vapour becomes a major constituent of the stratosphere, it can be readily photolysed by solar UV radiation. The hydrogen escapes to space, the oxygen goes back and reacts with the planet's surface, so the net result is loss of water. (Note that, in a low-O_2 atmosphere, water vapour can be photolysed in the troposphere as well, as pointed out in the previous section. However, this does *not* result in increased loss of hydrogen to space because the by-products of H_2O photolysis, H and OH, can recombine to form H_2O, which can then condense out as a liquid. That is why it is necessary to understand the technical details of the hydrogen escape process, which are discussed in the next section.)

The calculations just described can be used to estimate the position of the inner edge of the habitable zone, or HZ. According to the inverse square law, the solar flux at which water loss becomes rapid ($S/S_0 = 1.1$) corresponds to an orbital distance of $\sim$0.95 AU. This estimate for the inner edge of the HZ is precisely the same as that estimated by Hart (1978). The agreement is coincidental, however, as the two climate models were very different, and because Hart's model experienced a true runaway greenhouse (complete evaporation of the oceans) at this flux, whereas the Kasting (1988) model experienced what I termed a 'moist' greenhouse, in which a liquid ocean was still present even as the water was being lost. The moist greenhouse model is favoured for early Venus if Venus received a large initial complement of water and if clouds (which were neglected in these calculations) cooled the climate, as seems likely. This model can also better explain the loss of the last $\sim$10 per cent of Venus' water because the presence of an ocean should have kept most of Venus' CO_2 sequestered in carbonate rocks, thinning the atmosphere and allowing the tropopause cold trap to become established at a later phase in Venus' climate history.

8.2.2. *The climate of early Mars*

Mars presents the opposite problem from a planetary habitability standpoint. Mars orbits at 1.524 AU, so the solar flux is only 43 per cent of the terrestrial value. All other things being equal, this should make the martian climate much colder than that of Earth. And it is. As mentioned earlier, Mars' mean surface temperature is 218 K, or -55 °C, way below the freezing point of water. (Seawater, being salty, freezes at -2 °C. Even dense brines freeze at temperatures of -30 °C.) To be sure, climate models predict that surface

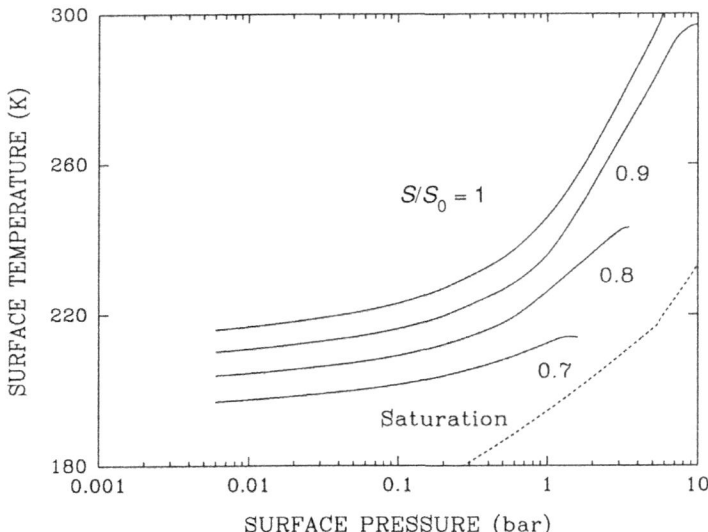

FIGURE 8.7. Solid curves represent mean surface temperature of Mars for different values of its relative present-day solar flux, S/S_0, and for different surface pressures. A CO_2–H_2O atmosphere was assumed, and the tropospheric relative humidity was set to unity (to produce an upper limit on surface temperature). From Kasting 1991.

temperatures do climb above 0 °C at noontime near the equator; however, any water ice that is present probably sublimates, rather than melting, because the surface pressure (6–8 mbar) is not much above the triple point pressure for water, 6.1 mbar. Kahn (1985) has provided a nice discussion of the physics of this process. (Indeed, Kahn suggests that the surface pressure is regulated at a point where transient liquid water just fails to exist, preventing further conversion of atmospheric CO_2 into carbonates.)

The interesting thing about Mars, from a climatologist's standpoint, is that Mars does not appear to have been cold throughout its history. Fluvial channels have been observed on Mars' surface by various spacecraft sent there since the early 1970s. Some of the most recent photographs (Malin & Edgett 2000; Mangold *et al.* 2004) have been taken at high resolution and provide convincing evidence for 'persistent' stream flow and relatively warm palaeoclimates. Surface observations by the *Mars Exploration Rovers* have revealed minerals, such as haematite 'blueberries', that must have formed in contact with liquid water (Squyres *et al.* 2004).

What makes these observations particularly puzzling is that most of the fluvial features are found on heavily cratered parts of the martian surface. From studies of lunar cratering (which include radiometrically dated Moon rocks), we believe that most of the cratering in the inner Solar System occurred at or prior to 3.8 Ga, during the time interval referred to as the 'heavy bombardment period'. This makes the problem of explaining Mars' early warmth even more difficult, as the Sun was less than 75 per cent of its current brightness at that time. Consequently, climatologists (including myself) have struggled to find a self-consistent mechanism for keeping early Mars warm. Gaseous CO_2 and H_2O by themselves are not adequate to do the job. Figure 8.7, from Kasting (1991), shows surface temperatures calculated by a one-dimensional climate model for various CO_2 partial pressures and solar fluxes. Mars' surface pressure today is 0.006 bar, and its atmosphere is nearly pure CO_2. According to this model, for present Mars ($S/S_{0\,\mathrm{Mars}} = 1$), a CO_2

partial pressure of ~3 bar would be sufficient to bring the mean surface temperature up above freezing. If $S/S_{0\,\mathrm{Mars}} < 0.85$, however, no amount of CO_2 would be sufficient to do this. The reasons are twofold: (1) as temperatures become cooler, condensation of CO_2 within the troposphere reduces the convective lapse rate, thereby lowering the greenhouse effect, and (2) high CO_2 partial pressures lead to increased Rayleigh backscattering of incoming solar radiation, thereby raising the planetary albedo. The combination of these two factors makes it impossible to warm early Mars using this mechanism. For $S/S_{0\,\mathrm{Mars}} = 0.75$ (the value at 3.8 Ga), the maximum surface temperature achieved in this model is 225 K – nearly 50 degrees below the freezing point.

As one might expect, there are various ways out of this dilemma. Real planets are of course three-dimensional, not one-dimensional, so it is possible that parts of the martian surface could have been above freezing even if the mean surface temperature was low. So far, no one has repeated these calculations with a three-dimensional climate model. I predict, however, that this will *not* solve the problem. Indeed, it is likely to exacerbate it instead. CO_2 condensation in the cold polar regions becomes an important consideration. And ice albedo feedback, which was not included in the one-dimensional calculations, may also become a factor. In part, this depends on the amount of water with which Mars was initially endowed. If it was largely covered by oceans, like Earth, then ice albedo feedback would be a huge problem. Recent 1.5-dimensional climate modelling suggests that mean planetary surface temperatures below about $-5\,°C$ are unsustainable in these circumstances; hence, the climate of a wet planet will either be Earthlike, or will resemble Snowball Earth. This result holds true regardless of planetary obliquity (Kasting 1991). Mars' obliquity is thought to vary chaotically from 0–60° as a consequence of gravitational interactions with the other planets (Laskar & Robutel 1993; Touma & Wisdom 1993).

Other solutions to the problem of the martian palaeoclimate also exist. Forget & Pierrehumbert (1997) suggested that CO_2 clouds themselves may have warmed the climate by creating a 'scattering' greenhouse effect. This might work; however, it requires nearly 100 per cent cloud cover, which is unlikely for condensation clouds such as these. Venus and Titan both have nearly 100 per cent cloud cover, but in both cases the clouds are formed photochemically from some other non-condensable gas (SO_2 for Venus, CH_4 for Titan). Earth's H_2O clouds, and martian CO_2 clouds, form primarily on updrafts and are thus unlikely to provide complete surface coverage. (On Earth, clouds obscure about 60 per cent of the surface.)

Another possibility is methane. We have seen earlier that CH_4 is thought to have played a major role in Earth's palaeoclimate. Perhaps it did the same thing for early Mars. Most of Earth's CH_4 is biological, however. Hence, high CH_4 concentrations on early Mars may imply the presence of life. We have begun to investigate this possibility using the same climate model used to generate Figure 8.4; however, we have not published the results because we are trying to correct problems with CH_4 absorption in the near-IR region. Some support for the idea that CH_4 was important on early Mars comes from recent reports that there may be ~10 ppbv (parts per billion by volume) of CH_4 in the *present* martian atmosphere (Mumma *et al.* 2003; Formisano *et al.* 2004; Krasnopolsky *et al.* 2004). These measurements are all on the verge of spectroscopic believability, so one should retain some scepticism about this result. If it is correct, though, it may imply that there are methanogens (or their martian equivalents) living at depth within some subsurface aquifer. This is not inconceivable. Putting a high-resolution spectrometer into orbit around Mars to further test this possibility is an excellent idea.

In any case, even though we cannot claim to understand the climate of early Mars, we can use the results shown in Figure 8.7 to obtain a (conservative) estimate for the outer edge of the HZ. If we restrict ourselves to greenhouse warming by gas-phase CO_2

and H_2O, the lowest solar flux at which Mars' mean surface temperature can be raised above freezing corresponds to $S/S_{0\oplus} = 0.43 \times 0.85 = 0.37$. This, in turn, corresponds to an orbital distance of $1 \text{ AU} \times (1/0.37)^{0.5} = 1.64$ AU. The presence of widespread, optically thick CO_2 ice clouds could conceivably raise this value to ~ 2.0 AU (Forget & Pierrehumbert 1997; Mischna et al. 2000).

Why then is Mars not presently habitable? This question is not too difficult to answer. Mars is much smaller than Earth (~ 0.1 Earth mass), and consequently it cooled off much faster. Thus, there is by this time little or no active volcanism or plate tectonics to recycle carbonate rocks back into gaseous CO_2. An Earth-sized planet at Mars' orbital distance would probably be habitable, but Mars itself is small, dry and cold. Clearly, other factors besides orbital distance are required to ensure planetary habitability.

Finally, one can also use these results to estimate the width of the *continuously habitable zone* (CHZ) around the Sun. The CHZ is the region that remains habitable over some specified time period, usually taken as the lifetime of the Solar System, 4.6 Ga. As the Sun was only 70 per cent as bright at the beginning of this time, the outer edge of the HZ would have been in at $1.64 \text{ AU} \times 0.7^{0.5} = 1.37$ AU (or, if one uses the more optimistic estimate of 2.0 AU for the outer edge of the present HZ, the CHZ outer edge would be at 1.67 AU). This is equivalent to saying that early Mars could have been warmed by CO_2 clouds if they were sufficiently widespread.

8.2.3. Habitable zones around main sequence stars

One can put together calculations like those discussed in the previous two subsections to estimate the width of the habitable zone around the Sun and other main sequence stars. The results of doing so are summarized in Figure 8.8, which shows the zero-age-main-sequence HZ around stars in the mass range of about 0.1–2.5 $M_\odot$. One can only make such a plot for a particular time in the star's history. Stars brighten with age, massive stars faster than smaller ones, so their HZs migrate outward at different rates as time goes by.

The basic results shown in Figure 8.8 are intuitive. The HZ moves outward for brighter, bluer stars and inward for dimmer, redder stars. It remains roughly the same width on a log scale, however. Thus, if planets around other stars are spaced logarithmically, as they are in our own Solar System, the chances of finding one within the HZ are reasonably good (~ 50 per cent) (Kasting et al. 1993). Problems arise, though, for stars too different from the Sun. Hot, blue stars have short main sequence lifetimes and they emit much of their energy at UV wavelengths. The UV radiation could pose problems for life, particularly prior to the rise of atmospheric O_2. Once O_2 becomes abundant in a planet's atmosphere, *even planets around F-stars would be well shielded from UV because of the development of a thick ozone layer* (Kasting et al. 1997; Segura et al. 2003). For planets around dim red stars, two other problems arise. First, the HZ moves within the star's tidal locking radius. Thus, a planet within the HZ would be in danger of developing synchronous rotation (one side always facing towards its parent star), just as the Moon rotates synchronously around the Earth. This could prove fatal towards habitability, as the atmosphere and ocean could freeze out on the night side to form a giant ice cap. However, as pointed out by Joshi et al. (1997), planets that are relatively far out in their HZs should build up dense CO_2 atmospheres that could transfer heat effectively from their daysides to their nightsides. Thus, some M-star planets could remain habitable. Another way out of this problem is illustrated by the behaviour of the planet Mercury in our own Solar System. Mercury is within the Sun's tidal locking radius (Figure 8.8); however, it does *not* rotate synchronously. Rather, it was on its way to doing so when it became trapped in a 3:2 spin–orbit resonance, in which it remains today. (Mercury's

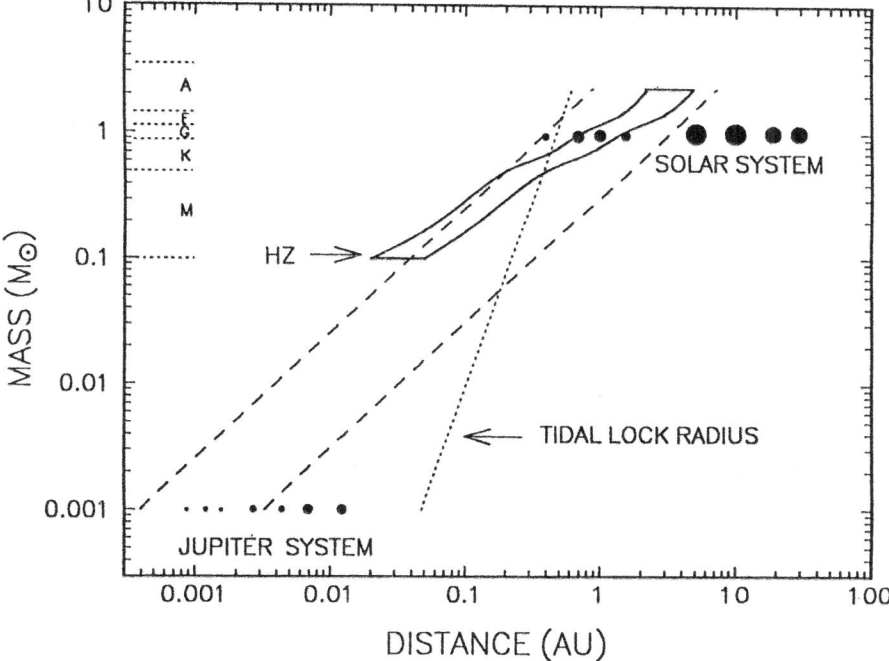

FIGURE 8.8. Diagram showing the boundaries of the liquid water habitable zone for stars of different masses at the time when they first enter the main sequence. The nine planets in our Solar System are shown. The dashed curve represents the 4.5-Gyr tidal locking radius, i.e. the distance within which a planet's spin would become tidally locked over that time interval. From Kasting *et al.* 1993.

mass distribution is slightly asymmetric, so its lowest energy state is found when the axis corresponding to its smallest moment of inertia is pointed towards the Sun when the planet is at perihelion.)

8.3. Abiotic oxygen levels on the Earth and Earthlike planets

Climate is an important factor in determining whether a planet might be habitable, but it is not the only one. For multicellular, eukaryotic organisms like plants, animals and humans, molecular oxygen (O_2) is another fundamental requirement. (*Eukaryotes* are organisms whose cells have nuclei; *prokaryotes*, including Bacteria and Archaea, are single-celled organisms whose cells lack nuclei.) O_2 is used by eukaryotes, and by some prokaryotes as well, for respiration, in which it is combined with organic matter to produce energy for metabolism.

O_2 is important for another reason as well. Along with its photochemically produced product, O_3 (ozone), O_2 is perhaps the best biomarker gas that might be used for remote life detection. O_2 has a strong absorption band at 0.76 µm, just beyond visible wavelengths, so it could potentially be observed by a space-based coronagraph such as NASA's *TPF-C* mission. Furthermore, nearly all of the O_2 in Earth's atmosphere is produced biologically by the process of photosynthesis,

$$CO_2 + H_2O \rightarrow CH_2O + O_2. \tag{R3}$$

Hence, under most circumstances, the detection of O_2 in the atmosphere of an extrasolar planet would be synonymous with the detection of life. There are, however, certain easily imaginable situations under which free O_2 could be produced abiotically in a planet's atmosphere. The most obvious of these is a post-runaway greenhouse planet, like early Venus, that started its life with lots of water, then lost it through photodissociation and escape of hydrogen to space (Kasting 1997). A somewhat less obvious case is a frozen planet, like Mars but a little bit bigger, that could retain its atmosphere effectively but that was too small to maintain a volcanic source of reduced gases (Kasting 1997). Such a planet could lose hydrogen relatively slowly but still build up O_2 because its oxygen sinks would be small.

In this section, I outline the basic processes governing hydrogen escape, and I describe how one can estimate atmospheric O_2 levels on a planet like the prebiotic Earth.

8.3.1. A historical perspective on the abiotic O_2 problem

It may be worthwhile beginning with a brief historical perspective on the problem of computing O_2 levels on the prebiotic Earth. This problem has been of interest to both geologists and biologists for many years because of its possible relevance to the origin of life and to early biological evolution. The first scientists to make a serious attempt to solve this problem were Berkner and Marshall (1964, 1965, 1966, 1967). These authors correctly deduced that a net abiotic source of O_2 could be provided by photodissociation of H_2O followed by escape of hydrogen to space. They also realized that both O_2 and H_2O are photolysed at approximately the same wavelengths, shortward of $\sim$240 nm. Most of the H_2O is confined to the troposphere, the lowest 10–15 km of Earth's atmosphere, whereas O_2 is well mixed up to $\sim$100 km altitude. Thus, Berkner and Marshall reasoned that O_2 should build up by this process until it became abundant enough to shield H_2O from photolysis. In their model, this occurred at an O_2 concentration of 10^{-4}–10^{-3} PAL. (PAL means 'times the present atmospheric level'.)

Following their work, Brinkmann (1969) came along and published his own solution to this problem. Despite adding an additional, crustal loss process for O_2, he predicted abiotic O_2 concentrations as high as 0.27 PAL. If this result were correct, then it would be hopeless to use O_2 as a biomarker. In constructing his model, however, Brinkmann assumed (for reasons that remain unclear) that precisely 1/10 of the H atoms produced by H_2O photolysis eventually escape to space. As demonstrated below, this fraction is enormously larger than the fraction of H atoms that actually escape. Hence, the key to solving the abiotic O_2 problem is to understand the factors that control hydrogen escape. The next section describes what we know about this process.

8.3.2. Escape of hydrogen to space / the diffusion-limited flux

Continuing our historical perspective for the moment, the first researcher to correctly elucidate the factors controlling hydrogen escape from Earth was Hunten (1973). Prior to Hunten, Sir James Jeans had shown that hydrogen (and helium) could be lost to space by evaporation from the *exobase*. The exobase, at $\sim$500 km altitude, is the height at which the atmosphere becomes collisionless. H atoms (or H_2 molecules) that are moving upwards at this height at speeds exceeding the escape velocity, 11 km s^{-1}, have a good chance of escaping. Mathematical descriptions of this process can be found in Walker (1977), Chamberlain & Hunten (1987) and in many other upper atmosphere texts. Because Earth's upper atmosphere is hot (1000 K at solar minimum, 2500 K at solar maximum), H atoms in the high-energy tail of the Maxwell–Bolzmann velocity distribution generally have enough energy to do this. This thermal escape process is often termed 'Jeans escape' in honour of the person who first described it.

What Hunten discovered (from studies of Saturn's moon Titan) was that hydrogen escape can also be limited at lower altitudes. The escaping hydrogen must move by diffusion through the non-escaping background gas, which for Earth consists primarily of N_2 and O_2. The static background gas slows the escape of hydrogen by creating friction. Mathematically, Hunten showed (but see Walker 1977 for a lucid discussion) that the diffusion is most limiting at the *homopause*, near 90–100 km altitude. Above the homopause, gases move relative to each other by the process of molecular, or Fickian, diffusion. Below the homopause, transport of gases is dominated by turbulence and large-scale motions, collectively parameterized as 'eddy diffusion'. The maximum upwards flux of hydrogen at the homopause is proportional to the total mixing ratio of hydrogen in all of its chemical forms, $f_{tot}(H)$:

$$\phi_{lim}(H) \cong 2.5 \times 10^{13} f_{tot}(H) \quad \text{molecules cm}^{-2} \text{s}^{-1}. \tag{8.2}$$

Here, the total hydrogen mixing ratio is given by the sum of the mixing ratios of each H-bearing species, weighted by the number of H atoms it contains

$$f_{tot}(H) = f(H) + 2f(H_2) + 2f(H_2O) + 4f(CH_4) + \cdots. \tag{8.3}$$

The term 'mixing ratio' denotes fractional amount by volume, equivalent to 'mole fraction'. The constant factor in Eq. (8.2) depends weakly on the mixture of chemical forms of hydrogen at the homopause. The leading factor is $\sim 2 \times 10^{13}$ for atomic H and 4×10^{13} for H_2, and the dominant H-bearing species at Earth's homopause are H and H_2.

What makes Eqs. (8.2) and (8.3) useful is that the total hydrogen mixing ratio is nearly constant between the tropopause (~ 10 km) and the homopause. This is intuitively clear, as photochemical reactions can neither create nor destroy hydrogen atoms; rather, they merely convert one form of H-bearing gas into another. Below the tropopause, H_2O can condense and so the total hydrogen mixing ratio is *not* constant with altitude. At the tropopause, the dominant H-bearing species are H_2O (~ 3 ppmv) and CH_4 (~ 1.7 ppmv). Because CH_4 contains twice as many H atoms as does H_2O, both gases make approximately equal contributions to the escape rate. Thus, the escape rate of hydrogen from the modern Earth is approximately

$$\begin{aligned}\phi_{lim}(H) &\cong 2.5 \times 10^{13} \cdot [2(3 \times 10^{-6}) + 4(1.7 \times 10^{-6})] \\ &= 3.2 \times 10^8 \, \text{H atom cm}^{-2} \, \text{s}^{-1}.\end{aligned} \tag{8.4}$$

This number has been compared with H escape rates estimated from spacecraft observations (e.g. Liu & Donahue 1974) and has been shown to be in good agreement. At solar minimum, the Jeans escape rate is actually somewhat slower than this; hence, hydrogen escape could in theory be limited at the exobase rather than at the homopause. In practice, however, there are also several non-thermal hydrogen escape processes, e.g. charge exchange with hot H^+ ions in Earth's magnetosphere, that allow H atoms to escape. Thus, the *diffusion-limited flux* predicted by Eqs. (8.2) and (8.3) provides a good estimate to how fast hydrogen escapes from the modern Earth.

Already one should be able to tell from this discussion where both Berkner & Marshall and Brinkmann went wrong. The escape rate of hydrogen into space is *not* simply related to the H_2O photolysis rate. In the low-O_2 photochemical models described below, H_2O is photolysed at an extremely rapid rate in the troposphere, 10^{12}–10^{13} cm^{-2} s^{-1}; however, escape rates are 2–3 orders of magnitude lower than this. It is the total mixing ratio of hydrogen in the stratosphere that matters, not how fast H_2O is photolysed. In the extreme case of a very quiescent M-star, where no short-wavelength UV photons were available, photolysis of H_2O might itself become the limiting factor. However, for most actual stars, the processes described above are the ones that should be considered.

8.3.3. *Balancing the atmospheric hydrogen budget: first-order calculation*

Once one understands the factors that control hydrogen escape, it now becomes possible to reliably estimate the atmospheric O_2 concentration on an abiotic, or prebiotic, Earth. The trick is to first balance the atmospheric hydrogen budget. Balancing the atmospheric hydrogen budget is equivalent to balancing the oxygen budget – it is just easier to do because most of the inputs to and outputs from the atmosphere are in the form of H_2. The first researcher to do this calculation was Walker (1977). Photolysis of H_2O followed by escape of hydrogen to space is the source for O_2. Sinks for O_2 are provided by oxidation of reduced volcanic gases and of reduced minerals at Earth's surface. The volcanic gases are more readily oxidized than is the surface; hence, they provide the most potent O_2 sink.

How can one estimate the flux of reduced volcanic gases? This is not an easy task. Instrumenting volcanoes globally with gas detectors would be completely unfeasible. However, there are ways around this problem. One can estimate these numbers indirectly by analyzing the global carbon cycle. In particular, one must look at the inorganic, or carbonate–silicate cycle. (The organic carbon cycle, which is both faster and more familiar to most people than the inorganic cycle, is irrelevant to this discussion.) Walker did this in his 1977 book. The most reliable and up-to-date numbers, however, come from Holland (1978, 1984, 2002). Holland's approach is to start from dissolved bicarbonate (HCO_3^-) concentrations in rivers. This bicarbonate derives from atmospheric CO_2 that was used in weathering of surface rocks, mostly silicates and carbonates. From considerations of mass balance, one can show that over long timescales the rate of silicate weathering must equal the rate at which CO_2 is outgassed from volcanoes. By combining the measured bicarbonate concentrations in river water with the (known) global flux of river water into the oceans, one can derive a number for the total CO_2 loss rate, $\sim 6 \times 10^{12}$ mol yr^{-1}. If the global carbonate–silicate cycle is currently in a steady state, then the CO_2 outgassing rate must be equal to this value. (Note that the global *organic* carbon cycle is clearly not in a steady state, largely as a consequence of the burning of fossil fuels. This is causing atmospheric CO_2 levels to rise and may well be causing the Earth to warm at the same time. Thus far, though, the temperature increase has been relatively small, <1 °C, and so its effect on weathering rates and the carbonate–silicate cycle should be minimal.)

Having obtained the CO_2 outgassing rate, one then applies two additional logical steps to get the reduced gas flux. From observations we know that the ratio of H_2O to CO_2 in volcanic gases varies between ~ 10 and 50 (Holland 2002). We can also predict theoretically that the ratio of H_2 to H_2O in volcanic gases is ~ 0.02 (Holland 2002). This prediction comes from assuming thermodynamic equilibrium between these two gases:

$$H_2 + 1/2\ O_2 \leftrightarrow H_2O, \tag{R4}$$

and then setting the oxygen fugacity equal to its value in equilibrium with typical silicate melts. (The term 'fugacity' is equivalent to 'partial pressure' and is used to describe the activity of a gas in equilibrium with a given mineral assemblage.) The ratio of 0.02 comes from using an oxygen fugacity near the QFM (quartz–fayalite–magnetite) synthetic buffer, which is thought to be representative of most of Earth's upper mantle. When one combines all these numbers together, one gets a volcanic H_2 flux,

$$\phi_{\text{out}}(H) \cong 5 \times 10^{12} \text{ mol yr}^{-1}, \quad \text{or} \quad 3.7 \times 10^{10} \text{ H atom cm}^{-2} \text{ s}^{-1}$$

(Holland 2002). Other reduced gases such as CO and SO_2 are also present in volcanic emissions, but according to Holland's analysis H_2 is by far the most abundant reduced species.

We are now in a position to say something about abiotic O_2 concentrations. H_2 released from volcanoes would have been a sink for O_2 via reaction (R4). The actual reaction would not be an equilibrium one, as shown in (R4). Rather, it would have been catalysed by the photolysis byproducts of H_2O, e.g.

$$H_2O + h\nu \rightarrow H + OH$$
$$H_2 + OH \rightarrow H_2O + H \quad (\times 2)$$
$$H + O_2 + M \rightarrow HO_2 + M$$
$$H + HO_2 \rightarrow 2OH$$
$$\underline{H + OH + M \rightarrow H_2O + M}$$
$$\text{Net } 2H_2 + O_2 \rightarrow 2H_2O. \tag{R5}$$

Meanwhile, the net abiotic source for O_2 would have come from photolysis of H_2O followed by escape of H to space. To estimate this rate, go back to Eq. (8.4) above. However, this time, we want to count only those H atoms that are coming from H_2O, not those coming from CH_4. Given that there are 3 ppmv of H_2O in the lower stratosphere, the H escape rate from water vapour is

$$\phi_{\lim}(H) \cong 2.5 \times 10^{13} \cdot 2(3) \times 10^{-6}$$
$$= 1.5 \times 10^8 \text{ H atom cm}^{-2} \text{s}^{-1}. \tag{8.5}$$

This number is about 250 times smaller than the hydrogen outgassing rate estimated above. (To convert these numbers to O_2 production and loss rates, one needs to divide each of them by four. However, their ratio remains the same.) Hence, it is clear that, using numbers for the modern Earth, the abiotic sink for O_2 is much larger than the abiotic source. This implies that the H_2 that is outgassed from volcanoes *cannot* be used up by reacting with the oxygen left behind from photolysis of water vapour. Rather, that H_2 must either somehow combine with CO_2 to form organic matter that can be buried in sediments, or it must escape to space.

On the prebiotic Earth, with no organisms available to catalyse the reaction of H_2 with CO_2, the rate of production of organic matter was probably small. Thus, most of the outgassed H_2 should have escaped to space. We can obtain a lower bound for the H_2 concentration in the atmosphere by assuming that H_2 escaped at its maximum, diffusion-limited rate. To do this, we employ Eqs. (8.2) and (8.3) once again, but we can now neglect all contributions to the total hydrogen mixing ratio except for that from H_2, i.e. we set $f_{\text{tot}}(H) \cong 2f(H_2)$ and then equate the escape rate with the outgassing rate, yielding

$$2.5 \times 10^{13} \cdot 2f(H_2) = \phi_{\text{out}}(H) = 3.7 \times 10^{10} \text{ H atom cm}^{-2} \text{s}^{-1},$$

or

$$f(H_2) = 3.7 \times 10^{10}/(5 \times 10^{13}) \cong 7 \times 10^{-4}. \tag{8.6}$$

These numbers are all approximate, so for convenience we will just round this to $f(H_2) \cong 10^{-3}$. Thus, at a minimum, the prebiotic atmosphere should have contained about 0.1 per cent H_2. This number could have been higher if volcanic outgassing was faster in the past (as seems likely) or if hydrogen escaped to space at less than the diffusion-limited rate. Recent numerical calculations of hydrodynamic escape rates from H_2-rich atmospheres (Tian *et al.* 2005) suggest that the escape should have been relatively slow, in which case H_2 could have built up to concentrations of several tens of per cent. This uncertainty in the actual H_2 concentration is acceptable in the sense that we are mostly interested in calculating an upper limit on the abiotic O_2 concentration. *If we can show that this upper limit is below measureable thresholds, then we can still have confidence that an actual detection of O_2 in a planet's atmosphere would be indicative of life.*

8.3.4. *More rigorous calculation of the atmospheric hydrogen budget*

For most purposes, the estimation of atmospheric H_2 concentrations outlined in the previous section is sufficient, as volcanic outgassing and escape to space appear to be the dominant terms in the atmospheric hydrogen budget. There are, however, other processes that should affect the H_2 concentration to some extent. These bear on the issue of possible false positives for (biotic) O_2; hence, a brief discussion of them is included here.

To begin, one should realize that when we talk about balancing either the atmospheric hydrogen budget or the oxygen budget, we are really talking about balancing the atmospheric redox budget. Or, to put it another way, we are balancing the flux of available electrons in the system. This can be illustrated by considering the effect of volcanic outgassing of reduced gases other than H_2. Take carbon monoxide, CO, as an example. According to Holland (2002), the volcanic flux of CO is ~5 per cent that of H_2. Most of the CO that is outgassed is eventually oxidized to CO_2. The direct oxidation reaction is: $CO + OH \rightarrow CO_2 + H$. The required oxygen atom comes from H_2O originally, though, so the reaction that is relevant to the redox budget is:

$$CO + H_2O \rightarrow CO_2 + H_2. \tag{R6}$$

Hence, each mole of CO that is outgassed is equivalent to one mole of H_2. Similarly, one can show that each mole of outgassed CH_4 is equivalent to four moles of H_2:

$$CH_4 + 2H_2O \rightarrow CO_2 + 4H_2. \tag{R7}$$

So outgassed CH_4 should enter into the redox budget with a stoichiometric coefficient of 4. (For this more detailed calculation it is convenient to keep track of the hydrogen budget in terms of H_2 molecules, rather than H atoms, as was done in the previous section.)

In addition to volcanic outgassing, the atmosphere can also gain or lose reducing power (hydrogen) by rainout of soluble species that are produced photochemically within the atmosphere. For example, one such species is formaldehyde, H_2CO. If we start from H_2 and CO_2, then the relevant redox reaction is

$$CO_2 + 2H_2 \rightarrow H_2CO + H_2O. \tag{R8}$$

Hence, for each mole of H_2CO that rains out, the atmosphere loses two moles of H_2. We are presuming here that the H_2CO goes into the ocean and is lost in some manner, e.g. by polymerizing to form sugars that precipitate out to form sediments. (Obviously, in doing such a 'rigorous' calculation, one needs to worry about what happens to the species once they enter the ocean.)

Rainout of oxidized species has just the opposite effect on the atmospheric hydrogen budget. For example, hydrogen peroxide, H_2O_2, is formed photochemically from reactions such as: $OH + OH + M \rightarrow H_2O_2 + M$. If the OH comes originally from H_2O, the net reaction is

$$2H_2O \rightarrow H_2O_2 + H_2. \tag{R9}$$

Hence, for each mole of H_2O_2 that rains out, the atmosphere gains one mole of H_2. In this case it is easier to think of what happens to the gas once it enters the ocean. The deep oceans, at least, should have contained significant concentrations of reduced (ferrous) iron, so the H_2O_2 would be consumed by oxidizing ferrous iron to ferric iron.

It should be clear from the foregoing discussion that one needs to have some reference redox state from which to calculate changes in H_2. We have already taken H_2O as the reference state for H-bearing gases and CO_2 as the reference state for C-bearing gases. This is done for convenience: if H_2O and CO_2 are defined as 'neutral' reference states, then one does not need to keep track of the flux of these gases into and out of the atmosphere (which would, after all, be virtually impossible to do!). For convenience, we also define N_2 and SO_2, respectively, as the reference states for N-bearing and S-bearing gases. One can then assign stoichiometric coefficients to each gas (or particle). As a final example, production of elemental sulfur particles, S_8, can be written as

$$8SO_2 + 16H_2 \rightarrow S_8 + 16H_2O. \tag{R10}$$

So the stoichiometric coefficient of S_8 in the redox budget is 16.

Having made these assignments, we can now express the redox budget in the form

$$\Phi_{out}(H_2) + \Phi_{out}(CO) + 4\Phi_{out}(CH_4) + 3/2\Phi_{out}(NH_3) + 3\Phi_{out}(H_2S) + \Phi_{rain}(ox)$$
$$= \Phi_{esc}(H_2) + \Phi_{rain}(red). \tag{8.7}$$

Here, $\Phi_{out}(i)$ represents the volcanic outgassing flux of species i, $\Phi_{rain}(ox)$ and $\Phi_{rain}(red)$ are the rainout rates of oxidized and reduced species, respectively, and $\Phi_{esc}(H_2)$ is the hydrogen escape rate to space, here expressed in terms of H_2 molecules (or moles). In a photochemical model, e.g. Pavlov et al. (2001), the rainout terms are each composed of two separate terms, one representing scavenging of gases by raindrops and the other representing direct surface deposition of gases into the ocean.

In practice, the rainout terms and the additional outgassing fluxes in Eq. (8.7) are typically only ~10 per cent of the outgassing and escape flux of H_2. That is why we were able to neglect them in the discussion of the previous subsection. However, there is one circumstance in which these terms could become quite important. Consider a planet that is like Earth in terms of climate and water abundance, but which is too small to maintain a substantial volcanic outgassing flux of reduced gases. (Mars, for example, satisfies this latter condition but not the former.) The only term remaining on the left hand side of Eq. (8.8), then, is $\Phi_{rain}(ox)$. If one neglected this term, then one would predict that escape of H_2 from H_2O photolysis could lead to essentially unlimited buildup of O_2. On this hypothetical Earthlike planet, however, rainout of oxidants such as H_2O_2 and H_2SO_4 will balance the hydrogen budget at an H_2 concentration of $\sim 10^{-4}$ (Kasting et al. 1984), ensuring that the abiotic O_2 level remains small. Essentially, the rainout of oxidants provides an efficient mechanism for oxidizing a planet's surface. Models that do not include this term, e.g. Selsis et al. (2002), predict unrealistically large abiotic O_2 concentrations on planets that lack volcanism.

In actuality, of course, Mars does *not* have an Earthlike climate, and so rainout of oxidants does not occur. Why, then, does Mars not build up O_2 indefinitely? The answer has to do with Mars' small size and correspondingly low gravity (about 1/3 that of Earth). Photochemical processes in Mars' upper atmosphere, e.g. dissociative recombination of O_2^+ ($O_2^+ + e \to O + O$), produce non-thermal O atoms that have sufficient energy to escape (McElroy et al. 1977). This limits Mars' O_2 concentration to $\sim$0.1 per cent. If Mars was, say, twice as massive as it is (but still not big enough to sustain volcanism), these non-thermal loss processes for O would be ineffective, and O_2 might well accumulate to much higher concentrations. This type of pathogenic situation could lead to a false positive detection of life by *TPF* or *Darwin*. Thus, *we should exercise care in interpreting any O_2 signal from a planet lying beyond the outer edge of the liquid water habitable zone.*

8.3.5. Abiotic O_2 levels

After all this discussion we are finally ready to present what might be a realistic estimate of the O_2 concentration in an abiotic, or prebiotic, atmosphere. As mentioned previously, Walker (1977) was the first to do this calculation. Walker calculated only the surface O_2 concentration. He got a number near 10^4 O_2 molecules cm^{-3}, corresponding to a mixing ratio of $\sim 10^{-15}$. This is slightly lower than my own estimate for the surface O_2 mixing ratio, $\sim 10^{-13}$. If we are interested in using O_2 as a biomarker in remote planetary atmospheres, however, we need to know its vertical profile as well. To calculate this, one needs to have a photochemical model that has resolution in the vertical. In such a model, one calculates photodissociation rates and other photochemical reaction rates as a function of altitude. One also needs to account for vertical mixing of gases by both eddy and molecular diffusion, as discussed earlier in this section. Typically, eddy mixing is assumed to occur at present mixing rates (which are empirically determined). This, of course, is a possible source of error when considering atmospheres other than our own, and so sensitivity studies must be performed in order to determine how changes in vertical mixing might affect the calculations.

Typical results from such a one-dimensional photochemical model calculation are shown in Figure 8.9. The assumed H_2 mixing ratio is $\sim 10^{-3}$, following the discussion in the previous subsections. A 1-bar atmosphere consisting of 80 per cent N_2 and 20 per cent CO_2 has been assumed. This is roughly enough CO_2 to compensate for the 25 per cent fainter young Sun back around 4 Ga. Although it cannot be seen from Figure 8.9, it is indeed the case that the ground-level O_2 concentration is very low, $\sim 10^{-13}$ PAL (present atmospheric level). The O_2 mixing ratio increases rapidly with altitude, however, up to a peak value of $\sim 10^{-3}$ at 60 km. This high-altitude peak in O_2 concentration is a consequence of photolysis of CO_2 followed by recombination of O atoms to form O_2:

$$CO_2 + h\nu \to CO + O, \quad \text{(R11)}$$
$$O + O + M \to O_2 + M. \quad \text{(R12)}$$

Here, 'M' represents a third molecule necessary to carry off the excess energy of the collision. Note that these reactions are *not* a net source of O_2 because the CO atoms produced by reaction (R1) flow downwards into the lower atmosphere where they combine with O_2 to reform CO_2. This reaction, like others discussed earlier, is mediated by the by-products of water vapour photolysis. Note also that despite the relatively high O_2 concentration in the upper stratosphere, the O_2 column depth is $<10^{-6}$ times the present O_2 column depth in Earth's atmosphere. *This is too small to be detected by a*

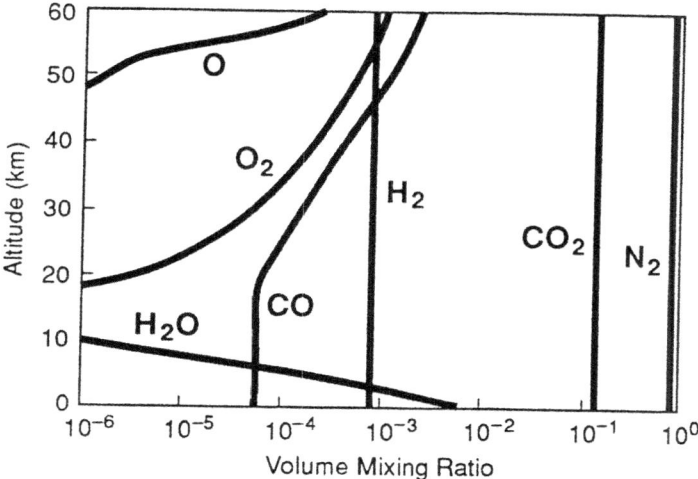

FIGURE 8.9. Vertical profiles of major gases in a weakly reduced prebiotic atmosphere. The assumed surface pressure is 1 bar. The H_2 mixing ratio is determined by balancing the volcanic outgassing rate with the diffusion-limited escape rate of hydrogen to space. From Kasting 1993.

low-resolution spectrometer and, hence, would yield a negative detection of photosynthetic life, consistent with the assumptions that went into the model.

8.4. The rise of oxygen and ozone

At some time close to 2.3 Ga, Earth's atmosphere underwent a dramatic change. In place of the H_2, and then later CH_4, that had dominated its redox balance, significant concentrations of free O_2 began to appear in the atmosphere. As pointed out in Section 8.2, the accompanying decrease in CH_4 concentrations may have thrown the climate into a deep-freeze and led to the first Snowball Earth glaciation(s).

8.4.1. A few thoughts on the rise of atmospheric O_2

This chapter is long enough already, and so I will not launch into a detailed discussion of the rise of O_2. There is no reason to do so here, particularly as this topic has been recently and extensively reviewed elsewhere (Holland 1994, 2002; Kasting & Catling 2003; Catling & Kasting 2006). Suffice it to say that a variety of geological evidence, first compiled by Cloud (1972), suggests that 2.3 Ga marked a sharp transition from basically anoxic to O_2-rich conditions. The evidence has been greatly buttressed in the last few years by new data on sulphur isotopes, which show evidence for so-called 'mass-independent' fractionation prior to this time (Farquhar et al. 2000, 2001; Kasting 2001; Pavlov & Kasting 2002; Ono et al. 2003). Unless new S-isotope data are uncovered that contradict the observed pattern, it seems safe to assume that the timing of the initial O_2 rise is now well determined.

This does *not*, however, imply that the rise of atmospheric O_2 is well understood. A number of interesting questions remain. One of the key questions is why did O_2 levels increase at 2.3 Ga when the organisms thought to be responsible for producing it, *cyanobacteria* (formerly called 'blue-green algae'), appear to have arisen at least 500 million years earlier (Brocks et al. 1999)? Carbon isotope data imply that production of O_2 was occurring at rates comparable to today (Kump et al. 2001). Thus, the sinks for O_2

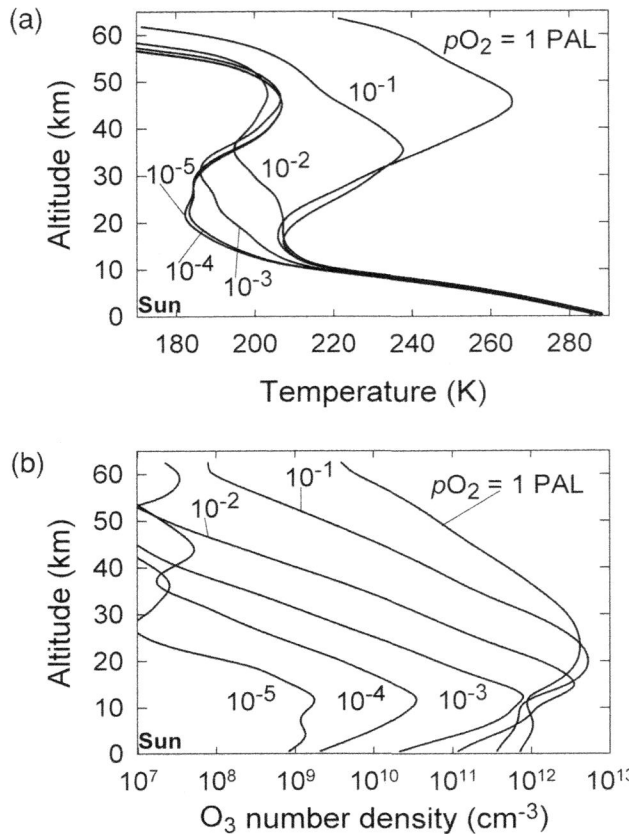

FIGURE 8.10. Vertical profiles of temperature (panel a) and ozone number density (panel b) as a function of atmospheric O_2 level for an Earthlike planet orbiting our Sun. 'PAL' means 'times the present atmospheric level'. From Segura et al. 2003.

must have been larger in the Late Archaean; however, the details of this explanation are not understood.

Another question is how high did O_2 levels go immediately following its initial rise, and what has been the subsequent time history of atmospheric O_2 concentrations? Canfield (1998), Canfield et al. (2000) and others (Anbar & Knoll 2002) have argued that the deep oceans remained anoxic (and sulphidic) during most of the Proterozoic. This suggests that atmospheric O_2 levels remained significantly lower than today, at least until 600–800 million years ago. As mentioned at the end of Section 8.2, such a scenario might imply that CH_4 remained an important greenhouse gas throughout this part of Earth history, too.

8.4.2. The rise of ozone

The rise of atmospheric O_2 should have been accompanied by a corresponding rise in atmospheric ozone, O_3. Ozone is interesting for at least two reasons: (1) it shields Earth's surface from damaging solar UV radiation in the 200–300 nm wavelength range; and (2) it can be an effective biomarker gas by way of its 9.6 μm band in the thermal

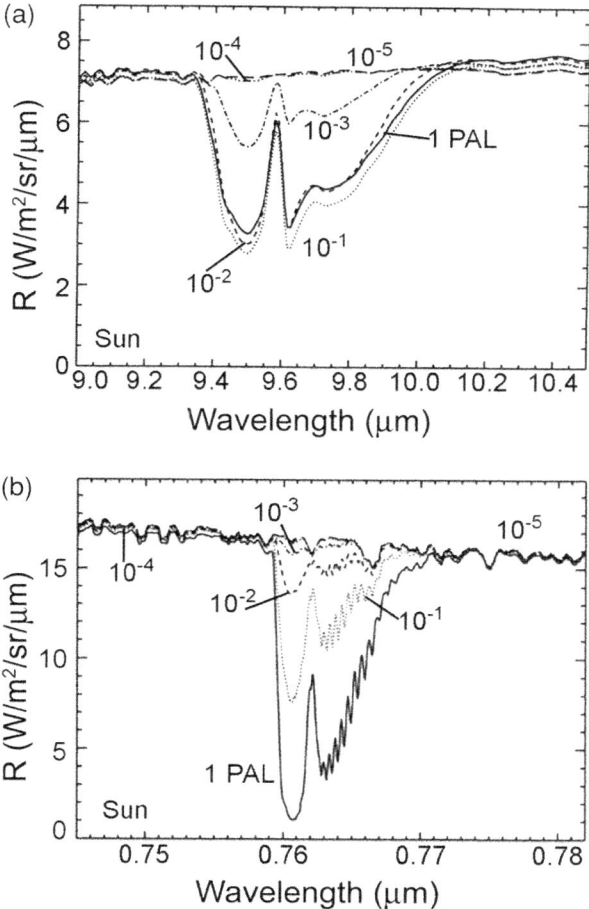

FIGURE 8.11. Calculated shape of the ozone 9.6-μm absorption band (a) and the O_2 0.76-μm band (b) for the atmospheric models shown in Figure 8.10. From Segura et al. 2003.

infrared. The calculation of ozone concentrations as a function of O_2 has been done many times, starting with Berkner & Marshall (1966). The most recent calculations have been done with coupled one-dimensional photochemical and radiative-convective models. An example is shown in Figure 8.10. As O_2 levels are decreased from the present value, the peak O_3 density decreases and moves downward in the stratosphere. This causes stratospheric temperatures to decrease, as absorption of solar UV radiation by ozone is the process that keeps Earth's stratosphere warm in the first place. The interesting result of doing this calculation, shown in Figure 8.11a, is that the strength of the O_3 9.6-μm band actually remains more or less constant down to 10^{-2} PAL of O_2. By contrast, the 0.76-μm band is nearly gone at this O_2 level (Figure 8.11b). The decreased column depth of ozone at lower O_2 levels is compensated by the decrease in temperature, resulting in a greater contrast between the hot surface and the cold stratosphere. As others have pointed out (Leger et al. 1993), this makes ozone a particularly sensitive indicator of atmospheric O_2. However, it also means that it is difficult or impossible to distinguish between various high levels of O_2 from observations of the ozone 9.6-μm band.

8.5. Conclusion

This review has been of necessity incomplete. To do justice to the topic of atmospheric evolution and planetary habitability would require writing a full book – something that I hope to accomplish within the next few years, but that has not yet been done. Here, though, I have tried to make several fundamental points.

(1) Earth's climate has been stabilized over long timescales by negative feedbacks involving both CO_2 and CH_4. The CO_2 feedback is abiotic and would operate on any Earthlike planet that had both liquid water and volcanism. The CH_4 feedback is biological and would only operate on planets, like Earth, that were inhabited with methane-producing organisms.

(2) A consequence of these stabilizing feedbacks is that the liquid water habitable zone around the Sun, and around stars not too dissimilar from the Sun, is relatively wide. Thus, from the standpoint of climate, the chances of finding other Earthlike planets around other stars are relatively good.

(3) Earth's prebiotic atmosphere was mildly reducing, with H_2 mixing ratios of 10^{-3} or higher and with O_2 concentrations too small to be observed by a low-resolution spectrometer like that envisioned for *TPF*. O_2 (and O_3) are therefore good potential biomarkers on most conceivable Earthlike planets. Planets just outside the boundaries of the habitable zone, though – a runaway greenhouse planet like early Venus, or a slightly larger version of present-day Mars – could conceivably develop O_2-rich atmospheres in the absence of life.

(4) O_2 and O_3 both rose to observable concentrations in Earth's atmosphere about 2.3 billion years ago. The timing of the initial O_2 rise is now well established but the precise reason why it rose when it did is not fully understood. Whether or not O_2 will be present on other Earthlike planets is a question that can only be settled by future space-based observations by projects such as NASA's *Terrestrial Planet Finder* and ESA's *Darwin* mission. Hopefully, the young generation of researchers at this winter school (and some of us older generation types, too!) will see this happen within their scientific careers and will contribute to the success of these missions.

REFERENCES

ANBAR, A. D. & KNOLL, A. H., 2002, *Science* **297**, 1137.
BERKNER, L. V. & MARSHALL, L. L., 1964, *Disc. Faraday Soc.* **34**, 122.
BERKNER, L. V. & MARSHALL, L. L., 1965, *J. Atmos. Sci.* **22**, 22.
BERKNER, L. V. & MARSHALL, L. L., 1966, *J. Atmos. Sci.* **23**, 133.
BERKNER, L. V. & MARSHALL, L. L., 1967, *Adv. Geophys.* **12**, 309.
BOOTHROYD, A. I., SACKMANN, I.-J. & FOWLER, W. A., 1991, *ApJ* **377**, 318.
BRINKMAN, R. T., 1969, *J. Geophys. Res.* **74**, 5355.
BROCKS, J. J., LOGAN, G. A., BUICK, R. & SUMMONS, R. E., 1999, *Science* **285**, 1033.
BUDYKO, M. I., 1969, *Tellus* **21**, 611.
CALDEIRA, K. & KASTING, J. F., 1992, *Nature* **359**, 226.
CANFIELD, D. E., 1998, *Nature* **396**, 450.
CANFIELD, D. E., HABICHT, K. S. & THAMDRUP, B., 2000, *Science* **288**, 658.
CATLING, D. & KASTING, J. F., 2007, in *Planets and Life,* Chapter 5, eds. W. T. Sullivan & J. Baross, Cambridge: Cambridge University Press.
CHAMBERLAIN, J. W. & HUNTEN, D. M., 1987, *Theory of Planetary Atmospheres,* Orlando: Academic Press.

CLOUD, P. E., 1972, *Amer. J. Sci.* **272**, 537.
CROWELL, J. C., 1999, *Pre-Mesozoic Ice Ages: Their Bearing on Understanding the Climate System,* Boulder, CO: Geological Society of America.
DONAHUE, T. M., HOFFMAN, J. H. & HODGES, R. R. JR., 1982, *Science* **216**, 630.
EVANS, D. A., BEUKES, N. J. & KIRSHVINK, J. L. 1977, *Nature* **386**, 262.
FARQUHAR, J., BAO, H. & THIEMANS, M., 2000, *Science* **289**, 756.
FARQUHAR, J., SAVARINO, J., AIRIEAU, S. & THIEMENS, M. H., 2001, *J. Geophys. Res.* **106**, 1.
FORGET, F. & PIERREHUMBERT, R. T., 1997, *Science* **278**, 1273.
FORMISANO, V., ATREYA, S., ENCRENAZ, T., IGNATIEV, N. & GIURANNA, M., 2004, *Science* **306**, 1758.
GOUGH, D. O., 1981, *Solar Phys.* **74**, 21.
GRAEDEL, T. E., SACKMANN, I.-J. & BOOTHROYD, A. I., 1991, *Geophys. Res. Lett.* **18**, 1881.
HART, M. H., 1978, *Icarus* **33**, 23.
HART, M. H., 1979, *Icarus* **37**, 351.
HOFFMAN, P. F., KAUFMAN, A. J., HALVERSON, G. P. & SCHRAG, D. P., 1998, *Science* **281**, 1342.
HOLLAND, H. D., 1978, *The Chemistry of the Atmosphere and Oceans,* New York: Wiley.
HOLLAND, H. D., 1984, *The Chemical Evolution of the Atmosphere and Oceans,* Princeton: Princeton University Press.
HOLLAND, H. D., 1994, in *Early Life on Earth,* ed. S. Bengtson, New York: Columbia University Press, p. 237.
HOLLAND, H. D., 2002, *Geochim. Cosmochim. Acta* **66**, 3811.
HUNTEN, D. M., 1973, *J. Atmos. Sci.* **30**, 1481.
INGERSOLL, A. P., 1969, *J. Atmos. Sci.* **26**, 1191.
JOSHI, M. M., HABERLE, R. M. & REYNOLDS, R. T., 1997, *Icarus* **129**, 450.
KAHN, R., 1985, *Icarus* **62**, 175.
KASTING, J. F., 1987, *Precambrian Res.* **34**, 205.
KASTING, J. F., 1988, *Icarus* **74**, 472.
KASTING, J. F., 1991, *Icarus* **94**, 1.
KASTING, J. F., 1993, *Science* **259**, 920.
KASTING, J. F., 1997, *Orig. Life* **27**, 291.
KASTING, J. F., 2001, *Science* **293**, 819.
KASTING, J. F., 2002, in *Geosphere–Biosphere Interactions and Climate,* eds. L. O. Bengtsson & C. U. Hammer, Cambridge: Cambridge University Press.
KASTING, J. F. & ACKERMAN, T. P., 1986, *Science* **234**, 1383.
KASTING, J. F. & CATLING, D., 2003, *Ann. Rev. Astron. Astrophys.* **41**, 429.
KASTING, J. F. & POLLACK, J. B., 1983, *Icarus* **53**, 479.
KASTING, J. F., POLLACK, J. B. & CRISP, D., 1984, *J. Atmos. Chem.* **1**, 403.
KASTING, J. F., TOON, O. B. & POLLACK, J. B., 1988, *Sci. Am.* **256**, 90.
KASTING, J. F., WHITMIRE, D. P. & REYNOLDS, R. T., 1993, *Icarus* **101**, 108.
KASTING, J. F., WHITTET, D. C. B. & SHELDON, W. R., 1997, *Orig. Life Evol. Biosph.* **27**, 413.
KIRSCHVINK, J. L., 1992, in *The Proterozoic Biosphere: a Multidisciplinary Study,* eds. J. W. Schopf & C. Klein, Cambridge: Cambridge University Press.
KRASNOPOLSKY, V. A., MAILLARD, J. P. & OWEN, T. C., 2004, *Icarus* **172**, 537.
KUMP, L. R., KASTING, J. F. & BARLEY, M. E., 2001, *Geol. Geochem. Geophys.* (on-line) **2**.
LASKAR, J. & ROBUTEL, P., 1993, *Nature* **361**, 608.

LEGER, A., PIRRE, M. & MARCEAU, F. J., 1993, *A&A* **277**, 309.

LIU, S. C. & DONAHUE, T. M., 1974, *J. Atmos. Sci.* **31**, 1118.

MALIN, M. C. & EDGETT, K. S., 2000, *Science* **288**, 2330.

MANGOLD, N., QUANTIN, C., ANSAN, V., DELACOURT, C. & ALLEMAND, P., 2004, *Science* **305**, 78.

MCELROY, M. B., KONG, T. Y. & YUNG, Y. L., 1977, *J. Geophys. Res.* **82**, 4379.

MCKAY, C. P., POLLACK, J. B. & COURTIN, R., 1991, *Science* **253**, 1118.

MISCHNA, M. M., KASTING, J. F., PAVLOV, A. A. & FREEDMAN, R., 2000, *Icarus* **145**, 546.

MORBIDELLI, A., CHAMBERS, J., LUNINE, J. I., et al., 2000, *Meteor. Planet. Sci.* **35**, 1309.

MUMMA, M. J., NOVAK, R. E., DISANTI, M. A. & BONEV, B. P., 2003, *Bull. Amer. Astron. Soc.* **35**, 937.

ONO, S., EIGENBRODE, J. L., PAVLOV, A. A., et al., 2003, *Earth Planet. Sci. Lett.* **213**, 15.

PAVLOV, A. A., HURTGEN, M. T., KASTING, J. F. & ARTHUR, M. A., 2003, *Geology* **31**, 87.

PAVLOV, A. A. & KASTING, J. F., 2002, *Astrobiology* **2**, 27.

PAVLOV, A. A., KASTING, J. F., BROWN, L. L., RAGES, K. A. & FREEDMAN, R., 2000, *J. Geophys. Res.* **105**, 11981.

PAVLOV, A. A., KASTING, J. F. & BROWN, L. L., 2001, *J. Geophys. Res.* **106**, 23267.

PIERREHUMBERT, R. T., 2004, *Nature* **429**, 646.

POLLARD, D. & KASTING, J. F., 2005, *J. Geophys. Res.* **110**, C07010.

RASOOL, S. I. & DEBERGH, C., 1970, *Nature* **226**, 1037.

ROSCOE, S. M., 1969, *Geol. Surv. Can. Pap.* **68–40**, 205.

ROSCOE, S. M., 1973, *Geol. Soc. Can. Spec. Pap.* **12**, 31.

RYE, R., KUO, P. H. & HOLLAND, H. D., 1995, *Nature* **378**, 603.

SEGURA, A., KRELOVE, K., KASTING, J. F., et al., 2003, *Astrobiology* **3**, 689.

SELLERS, W. D., 1969, *J. Appl. Meteor.* **8**, 392.

SELSIS, F., DESPOIS, D. & PARISOT, J.-P., 2002, *A&A* **388**, 98.

SLEEP, N. H. & ZAHNLE, K., 2001, *J. Geophys. Res.* **106**, 1373.

SQUYRES, S. W., ARVIDSON, R. E., BELL, J. F., et. al., 2004, *Science* **306**, 1698.

TIAN, F., TOON, O. B., PAVLOV, A. A. & DE STERCK, H. A., 2005, *Science* **308**, 1014.

TOUMA, J. & WISDOM, J., 1993, *Science* **259**, 1294.

VALLEY, J. W., PECK, W. H., KING, E. M. & WILDE, S. A., 2002, *Geology* **30**, 351.

WALKER, J. C. G., 1977, *Evolution of the Atmosphere*, New York: Macmillan.

WALKER, J. C. G., 1985, *Orig. Life* **16**, 117.

WALKER, J. C. G., HAYS, P. B. & KASTING, J. F., 1981, *J. Geophys. Res.* **86**, 9776.

WOESE, C. R. & FOX, G. E., 1977, *Proc. Natl. Acad. Sci. USA* **74**, 5088.

WOOD, B. E., MULLER, H.-R., ZANK, G. P. & LINSKY, J. L., 2002, *ApJ* **574**, 412.

YOUNG, G. M., 1991, *Stratigraphy, Sedimentology, and Tectonic Setting of the Huronian Supergroup*, Toronto: Geological Association of Canada.

ZAHNLE, K. J., 1986, *J. Geophys. Res.* **91**, 2819.

ZAHNLE, K. J., KASTING, J. F. & POLLACK, J. B., 1988, *Icarus* **74**, 62.

9. Biomarkers of extrasolar planets and their observability

FRANCK SELSIS, JIMMY PAILLET AND FRANCE ALLARD

The first space-borne instruments able to detect and characterize extrasolar terrestrial planets, *Darwin* (ESA) and *TPF* (*Terrestrial Planet Finder*, NASA), should be launched at the end of the next decade. Beyond the challenge of planet detection itself, the ability to measure mid-infrared (*Darwin, TPF-I*) and visible (*TPF-C*) spectra at low resolution will allow us to characterize the exoplanets discovered. The spectral analysis of these planets will extend the field of planetary science beyond the Solar System to the nearby Universe. It will give access to certain planetary properties (albedo, brightness, temperature, radius) and reveal the presence of atmospheric compounds, which, together with the radiative budget of the planet, will provide the keys to understanding how the climate system works on these worlds. If terrestrial planets are sufficiently abundant, these missions will collect data for numerous planetary systems of different ages and orbiting different types of stars. Theories for the formation, evolution and habitability of the terrestrial planets will at last face the test of observation. The most fascinating perspective offered by these space observatories is the ability to detect spectral signatures indicating biological activity. In this chapter, we review and discuss the concept of extrasolar *biosignatures* or *biomarkers*. We focus mainly on the identification of oxygen-rich atmospheres through the detection of O_2 and O_3 features, addressing also the case of other possible biomarkers and indicators of habitability.

9.1. Introduction: the search for habitable worlds

The search for habitable terrestrial planets raises considerable scientific and philosophical interest. However, it is technically much more difficult than the detection of giant, short period planets. Terrestrial planets at habitable orbital distance do not sufficiently perturb the trajectory of their parent star to produce an indirect detectable feature, and their brightness, 10^6 to 10^{10} times lower than the stellar one, is diluted within the diffraction pattern of the star. Among the projects aiming at the detection of terrestrial planets, the first to come will search for planetary transits, observing continuously for months or years selected dense fields of stars. These space missions are *COROT* (CNES, Rouan *et al.* 1998), *Kepler* (NASA, Borucki *et al.* 1997) and *Eddington* if re-scheduled by ESA (Roxburgh and Favata 2003). These missions will give us statistics on the abundance of terrestrial planets, their size, period and orbital distance. The first projects for the direct detection of terrestrial planets, *Darwin* (ESA, Léger *et al.* 1996, Volonte *et al.* 2000) and *TPF* (Terrestrial Planet Finder, NASA, Beichman *et al.* 1999), are not expected before 2014. *Darwin* is an infrared observatory (7–20 μm) based on the 'nulling interferometry' technique. This method, combining the light from several free-flying mirrors, allows one to dim the light from the central star and to increase the contrast between the star and possible planets (Bracewell 1978; Ollivier 1999). The *TPF* program consists of two distinct missions. With launch scheduled in 2014, *TPF-C* will be an optical (0.5–0.8 μm) coronograph using an elliptic (4 m × 6 m) telescope, while a second component, *TPF-I*,

Extrasolar Planets, eds. Hans Deeg, Juan Antonio Belmonte and Antonio Aparicio.
Published by Cambridge University Press.
© Cambridge University Press 2007.

will be an infrared nulling interferometer similar to *Darwin*. *TPF* and *Darwin* will expand the field of planetary science to other systems by providing some characterization of the detected planets. We can expect to learn about their size, temperature, climate and composition, and thus about their habitability. This would already be a major leap in science, justifying by itself and by far these ambitious instruments, but the perspective to search for biosignatures gives these telescopes an even more fascinating dimension.

The emergence and evolution of life on Earth was made possible by the chemical composition, size, mass and orbital properties of our planet, as described in Chapter 8. Furthermore, the mass, diversity and productivity of the biosphere reached such a level that life became a key agent of the global geochemical cycles, making Earth a different world. As a consequence, the Earth observed from space exhibits some atmospheric and surface properties that would not be found in the absence of life. These *biosignatures* (or *biomarkers*) are extremely numerous and the presence of life would be obvious for an observer based on a nearby planetary system and using a 'supertelescope' providing unlimited spatial and spectral resolutions. However, observed from a few parsecs away and with the best instruments we should be able to build during the next decade, the image of the Earth shrinks into a single pixel, whose light can only be dispersed with a very low resolution. Although the immense majority of the signs of life are lost in this situation, we will see in this chapter that some spectral features (mainly due to our oxygen-rich atmosphere) would still reveal the existence of the biosphere. This means that life on Earth could be suspected or detected by a distant observer possessing the technology we are now developing. Therefore, not only are we about to design telescopes to detect terrestrial exoplanets around the closest stars, but also to search for inhabited ones.

9.2. The search for spectroscopic biosignatures

As soon as photons coming from a planet can be distinguished from those coming from a star, a spectral analysis is feasible within the available signal-to-noise ratio and sensitivity. The physical and chemical properties of the planets and their atmosphere can be studied. The spectroscopy of extrasolar planetary atmospheres is a very young science that has given its very first results only recently. Some atmospheric compounds (Na, H, C, O) were detected in the upper atmosphere of a hot jupiter (HD 209458b) in absorption during a transit. This was the first spectral information gained on the atmosphere of an exoplanet (Charbonneau *et al.* 2002; Vidal-Madjar *et al.* 2004a,b). In 2005, the space telescope *Spitzer* was able to detect the infrared emission from two transiting hot jupiters (HD 209458b and TrES-1) by measuring the decrease in infrared luminosity occurring when the planet is eclipsed by its star (Deming *et al.* 2005; Charbonneau *et al.* 2005).

As life on Earth has strongly modified the planet (atmosphere, ocean, surface), can we use this fact to distinguish spectroscopically the presence of a similar ecosystem on another planet? In the particular case of Earth, O_2 is fully produced by the biosphere, with less than 1 ppm coming from abiotic processes (Walker 1977). Cyanobacteria and plants are responsible for this production by using the solar photons to extract hydrogen from water and using it to produce organic molecules from CO_2. This metabolism, called oxygenic photosynthesis, can be summarized as follows:

$$2H_2O^* + CO_2 + \text{photons} \leftrightarrow CH_2O + H_2O + O_2^*$$

(* indicates that the O atoms in O_2 originate from water).

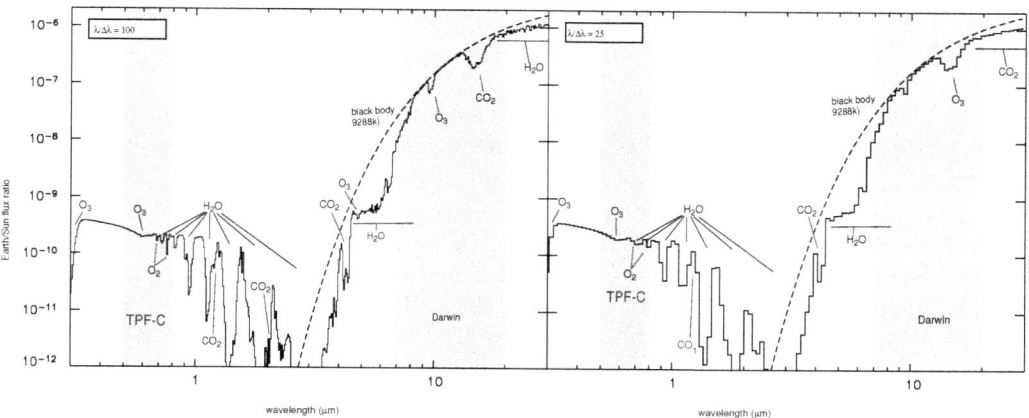

FIGURE 9.1. Synthetic spectra of the Earth from UV to IR. These graphs show the spectrum of the Earth at two different resolution powers ($R = 100$ and 25) computed with the model PHOENIX (Paillet et al., in prep.). The intensity is given as a fraction of the solar intensity. The features produced by atmospheric species and the spectroscopic ranges of Darwin and TPF-C are indicated. Note the disappearance of the 0.76 µm O_2 band at $R = 25$.

The reverse reaction, using the oxygen to oxidize the organics produced by photosynthesis, can occur abiotically when organics are exposed to free oxygen, or biotically by eukaryotes breathing O_2 and consuming organics. Because of this balance, the net release of O_2 in the atmosphere is due to the burial of organics in sediments. Each reduced carbon that is buried frees an O_2 molecule in the atmosphere. This net release rate is also balanced by the exposure and weathering of fossilized carbon. The oxidation of reduced volcanic gases such as H_2 or H_2S also accounts for a significant fraction of the oxygen losses. While respiration recycles the 10^{18} kg of O_2 contained in the atmosphere in less than 10 000 yr, the geochemical recycling is much slower and takes about 10 Myr. On the other hand, the abiotic production rate of O_2 due to the photolysis of atmospheric H_2O followed by the escape of H to space is more than 100 times lower than the abiotic loss due oxidation of rocks and volcanic gases. This is due to the small abundance of H_2O (10^{-6}) above the tropopause and the diffusion-limited loss of hydrogen to space (see Chapter 8). At the present loss rate, it would take more than 1 Gyr to build up the present O_2 level *in the total absence of oxygen loss by oxidation*.

On the basis of these arguments, Owen (1980) suggested searching for O_2 as a tracer of life. In a famous paper, Sagan et al. (1993) analyzed a spectrum of the Earth taken by the Galileo probe, searching for signatures of life. They concluded that the large amount of O_2 and the simultaneous presence of CH_4 traces are suggestive of biology. Moreover, the detection of a widespread red-absorbing pigment with no likely mineral origin supports the hypothesis of biophotosynthesis. Recently Arnold et al. (2002) and Woolf et al. (2002) independently recorded spectra of earthshine on the Moon (the light reflected by the Earth, then by the Moon back to the Earth). They observed the signatures of oxygen and ozone, and searched for the specific albedo of vegetation. Instead of searching directly for O_2, Angel et al. (1986) suggested considering O_3 and its mid-infrared band at 9.6 µm, which produces a strong feature in the Earth's emission. At this wavelength, the star/planet brightness contrast is 1000 times more favourable than in the visible (see Figure 9.1). Léger et al. (1993) have investigated the O_3 feature as a tracer of O_2 in planetary atmospheres and the use of O_3 as a tracer of oxygen-rich atmospheres sustained by life. This concept is at the root of *Darwin*, as proposed to ESA by Léger et al. (1996), and of *TPF-I* (*Terrestrial Planet Finder*, Beichman et al. 1999).

9.3. Oxygen and ozone as biosignatures

In order to investigate the astrobiological possibilities of the *Darwin/TPF* missions, it is necessary to estimate the risk of false positive and false negative detection. A false positive case results from the detection of an abiotic feature that is wrongly attributed to some form of biological activity. On the other hand, a false negative case occurs when an inhabited planet does not present any of the features sought, whether because the dominant metabolisms do not produce the required biosignature or because this biosignature is undetectable or masked by some other process. In the following, we address the relevance of searching for spectroscopic features of O_2 and O_3 to detect extraterrestrial ecosystems. It has to be clearly stated here that the presence of a biosphere does not imply an oxygen-rich atmosphere. Earth's biosphere had existed for at least 1 Gyr and probably more in an anoxic atmosphere (see Chapter 8). The non-detection of O_2 or O_3 on an exoplanet cannot, then, be interpreted as the absence of life. But, as we will see, some causes can prevent us from identifying O_2 or O_3 in an oxygen-rich atmosphere. Such a case is what we consider a false negative.

9.3.1. O_3 as a tracer of O_2-rich atmospheres

Is ozone a good tracer of oxygen-rich atmospheres? The answer depends obviously on the wavelength range considered. The range giving the best contrast between the stellar and planetary emission is the mid-infrared, around 10 microns, where the thermal emission from terrestrial planets peaks. In this spectroscopic window, O_2 itself does not have a vibrational transition, while O_3 exhibits a strong band centred on 9.6 µm. This band (and also the H_2O bands) produces a strong feature in the Earth's infrared emission that makes our planet distinguishable from any other in the Solar System (see Figures 9.1 and 9.2).

Ozone is produced in the atmosphere by a unique chemical reaction:

$$O + O_2 + M \rightarrow O_3 + M,$$

where M is any compound. This reaction is not very efficient as it requires at the same time a high enough pressure (because of the third body), and oxygen atoms that are produced at lower pressures and at upper altitudes, where photolysis of O_2 by UV can occur.

In contrast, ozone can be efficiently destroyed by a large number of reactions. In the Earth's atmosphere, these are dominated by catalytic cycles involving trace species such as hydrogenous compounds (H, OH, HO_2), nitrogen oxides (NO_X) and chlorine compounds (ClO_X). These species have various origins and their amount depends on the nature and the intensity of the bio-productivity, the thermal profile of the atmosphere, human pollution and many other parameters. Without these compounds, an atmosphere made only of N_2 and O_2 would contain ten times more O_3. As the amount of O_3 in an O_2-rich atmosphere strongly depends on the abundance of trace species with various origins and complex behaviours, it is difficult to extrapolate the chemistry of our present atmosphere to any other planet that would host an oxygen-producing ecosystem. However, for a given atmospheric composition, the amount of O_3 is weakly sensitive to the total amount of O_2, as seen in Figure 9.3. The abundance of O_3 also varies with the spectral distribution of the incoming stellar radiation. O_3 production depends on

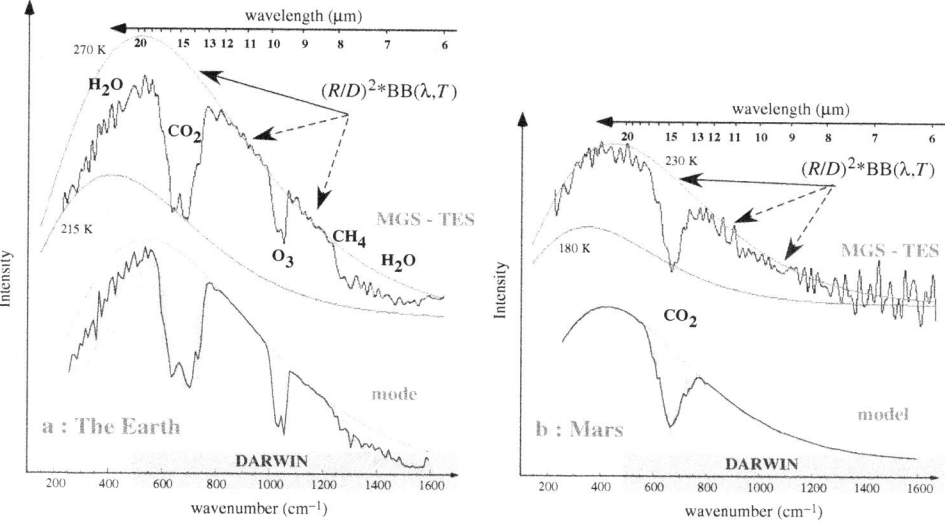

FIGURE 9.2. Thermal emission from the Earth and Mars. Infrared spectra for the Earth (a) and Mars (b) were measured by the TES instrument of the *Mars Global Surveyor*, and are compared to synthetic spectra obtained with a numerical model. The envelope of each spectrum is given by $(R/D)^2 \times$ BB (λ, T) where D is the distance to the planetary system, R the radius of the planet and BB is the Planck function.

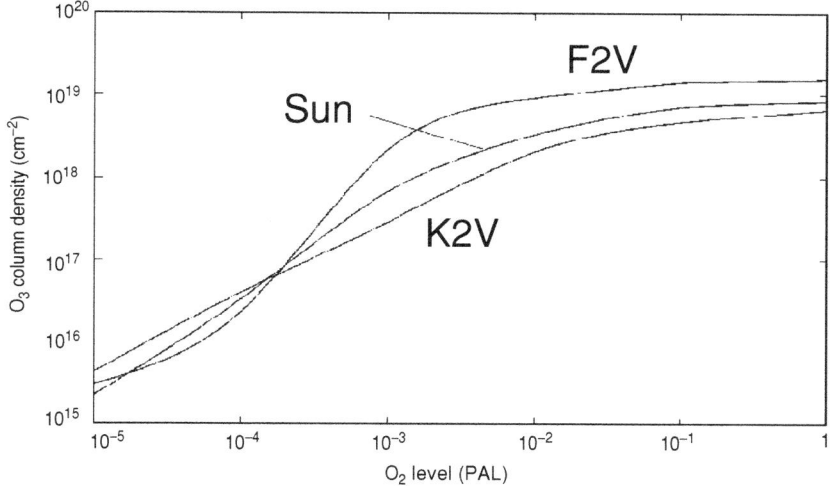

FIGURE 9.3. Ozone column density versus oxygen level. These curves show the relation between the O_3 column density and the level of O_2 in an N_2-rich 1-bar atmosphere, for a G (solar), K- and F-star. Published in Segura *et al.* (2003), courtesy of Antigona Segura and Mary Ann Liebert, Inc.

the availability of oxygen atoms and therefore on the photolysis of O_2 by the UV radiation. However, more UV radiation also enhances ozone destruction, directly by photolysis, but mostly indirectly due to the increased photochemical production of highly reactive radicals. For atmospheric compositions similar to that of the Earth, numerical simulations show a quantity of O_3 increasing with the UV flux (Selsis 2000; Segura *et al.* 2003).

This property could play a very important role in the surface habitability by providing an ozone shield tuned to the level of incoming lethal UV radiation.

The detectability of the O_3 feature, or its depth in the infrared spectrum, depends on the atmospheric profiles of ozone abundance and temperature. The ability to detect the 9.6 μm band of O_3, with a low resolution ($\lambda/\Delta\lambda < 25$: only 1 or 2 bins for the O_3 band) and with a signal-to-noise ratio less than 10, requires a deep feature. On Earth, the O_3 band is saturated and would still be for O_3 column densities 10 or even 100 times lower. Therefore, what produces the feature in the spectrum is the difference beween the brightness temperature of the continuum (given by the temperature of the surface and/or the clouds) and the temperature of the ozone layer (where the opacity of O_3 is 1). On average, in the case of the Earth the ozone layer is cooler than the surface and its band appears as an 'absorption' feature. However, it is possible (for instance over the Antarctic) to see the ozone band locally (as well as the 15 CO_2 μm band) above the continuum, appearing as an 'emission' feature. Also, as the clouds are cooler than the surface, cloudiness significantly decreases the depth of the atmospheric bands in the infrared.

What could happen in an extrasolar O_2-rich atmosphere? The complex coupling between radiation, photochemistry and temperature prevents us from generalizing the processes occurring in the Earth's atmosphere and requires detailed and self-consistent modelling. For instance, ten times less O_3 would produce less stratospheric warming and the altitude where the ozone opacity is 1 would be lower: these are two factors that would make the ozone feature deeper. By using a numerical code that simulates the photochemistry of a wide range of planetary atmospheres, we have simulated a replica of our planet orbiting different types of stars (Selsis 2000): an F-type star (more massive and hotter than the Sun) and a K-type star (smaller and cooler than the Sun). The orbital distance was chosen in order to give the planet the Earth's effective temperature (by receiving the same energetic flux): 1.8 and 0.5 AU respectively for the F- and the K-type star. Scaling the energetic flux allows us to consider habitable planets irradiated by a non-solar spectrum: the contribution of the UV range (150–400 nm, the most important for the photochemistry) is higher for the F-type star and lower for the K-type star (this is no longer true in the EUV range, below 150 nm, where low-mass stars, like K-type stars, are very active). A more detailed study has also been done recently by Segura et al. (2003). Let us summarize here some of the important results found. The planet orbiting the K-star has a thin O_3 layer, compared to that of the Earth, but still exhibits a deep O_3 absorption: indeed, the low UV flux is absorbed at lower altitudes than on Earth, which results in less efficient warming (because of the higher heat capacity of the dense atmospheric layers). The ozone layer is therefore much colder than the surface and this temperature contrast produces a strong feature in the thermal emission. The process works the other way around in the case of an F-type host star. Here, the ozone layer is denser and warmer than the terrestrial one, exhibiting temperatures close to the surface temperature. Thus, the resulting low temperature contrast produces only a weak and barely detectable feature in the infrared spectrum. This comparison shows that G- (solar) and K-type stars may offer better candidates for the search for the O_3 signature than F-type stars. This result is promising since G- and K-type stars are much more numerous than F-type stars, the latter being rare (only four are within 10 pc) and are affected by a short lifetime (less than 1 Gyr). Segura et al. (2003) have studied the photochemistry of an Earthlike atmosphere for different levels of O_2 and for different spectral types of central star (K, G and F). As in Selsis (2000), the photochemical modelling is coupled to a self-consistent retrieval of the temperature profile and the trace gases of importance for the photochemistry of O_3 (CH_4 and N_2O for instance) are assumed to be

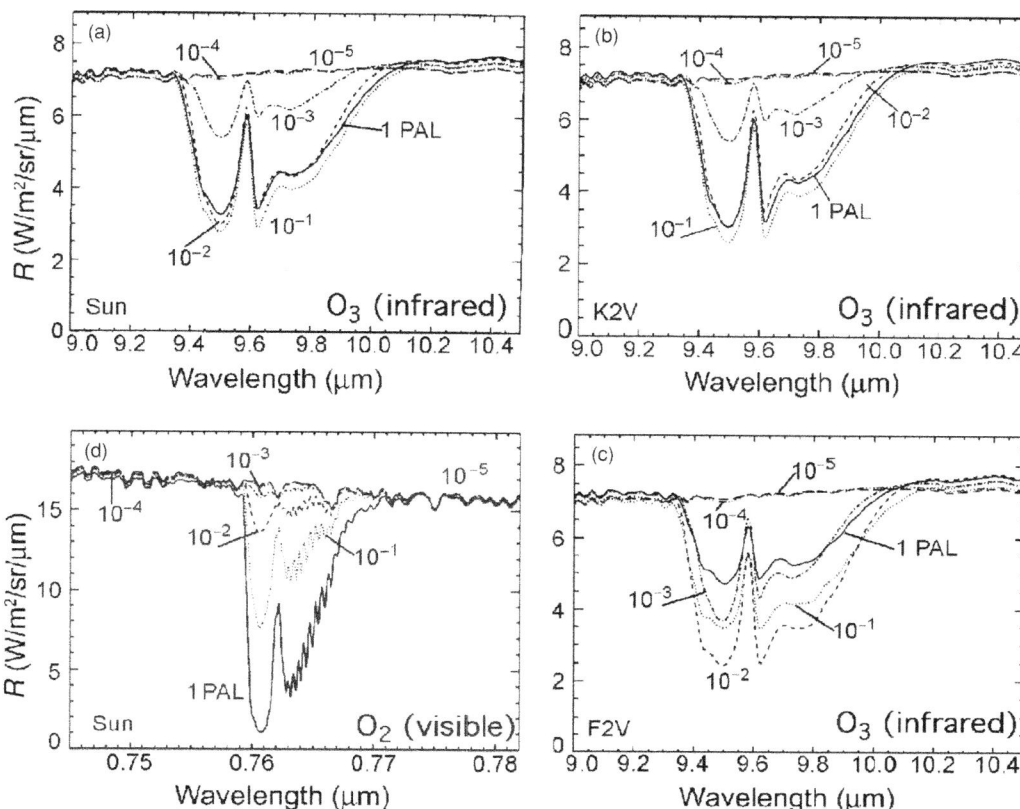

FIGURE 9.4. Variation of the spectral features of O_2 and O_3 with the O_2 level and spectral type of the central star replica. The spectral features produced by O_3 (9.6 μm – frames A, B and C) and O_2 (0.76 μm – frame D) are computed for an Earth replica orbiting a G-, K- or F-star for different O_2 levels (from 10^{-5} to 1 PAL). For the K- and G-stars, the planet–star distance is chosen in order to provide a similar mean surface temperature. The O_2 feature is weakly sensitive to the thermal profile and its shape does not depend on the spectral type. From Segura et al. (2003), courtesy of Antigona Segura and Mary Ann Liebert, Inc.

released at the same rate as on the present Earth. Figure 9.4 shows the relation between the ozone column density and the O_2 level obtained by these authors: one can see that, down to 10^{-3} times the present atmospheric level (PAL) of O_2, the amount of O_3 does not decrease by more than one order of magnitude. For all their simulations, they have computed the spectral features of O_2 (in the visible) and O_3 (in the infrared). For 1 PAL of O_2, the spectra obtained for K and F central stars are similiar to those obtained by Selsis (2000), but the result that really makes O_3 a good tracer of O_2-rich atmospheres is that the depth of the 9.6 μm is similar or even higher with lower amounts of O_2 (down to about 10^{-3} PAL).

Playing with a few parameters only, such as the stellar spectrum or the O_2 level, is a way to identify fundamental mechanisms in the formation of the O_3 signature but certainly not to predict the general atmospheric and spectral properties of exo-Earths. In order to understand the sensitivity to other parameters, new simulations should of course be performed by changing variables such as gravity, pressure and the background composition of the atmosphere (especially the CO_2 level, which depends theoretically on the orbital distance inside the habitable zone, see Chapter 8), biogenic emission of tropospheric gases such as CH_4 or N_2O, and orbital distance.

9.4. Detection of an O_2-rich atmosphere in the reflected spectrum

The main spectroscopic feature of molecular oxygen in the Earth's atmosphere is the 0.76 µm oxygen A line in the reflected spectrum. This 'line' is in fact a band whose structure is shaped by thousands of individual lines, as seen in Figure 9.4. In the atmosphere of the Earth, the mean opacity of this feature is about 0.5, which means that a lower abundance would produce a shallower absorption. The relation between the line depth and the O_2 level for a cloud-free atmosphere is discussed by Des Marais *et al.* (2002) and Segura *et al.* (2003). However, clouds have strong effects on the reflected spectrum, making the retrieval of the O_2 column density from remote observations extremely difficult. Clouds make the spectroscopic study technically easier by significantly enhancing the albedo and thus the level of the continuum thanks to Mie scattering, but they have two opposite effects that make the relation between the line depth and the O_2 abundance degenerate: they hide the lowest part of the atmosphere where most of the O_2 lies and they increase the optical path in the clouds by multiple scattering. Therefore, ability to infer the O_2 column density is not a clear advantage of the visible range over the infrared. The detection of the 0.76 µm O_2 line requires a minimum resolving power of $R = 55$ (and $R \approx 70$ for the detection of two or three additional lines, see Figure 9.6). The 'Chappuis' band of O_3 (a photodissociation continuum between 400 and 850 nm that peaks at 600 nm) is very wide but requires very high sensitivity. The earthshine spectra obtained by Woolf *et al.* (2002) and Arnold *et al.* (2002) clearly show how difficult the detection of this feature on extrasolar planets would be. Observations in the near-infrared are needed to detect the presence of CO_2, which is a crucial piece of information for both habitability and biosignatures. In Section 9.8 we discuss the benefits of having both infrared (thermal) and visible (reflected) spectra of the same planet.

9.4.1. *Abiotic production of O_2 and O_3*

The fact that, on the Earth, oxygen and indirectly ozone are byproducts of the biological activity does not mean that life is the only process able to enrich an atmosphere with these compounds. The question of the abiotic synthesis of biomarkers is crucial, though very few studies have yet been dedicated to it (Rosenqvist and Chassefiere 1995; Kasting 1995; Léger *et al.* 1999; Selsis *et al.* 2002). To qualitatively associate oxygen and life without investigating further possible abiotic sources is an error that has already been committed in the past. In the 1950s, although no precise data about the composition of the martian atmosphere was available, it was already understood that the red colour of the martian surface was due to oxidation. Some scientists inferred from this hypothesis (true) that oxygen was the main component of the atmosphere (wrong: there is only 0.1%), and that this oxygen should have a biological origin (Spencer Jones 1959). The presence of O_2 was interpreted as a confirmation of the presence of vegetation, which was suggested by some authors as an explaination of seasonal colour variations on the planet (now understood as seasonal dust storms). However, martian O_2 has an abiotic photochemical origin.

The martian atmosphere is the perfect laboratory in which to learn about the photochemical synthesis of O_2 and O_3. Indeed, if a low abiotic production of O_2 does exist on Earth, it is totally masked by the biological release. The main constituent of the martian atmosphere is carbon dioxide (CO_2: 95.3% – N_2: 2.7% – Ar: 1.6% – O_2:

0.13% – CO: 0.08% – H_2O: $\sim 0.01\%$ – O_3: $\sim 10^{-6}$ %). CO_2 is photolysed by UV radiation at wavelengths below 227 nm, producing carbon monoxide, CO, and atomic oxygen, O. The reaction between two oxygen atoms to form O_2 is much more efficient than the CO + O recombination into CO_2 (the reaction is 10^5–10^7 times faster at temperatures prevailing in the martian atmosphere: 145–270 K). This simple fact seems to be in contradiction to the stability of a CO_2 atmosphere, which should be converted into a mixture of O_2 and CO. Indeed, the conversion of a pure CO_2 atmosphere should be limited only by the screening of CO_2 by the produced O_2, which absorbs the UV radiation. Some photochemical simulations (Nair et al. 1994; Selsis et al. 2002) have shown that, in a martian atmosphere made purely of CO_2, O_2 would indeeed reach a level of about 3% (in the absence of other chemical loss of O_2 such as surface oxidation). If O_2 is a minor component of the martian atmosphere, this is due to the presence of another atmospheric compound playing a critical role in the photochemistry: water vapour. H_2O is only a trace gas on Mars: if condensed, the whole content of water vapour in the atmosphere would represent a 3 µm layer of water on the surface. But, despite its low abundance, H_2O strongly influences the photochemistry, and above all, the level of O_2 and O_3. Indeed, H_2O, like CO_2 and O_2, is photodissociated by solar UV radiation. H_2O photolysis produces hydrogen atoms and hydrogenated compounds (OH, HO_2), that catalyse the recombination of CO into CO_2. The following catalytic cycle is the main recombination route for CO_2 in the martian atmosphere:

$$H + O_2 + M \rightarrow HO_2 + M$$
$$O + HO_2 \rightarrow O_2 + OH$$
$$CO + OH \rightarrow CO_2 + H$$
$$\overline{CO + O \rightarrow CO_2}$$

The small amount of water vapour in the atmosphere of Mars is then enough to prevent the accumulation of O_2. Moreover, the photolysis of H_2O, directly or indirectly, produces oxidants much more reactive than O_2: OH, HO_2 and H_2O_2. The oxidation of the surface, and hence the loss of oxygen, is therefore much higher in their presence.

H_2O photolysis enhances the consumption of O_2, but it can also result in its production and accumulation when it is associated with a significant loss of hydrogen to space. Indeed, a fraction of the hydrogen atoms released by H_2O photolysis in the upper atmosphere have enough kinetic energy to exceed the escape velocity, inducing an oxidation of the atmospheric content. One can summarize this process as follows:

$$4 \times (H_2O + h\nu \rightarrow OH + H)$$
$$2 \times (OH + OH \rightarrow H_2O + O)$$
$$O - O + M \rightarrow O_2 + M$$
$$H \text{ escapes}$$
$$\overline{2H_2O \rightarrow O_2 + (4H)}$$

The hydrogenous radicals produced from H_2O photolysis also destroy ozone very efficiently. Again, catalytic cycles are involved in the destruction process and the small abundances of the byproducts of water vapour photodissociation can strongly affect the ozone content.

$$\mathrm{OH} + \mathrm{O} \rightarrow \mathrm{H} + \mathrm{O}_2$$
$$\mathrm{H} + \mathrm{O}_3 \rightarrow \mathrm{OH} + \mathrm{O}_2$$

$$\mathbf{O + O_3 \rightarrow 2O_2}$$
$$\mathrm{OH} + \mathrm{O}_3 \rightarrow \mathrm{HO}_2 + \mathrm{O}_2$$
$$\mathrm{HO}_2 + \mathrm{O} \rightarrow \mathrm{OH} + \mathrm{O}_2$$

$$\mathbf{O + O_3 \rightarrow 2O_2}$$
$$\mathrm{OH} + \mathrm{O}_3 \rightarrow \mathrm{HO}_2 + \mathrm{O}_2$$
$$\mathrm{HO}_2 + \mathrm{O}_3 \rightarrow \mathrm{OH} + 2\mathrm{O}_2$$

$$\mathbf{2O_3 \rightarrow 3O_2}$$

The above cycles are responsible for the diurnal variation of O_3 in the upper stratosphere and mesosphere of the Earth and are the main pathways for the destruction of O_3 on Mars. This explains why the abundances of water and ozone are anti-correlated on Mars, the maximum of ozone being above the poles in winter where the atmosphere is the driest. For a detailed modelling of the martian ozone photochemistry with a three-dimensional general circulation model see Lefèvre et al. (2004).

Mars is not a unique place in the Solar System where oxygen is produced abiotically (see Table 9.1): some icy satellites of Jupiter (Europa, Ganymede) and of Saturn (Rhea, Dione) exhibit a tenuous atmosphere ($\sim 10^{-10}$ bar) quasi-exclusively made of O_2 and O_3 (Hall et al. 1995; Noll et al. 1996, 1997). The associated column densities (density integrated over the altitude) are very low. Nevertheless, they reveal an abiotic continuous synthesis of these two molecules from the dissociation of water (ice) by UV and, mostly, charged particles trapped in giant planet magnetospheres. The hydrogen released from water escapes from the low gravity field of the satellites, and a gas enriched in oxygen remains, partly in the atmosphere and partly in bubbles within the ice, mainly made of O_2 and O_3 (Johnson and Jesser 1997).

In the case of Venus, one could expect to find atmospheric oxygen. Indeed, this dry planet is believed to have had a significant water reservoir at the beginning of its history, and the hydrogen escape generally suggested to explain the disappearance of water (Kasting 1988) should have left considerable amounts of residual oxygen. The loss of the hydrogen content of a 2700 m layer of water (equal to the terrestrial ocean) would have resulted in 220 bars of pure O_2. However, molecular oxygen has not yet been observed on Venus, and its abundance in the middle atmosphere is lower than 10^{-6}. The missing oxygen was probably consumed through the oxidation of the surface and volcanic emission. Also, the initial amount of water on Venus is unknown and could have been much smaller than on Earth, which is totally plausible considering the latest theories about the origin of water on terrestrial planets (Raymond et al. 2004). This illustrates the fact that the rate of O_2 production has to exceed the oxidation rates in order to result in an O_2 build-up on a geologically active planet. From this viewpoint, Venus is very different from Mars: the latter is a small planet that cooled rapidly, whose surface is mostly inert and fully oxidized, consuming the atmospheric oxidants only at a very slow rate.

The study of the photochemistry in the planetary atmospheres of the Solar System reveals some mechanisms for the synthesis of O_2 and O_3. The production rates and amounts of these abiotically synthesized molecules are very low: the column densities of

TABLE 9.1 Observations of O_2 and O_3 in the Solar System. For the martian atmosphere, equivalent mixing ratios are calculated with a mean total column density of 2.08×10^{23} cm^{-2} corresponding to a mean surface pressure of 5.6 mbar.

Species and references	Column density (cm^{-2})	Equivalent mixing ratio
	Mars	
O_2	($\times 10^{20}$)	
Barker (1972)	2.6 ± 0.2	0.0012
Carleton & Traub (1972)	2.8 ± 0.3	0.0013
Trauger & Lunine (1983)	2.3 ± 0.1	0.0011
Owen et al. (1977)		0.0013[a]
(Viking – in situ)		
O_3	($\times 10^{15}$)	($\times 10^{-8}$)
Espenak et al. (1991)	4.0 ± 1.3	1.9
Clancy et al. (1999)	5.4–10.8	2.6–5.2
	Venus	
O_2: Mills (1999)	–	$< 3 \times 10^{-6}$
	Icy satellites	
	Ganymede (G), Europa (E), Rhea (R), Dione (D)	
Atmospheric O_2	($\times 10^{14}$)	
Hall et al. (1998)	2.4 –14 (G)	–
	1–10 (E)	–
O_3 trapped in ice	($\times 10^{16}$)	
Noll et al. (1996)	4.5 (G)	–
Noll et al. (1997)	1–6 (R & D)	–

[a] The surface mixing ratio was measured by Viking at a pressure of 7.5 mbar.

O_2 and O_3 on Mars are respectively 6×10^{-5} and 5×10^{-4} times those of the Earth, this ratio being 10^{-10} in the case of the icy satellites (see Table 9.1). Nevertheless, in some different environments such processes could a priori lead to higher levels (as we saw in the hypothetical case of a dry martian atmosphere) on extrasolar planets. It is therefore crucial to find some quantitative and/or qualitative criteria, accessible to remote detection, which would allow us to distinguish between biological and abiotic origins. A first conclusion arises: a biosignature is a more complex concept than the simple detection of a given compound. The question 'Is there oxygen (or ozone) in this atmosphere?' has to be replaced by 'How much oxygen (or ozone) is there?' That, however, is not enough. The search for signs of life implies gathering as much information as possible in order to understand how the observed atmosphere works physically and chemically.

9.5. Reliability of biomarkers inside and outside the habitable zone

The following discussion requires a good understanding of the *habitable zone* (HZ) as defined in astrobiology. We thus strongly recommend the reading of Chapter 8. The boundaries of the habitable zone as a function of stellar type are given in Figure 9.5. Two points are of particular importance and are briefly summarized here.

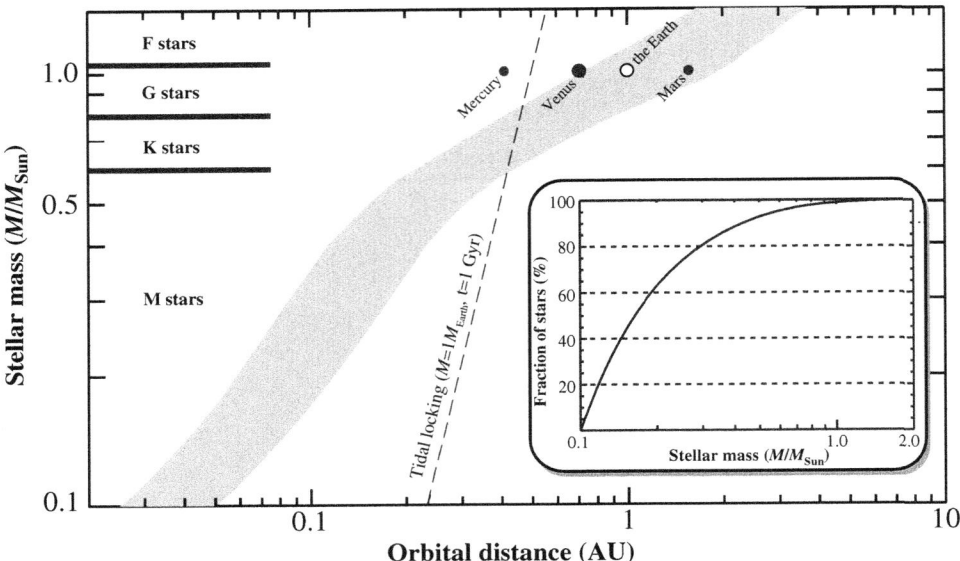

FIGURE 9.5. Boundaries of the continuously habitable zone as a function of the stellar mass. This graph gives the limit of the region where terrestrial planets can host a surface ocean for at least 1 Gyr as a function of the stellar mass. Stellar evolution models are from Baraffe et al. (1998) and surface temperature estimations from Kasting et al. (1993). The dashed line indicates the orbital distance at which planets of 1 $M_\oplus$ on circular orbits are tidally synchronized in 1 Gyr. The insert plot gives the mass distribution of stars in the [0.1–2] $M_\odot$ range. Figure by F. Selsis and J.-M. Grießmeier.

- On an Earthlike planet,[1] the level of CO_2 in the atmosphere depends on the orbital distance: CO_2 is a trace gas close to the inner edge of the HZ and a major compound in the outer part of the HZ.[2]
- Earthlike planets close to the inner edge are expected to have a water-rich atmosphere or to have lost their water reservoir to space.

The search for biomarkers should of course be focused on planets orbiting inside the HZ. As the HZ is defined for surface conditions only, chemo-lithotrophic life, whose metabolism does not depend on stellar light, can still exist outside the HZ, but such metabolisms (at least the ones we know on Earth) do not produce O_2, and other by-products such as methane would need to be sought. Also, because of their high productivity, metabolisms based on photosynthesis seem more likely to be able to transform a whole planetary environment and produce detectable features. This may be another reason to consider only the HZ. However, the limits of the HZ are known qualitatively, rather than quantitatively. This uncertainty is mainly due to the complex role of clouds but also

---

[1] Here we define Earthlike planets as planets with a size between 0.5 and 2 Earth-masses and an Earthlike chemical composition, especially with a similar water and volatile content. Such planets are assumed to develop plate tectonics and surface oceans when formed in the habitable zone.

[2] This might not be the case on ocean-planets, where the carbonate–silicate cycle has to be different, and for tidally-locked planets in the HZ of M-stars. This relation between the orbital distance and the CO_2 can be broken also by life itself if biogenic greenhouse gases dominate the climate regulation, as it may have been the case on Earth with biogenic methane before the rise of oxygen.

to three-dimensional climatic effects not yet included in current modelling. Thus, planets slightly beyond the computed HZ could still be habitable, while planets at habitable orbital distances might not be habitable because of their size or chemical composition. Because of these issues, we examine here the possible significance of an O_2/O_3 detection inside and slightly outside the habitable zone.

9.5.1. *From the star to the inner edge of the HZ*

In the inner part of a planetary system, Earthlike planets should evolve into Venus-like planets. In the initial stage, the fully vapourized ocean produces a dense H_2O atmosphere where intense H_2O photolysis and H escape take place. During this period, if oxygen is consumed at a lower rate than that at which hydrogen is lost, O_2 may build up in the atmosphere, producing a detectable feature in the reflected spectrum. Ozone, however, should not reach detectable levels because of its destruction by HO_X radicals produced by the photolysis of water vapour (Selsis *et al.* 2002). Moreover, if an ozone layer could form below the layers where H_2O is photolyzed, its signature would be masked by the dense H_2O atmosphere above: the infrared 9.6 μm O_3 would be screened by the H_2O–H_2O collision-induced absorption continuum and the visible Chappuis band by Rayleigh backscattering. Depending on the amount of water lost, this stage ends more or less rapidly, or lasts 'forever' if this reservoir is inexhaustible as it could be on ocean-planets (Léger *et al.* 2004). When all the water has been photodissociated and all the H is gone, there might still be a remaining O_2-rich atmosphere that is slowly consumed by the volcanic activity and surface recycling. In the absence of water vapour, an ozone layer could also form. What we learned from the study of Venus is that CO_2 can build up to very high levels in the absence of water. Above about 100 mbar of CO_2, the O_3 9.6 μm band can no longer be detected as it is screened by CO_2 bands (Selsis *et al.* 2002). In the reflected spectrum, both O_2 and O_3 could show up, but this should not lead to a wrong interpretation because such planets would not present any H_2O feature. Water is necessary to life as we know it but is also the source of oxygen for the production of O_2 by photosynthesis. As we will see in other cases also, considering biomarkers only when associated with H_2O features ensures a higher reliability.

9.5.2. *Inside the habitable zone of K-, G- and F-stars*

[3]Planets with a water reservoir and a biologically produced O_2-rich atmosphere, orbiting close to the inner edge of the HZ (0.84–0.95 AU for the present solar luminosity), would contain much more water vapour in their upper atmosphere than Earth (see Kasting 1988). The efficiency of the cold-trap that keeps the water in the troposphere would indeed be reduced by the thermal profile. In such conditions, no dense ozone layer is expected because of the high abundance of hydrogenous compounds (H, OH, HO_2, H_2O_2) produced by the photolysis of H_2O. Therefore the IR spectrum may not exhibit the O_3 band in this inner region of the HZ, while the reflected spectrum should still be similar to that of the Earth, with deeper H_2O features (cloudiness could significantly modify the spectrum, though).

Earthlike planets, where the carbonate–silicate cycle acts as on Earth, should be characterized inside the HZ by a CO_2 level determined by their orbital distance. Beyond

---

[3] The case of M-stars is not addressed here, although their abundance makes their case extremely interesting. The reason is that we do not yet understand the kind of atmosphere planets around M-stars may have. Such an atmosphere would have to be stable on a synchronously rotating planet and to survive intense erosion by X-rays, stellar wind and coronal mass ejections. (See Chapter 1 on this subject.)

about 1.2 AU, CO_2 partial pressures above 100 mbar are required to maintain a mean surface temperature above 273 K. At such levels, the CO_2 bands screen the O_3 9.6 µm feature, making the detection of an ozone layer impossible. Consequently, oxygenic photosynthesis can only be detected in a fraction of the HZ, approximately between 0.95 and 1.2 AU for the present solar luminosity. In the visible spectrum, no CO_2 band interferes with O_2, which remains detectable in the outer HZ, again making the two spectroscopic ranges highly complementary. It is difficult to know without detailed studies whether the Chappuis band of O_3 would be detectable in the reflected spectrum at high levels of CO_2. Indeed, an ozone layer at the same altitude as on Earth could be masked by Rayleigh scattering, but in a CO_2- and O_2-rich atmosphere the ozone layer might form at a higher altitude.

At the high CO_2 levels that should be found in the outer HZ (> 1.2 AU), the abiotic build-up of O_2 and O_3 is theoretically possible through the photochemical transformation of CO_2 into CO and O_2 (as occurs in the martian atmosphere). Selsis et al. (2002) have shown that this build-up can lead to a detection of O_2 in the visible in the following cases:
- $P_{CO_2} > 50$ mbar for a dry or icy planet, $P_{CO_2} > 1$ bar for a habitable planet (with surface liquid water),
- negligible release of volcanic reducing species,
- negligible loss of atmospheric oxidants on the surface or in the ocean.

Kasting & Catling (2003) recalled that these abiotic levels of O_2 and O_3 are not attainable in habitable planetary environments where tectonics, volcanism and liquid water are present. Indeed, atmospheric oxidants produced by photochemistry (such as H_2O_2) would react with volcanic gases and rocks, consuming oxygen and releasing hydrogen. However, we should still pay attention to some cases. First of all, a small-sized Mars-like planet without active volcanism or a carbonate–silicate cycle, and a CO_2 level below that required for habitability but higher than 100 mbar, would satisfy the conditions for an O_2 and O_3 build-up. Indeed, without surface recycling and liquid water, the surface is fully oxidized rapidly and stops consuming oxidants from the atmosphere (this is the case on Mars). The IR feature of ozone could not be detected because of CO_2 screening but the visible O_2 line could be. The risk of false-positive can be lowered if O_2 is considered as a biomarker only when associated with the spectroscopic features of H_2O. Also, if the size of the planet can be measured (for instance by the shape and intensity of the infrared emission, as shown in Figure 9.2) the observers should suspect that the planet has lost most of its internal heat and is no longer active or habitable.

Ocean-planets may also differ from Earthlike planets on this specific question. Indeed, on a planet with a very high water content, an H_2O-ice layer possibly deeper than a thousand kilometres lies beneath the ocean. No carbonate–silicate cycle or terrestrial volcanism should takes place in the absence of land weathering. The CO_2 level and the balance between the release of reducing species from the interior and the loss of hydrogen to space are extremely difficult to predict. These processes should, however, be studied with interest as they could lead to an accumulation of O_2 simultaneously with the signature of H_2O if the ocean does not contain reducing species able to trap the atmospheric excess of oxidants.

9.5.3. Beyond the outer edge of the habitable zone

On Earthlike planets (or ocean-planets) orbiting beyond the outer limit of the HZ, the surface is frozen. CO_2 can accumulate to very high levels, preventing the detection of any other species in the infrared. O_2 can be produced abiotically in the atmosphere from CO_2 photolysis and, once volcanic activity has declined, O_2 may reach a level at which

TABLE 9.2 Biogenic molecules with transition bands within the *Darwin* window [5–20 µm]. The estimated abundance needed to produce a signature detectable by *Darwin* in the spectrum and the current abundance in the Earth's present atmosphere are indicated. The resolution required for detection is given in Figure 9.6. (lo) and (mi) refer respectively to the lower atmosphere, <15 km, and to the middle or upper one, >15 km.

Species	Bands (µm)	Minimum abundance	Earth abundance
CH_4	7.5	10 ppm	2 ppm
NO	5.4	1 ppm	<1 ppb (lo)
			10 ppb (mi)
NO_2	6.2	10–100 ppb	1 ppb (lo)
			0.1 ppb (mi)
N_2O	17, 8.5, 7.8	1–10 ppm	0.3 ppm
NH_3	11–9, 6	1–10 ppm	0.01 ppb

it could be detected in the reflection spectrum. The icy surface would generate an H_2O abundance too small to produce a detection, especially if the lower atmosphere is hidden under CO_2 clouds.

9.6. Other biosignatures?

Oxygenic photosynthesis is only one among many possible metabolisms invented by life on Earth, and during half of its history our planet had an anoxic atmosphere. Other sources of carbon are used (CO, CH_4, organic molecules), as well as many sources of hydrogen or electron donors (H_2S, S, H_2, CH_4, organic molecules, NH_3, NO_2, Fe^{2+}, MN^{2+}, $SO4^{2-}$, to name only a few). The outputs of these various metabolisms are present in our atmosphere as trace materials, with abundances of about, or lower than, one ppm (part per million). Table 9.2 gives a list of the terrestrial atmospheric compounds produced by biological activity and potentially detectable in the thermal emission of a planet.

None of these compounds could be detected at their current terrestrial abundances with *Darwin/TPF*. However, these abundances could have been higher in the past and this was probably the case for CH_4. The methanogenic archaea that produce CH_4 by consuming CO_2 and H_2 are very primitive organisms that appeared prior to the oxygen producers. Before the build-up of atmospheric oxygen, which occurred around 2.3 Gyr ago, the photochemical lifetime of CH_4 in the atmosphere was much longer. Nowadays, the total production of CH_4 by the biosphere is 2×10^{14} g/yr (GEIA 2002), which sustains an atmospheric CH_4 abundance of 2×10^{-6} (2 ppm). In an atmosphere without oxygen (where O_2 would be replaced for instance by N_2, CO_2 or Ar) the same production would lead to an abundance between 100 and 1000 times larger and CH_4 would then be a major greenhouse gas, detectable in a planetary spectrum. In addition, methane is currently produced by anaerobic organisms that are confined in marginal environments (marsh, sediments, animal digestive systems, underwater hydrothermal sources; GEIA 2002). Before the rise of atmospheric O_2 on the Earth, these methanogens were certainly spread over a much wider biotope and the total CH_4 production was higher than it is today.

The Earth could thus have exhibited a strong biological CH_4 signature during an important period of its existence (Schindler and Kasting 2000). CH_4 is, then, an

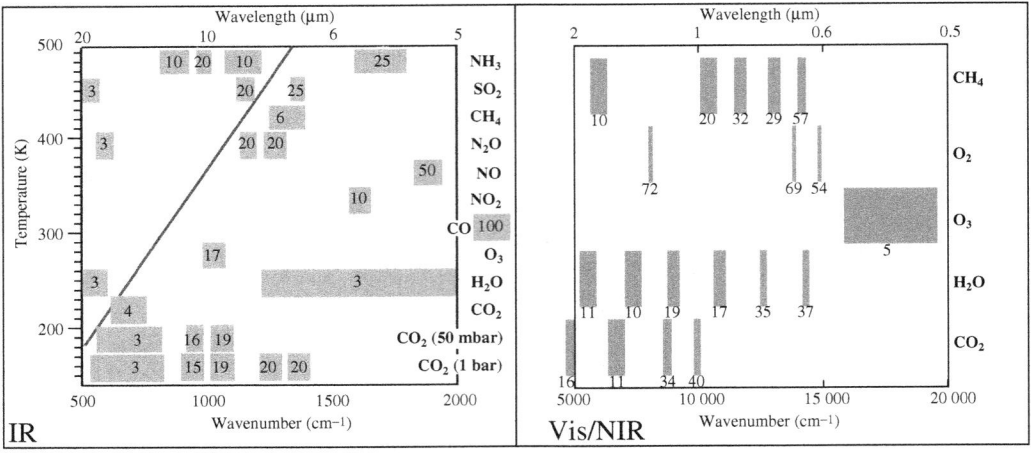

FIGURE 9.6. Potential spectroscopic signatures in the mid-infrared (5–20 μm) and in the visible/near-infrared (0.5–2 μm) ranges. The resolving power ($\lambda/\Delta\lambda$) required for detection is indicated for each molecular feature. Mid-IR: all these compounds are present (at least as trace gases) in the present-day Earth's atmosphere although only CO_2, H_2O and O_3 are effectively detectable at low resolution. The detection of the other species would imply abundances higher than their terrestrial values (see Table 1.2). The level required for detection and the consequent width of the bands have been estimated from synthetic spectra computed for various abundances. In the case of CO_2 some quantitative information can be inferred even at low resolution. The line gives the relation between the temperature and the wavelength at which black body photon emission peaks. From Selsis (2000). Visible/NIR: bands are indicated for present atmospheric abundances. Broader or new features (such as CH_4) can show up at higher levels but near-infrared is required for CO_2 detection. The Chappuis band of O_3 is a very broad but faint feature that requires a high signal-to-noise ratio. Adapted by Ollivier (2004) from spectra taken by Des Marais et al. (2002).

interesting biomarker, complementary to O_3. It is generally accepted that only a very small fraction of the CH_4 emitted into our present atmosphere comes from non-biological sources. This geothermal fraction, lower than 0.04% of the total emission, is produced in hydrothermal systems. The abiotic formation process is believed to rely on the oxidation of iron by water; this in turn releases H_2, which, in the presence of CO_2 and under specific temperature and pressure conditions, gives CH_4 (Holm and Andersson 1998). On the other hand, some authors suggest that this abiotic CH_4 flow could have been more important in the past or could exist on other planets (Pavlov et al. 2000). There is currently a controversy over the possible biological origin of methane that three independent groups claim to have detected on Mars (Krasnopolsky et al. 2004; Formisano et al. 2004; Mumma et al. 2004). It is important here to note that the very low abundance of CH_4 they found (10^{-8}, close to the detection limit) can be sustained by an extremely low emission rate (less than 10^5 molecules s^{-1} cm^{-2} on average) that can have non-abiotic sources. They give one example: methane can be produced by current or by ancient hydrothermal systems and now be released by the erosion of an old clathrate reservoir. Also, the large spatial variations of methane found by *Mars Express* are difficult to understand, whatever the origin of the CH_4, and cast some doubt on the interpretation of these measurements. Indeed, CH_4 has a photochemical lifetime of longer than 100 years and should be well mixed. The only explanation for this longitudinal gradient of CH_4 would therefore be a very recent and localized outburst, which seems unlikely. Interestingly, this debate illustrates (again) that, in such cases, the biological explanation is usually the first to be proposed. In the context of the search for biomarkers on exoplanets, it may be wise to keep in mind Carl Sagan's advice: 'Extraordinary claims require extraordinary evidence'.

Methane is also an abundant compound in the external and cold part of our Solar System: it is a major component of Titan's atmosphere and is part of the composition of giant gas planets. According to models (Prinn 1993), the carbon present in the protosolar nebula, before planet formation, was mainly in the form of CH_4 in the external region and in CO in its inner region. The limit beyond which CH_4 condenses, under the pressures and temperatures in this nebula, is at about 10 AU from the Sun, which explains why one finds CH_4 ice on Pluto, Triton and Kuiper belt objects. Cometary ices typically contain 0.5–1% of CH_4 (Crovisier 1994) and a large fraction of their mass is made of organic matter. Comet impacts on a terrestrial planet could thus be a significant source of CH_4 (Kress and McKay 2004), a part of which would be a byproduct of the organic matter after the impact. One percent of the mass of a Hale–Bopp-sized comet converted into methane after an impact would result in an atmospheric level of CH_4 equal to the present terrestrial one. This confirms that, in the absence of a more thorough study, it is not possible to consider CH_4 alone as an unambiguous biomarker. It remains an interesting indicator in the habitable zone, though. The simultaneous detection of O_2 (or O_3) and CH_4 was suggested as a biosignature by Lovelock (1975) and Sagan *et al.* (1993) and probably remains the most reliable biosignature in both the visible/NIR and IR ranges. However, the atmospheric levels of CH_4 required for a remote detection are higher than on the present-day Earth and might not be compatible with an O_2-rich atmosphere. Among the other components that have a direct or secondary biological origin, one finds several nitrogen compounds, mainly nitrogen protoxide, N_2O, for which no significant abiotic source is known. N_2O is produced by bacteria (known as nitrifying and denitrifying) on the ground and in the oceans, at a rate of 10^{13} g/yr (GEIA 2002). Its content in the atmosphere, although small (0.3 ppm), plays a major part in photochemistry. It is indeed the main source of nitrogen oxides (NO and NO_2), which are the main destructors of ozone in the middle and lower atmosphere. It is an optically very active gas in the infrared and, regarding the greenhouse effect, it is even more effective than CH_4, for a given mass. In a terrestrial thermal emission spectrum, the signature of N_2O at 7.8 μm is detectable, but a quantity at least ten times larger would be needed to detect N_2O on an exoplanet with *Darwin/TPF*. It is especially difficult to state whether such quantities are realistic and whether they could have occurred on the Earth, in the past – particularly before the rise of O_2. Segura *et al.* (2003) have calculated the level of N_2O for different O_2 levels and found that, although N_2O is a reduced species, its level decreases with O_2. This 'unexpected' result is due to the fact that a decrease in O_2 produces an increase in H_2O photolysis resulting in the production of more hydroxyl (OH) radicals responsible for the destruction of N_2O.

Atmospheric N_2O is oxidized and photolyzed in the upper atmosphere, where it produces NO and NO_2, which are themselves IR-active compounds. However, the minimum quantities for the detection of these two gases are much higher than their current terrestrial abundance and it is considered that they can also be produced by abiotic sources related to lightning (especially in volcanic plumes, Navarro-Gonzalez *et al.* 1998, and impacts, Prinn & Fegley 1987), at a rate larger than that coming indirectly from the biosphere.

Another nitrogen compound, ammonia (produced by nitrogen-fixing bacteria), is present at a trace level in our atmosphere. NH_3 has an extremely short lifetime because of our current oxidizing environment but also because of its photolysis by UV. This explains its very low abundance (0.01 ppb), which is 2×10^5 times lower than that of

CH$_4$ in spite of a production rate only five times lower (7.5 × 10^{13} g/yr; GEIA 2002). As for CH$_4$, there are abiotic sources of NH$_3$ in the outer Solar System. The necessary concentration for the detection of NH$_3$ (1–10 ppm) is so high that it would imply a large and continuous production source, which it would be difficult to assign to an abiotic process. It is also not obvious that any ecosystem would be able to produce such quantities of NH$_3$.

The spectral signature of chlorophyll (the red-edge seen at 700–750 nm and due to the absorption band of the pigment) has been suggested as a marker for vegetation on exoplanets (Arnold et al. 2002; Woolf et al. 2002). These authors have performed observations of earthshine as an analogue to the observation of an extrasolar Earthlike planet. However, the feature produced in an integrated Earth spectrum is very difficult to detect (and is still not clearly detected in the earthshine spectrum from our point of view) because of the small fraction of vegetation-covered lands and cloudiness. Paillet et al. (2007 in preparation) show that in order to produce a feature detected at SNR = 20 more than 20% of the cloud-free surface of a planet should be covered by Earthlike vegetation (equivalent to 40% on average for a uniform cloudiness of 50%). Such vegetation should have evolved through a similar pathway leading to the selection of chlorophyll as a pigment. Knowing that other pigments exist on Earth, and that some minerals can exhibit a similar spectral shape around 750 nm (Seager et al. 2005), the detection of the red-edge of the chlorophyll on exoplanets, despite its interest, should not be considered as a driver for the first launched planet finders.

9.7. Temperature and radius of the planets

Simultaneously with the information on the chemical composition of a planet's atmosphere, it would be very interesting to obtain information on the surface temperature from thermal emission as an indicator of its habitability. What can we know, remotely, about the temperature of an exoplanet? This question is far from easy to answer. In theory, spectroscopy can provide some detailed information on the thermal profile of a planetary atmosphere. This, however, requires a spectral resolution and sensitivity that are well beyond the performance of an instrument like *Darwin*. The question must thus be adapted to the observational means that will initially be available. First, it is possible to calculate the stellar energy, F_{star}, received at the measured orbital distance ($F_\odot$ = 1370 Wm^{-2} at 1 AU). This gives us, in fact, little information on the temperature of the planet, which also depends on its albedo. The albedo of terrestrial planets can range from about 5% (ocean) to 80% (ice). Also, even if the assumed albedo is close to the real one, the surface temperature is likely to be enhanced by greenhouse gases. If the radius of the planet is not known, there is no practical way to infer the temperature from the integrated IR flux alone. Indeed, a small but hot planet could emit the same energy as a large but colder planet.

However, with a low-resolution spectrum of the thermal emission ($\lambda/\Delta\lambda > 10$), the mean brightness temperature and the radius of the planet can be obtained by fitting the envelope of the thermal emission with a Planck function (see Figure 9.2). Ability to associate a brightness temperature with the spectrum relies on the existence and identification of spectral windows probing the surface or the same atmospheric levels (indicated by dashed arrows on the graph). Such identification is not trivial in the absence of any other information on the observed planet but there are some atmospheric windows that can be used in most cases, especially between 8 and 11 µm, as seen in Figure 9.6. This window would, however, become opaque at high H$_2$O partial pressures (in the inner part of the HZ, where a lot of water is vapourized) and at high CO_2 pressures (in the outer part of the HZ).

In the case of Venus, Earth and Mars, three terrestrial planets with a dense atmosphere, is the black-body temperature deduced from the IR spectrum a good approximation to the surface temperature? The answer is 'yes' in the case of Mars, where the surface emission is directly observed (except in the CO_2 band, see Figure 9.2). On Earth, the IR spectrum is a mixture of surface and cloud emission, the latter occurring at lower temperatures. The temperature given by the envelope of the spectrum is thus slightly lower, by about 10 K on average, than the average surface temperature. The temperature measured at one position in the orbit also depends on the observation geometry, as well as seasons, and differs from the mean temperature of the Earth ($\sim$288 K). In the extreme case of Venus, the spectrum envelope gives a temperature of 277 K, much lower than the 740 K of the surface. Observed in the mid-infrared, Venus could be considered as a habitable planet, with a mean temperature just above the triple point of water. The reason for this discrepancy is that the venusian atmosphere is completely opaque below 60 km because of the permanent cloud cover and the absorption continuum, induced at high pressure by CO_2–CO_2 collisions. With low-resolution spectral observations, it is difficult to know unambiguously whether or not the lower atmosphere contributes to the spectrum and hence whether the temperature reflects the surface conditions.

The presence of a cloud layer, or more generally of an opaque atmosphere, does not render the radius measurement impossible so long as the thermal emission comes from similar atmospheric levels or atmospheric levels at a similar temperature. The accuracy of the radius determination will depend on the quality of the fit (and thus on the sensitivity and resolution of the spectrum), the precision of the Sun–star distance (known with a precision of better than 2% for G- and K-stars within 20 pc of the Sun) and also the distribution of brightness temperatures over the planetary surface. Concerning the latter, we may consider a Lambertian sphere (like the Moon, Lawson *et al.* 2000) as a 'worst case' exhibiting high temperature contrasts. In such a case, the inferred radius would be at least 10% smaller than the real one, when the observation is made at full visual phase, and much smaller at lower phases. However, in this case, the variation of the thermal flux (and thus of the measured radius) with phase would reveal the absence of an atmosphere and the estimate could be readjusted. When the brightness temperature is stable along the orbit, the estimated radius is more reliable. Moreover, the radius can be measured at different points of the orbit and thus for different values of T_b, which should allow an estimate of the error made in its determination.

The determination of the radius and the brightness temperature, T_b, of the planet gives us the thermal flux, $\sigma T_b^4 R^2/d^2$, received by the observer (where d is the planet–observer distance). The observed brightness temperature averaged over a hemisphere depends on the relative positions of the star, the planet and the observer and varies with season, the inclination of the system and the phase of the planet. The thermal lightcurve (i.e. the integrated infrared emission measured at different positions in the orbit) exhibits variations due to the phase (whether the observer sees mainly the day side or the night side) and to the season. Important phase-related variations are due to a high day/night temperature contrast and imply a low greenhouse effect and the absence of a stable liquid ocean. Habitable planets can therefore be distinguished from airless or Mars-like planets by the amplitude of the observed variations of T_b (Gaidos and Williams 2004). Unfortunately, Venuslike atmospheres would also exhibit extremely low amplitudes. Low phase-related variations and a total absence of seasonal variations would be signature of an Earthlike planet with a $0°$ obliquity or a Venuslike planet.

The mean value of T_b estimated over an orbit can be used to estimate the Bond albedo, A, through the balance between the incoming stellar radiation and the outgoing IR emission:

$$F_{\text{star}}(1-A) = 4\sigma \langle T_b \rangle^4_{\text{orbit}},$$

where F_{star} is the flux of the central star at the planet's location.

In the visible ranges, the reflected flux allows us to measure the product $A \times R^2$, where R is the planetary radius. The first generation of optical instruments will be very far from the angular resolution required to directly measure an exoplanet radius. At present, such a measurement can only be performed when the planet transits its parent star and by an accurate photometric technique (see Chapter 3). This is probably the main weakness of the characterization in the visible range. Indeed, knowledge of the planetary radius is crucial for a general understanding of the physical and chemical processes occurring on the planet (tectonics, hydrogen loss to space). However, if the same target is observed in both the visible (*TPF-C*) and IR (*Darwin*), the albedo can be obtained once the radius is inferred from the IR spectrum. This measurement of the albedo can then be compared to the albedo estimated from the IR lightcurve.

9.8. The benefits from both infrared and visible observations

The possibility of obtaining spectral information from both the reflected light in the visible and the thermal emission in the IR allows us to characterize single planets in greater detail, and also to explore a wider domain of planet diversity. First of all, VIS + IR observations can confirm the presence of atmospheric compounds such as H_2O, CH_4 and O_3, having spectral features in both wavelength ranges. Oxygen-rich atmospheres detected through their O_3 band in the IR can be confirmed by the signature of O_2 in the visible. Some important species like CO_2 (a tracer of habitable planets), N_2O (a reliable biosignature) appear only in the IR range, while the reflected spectrum is the only one that can give information on the nature of the surface. This information on the surface composition (oceans, ice, rocks, the presence of clouds) can be obtained once the absolute level of the albedo is known, which requires a knowledge of the radius, obtained through the IR spectrum.

Being able to measure the outgoing visible and IR radiation and their variations along the orbit, to determine the albedo, and to identify greenhouse gases would allow us to understand the climate system at work on the observed world. The observed climate could be compared with the theoretical predictions, and in particular to the relation between the orbital distance and the atmospheric evolution: we could test for the existence of carbonate–silicate climate regulation on exoplanets, as well as the limits of the habitable zone.

Planets with significant cloudiness may be difficult to study in the infrared, while they provide better targets in the visible. This effect is illustrated in Figure 9.7: let us consider an atmospheric compound A with an optically thin transition in the visible (such as O_2) and a compound B with an optically thick transition in the infrared (e.g. O_3). The clouds hide the lower part of the column density of the absorber. In the visible, this lowers the effective opacity in the transition of A, but as the continuum level is enhanced by the reflection on the clouds, the absolute depth of A is more easily detected. In the infrared, the opacity between the top of the clouds and the top of the atmosphere remains very high but the difference between the brightness temperature of the continuum and at the bottom of the band is lowered by the presence of clouds, making the spectral feature of B more difficult to detect.

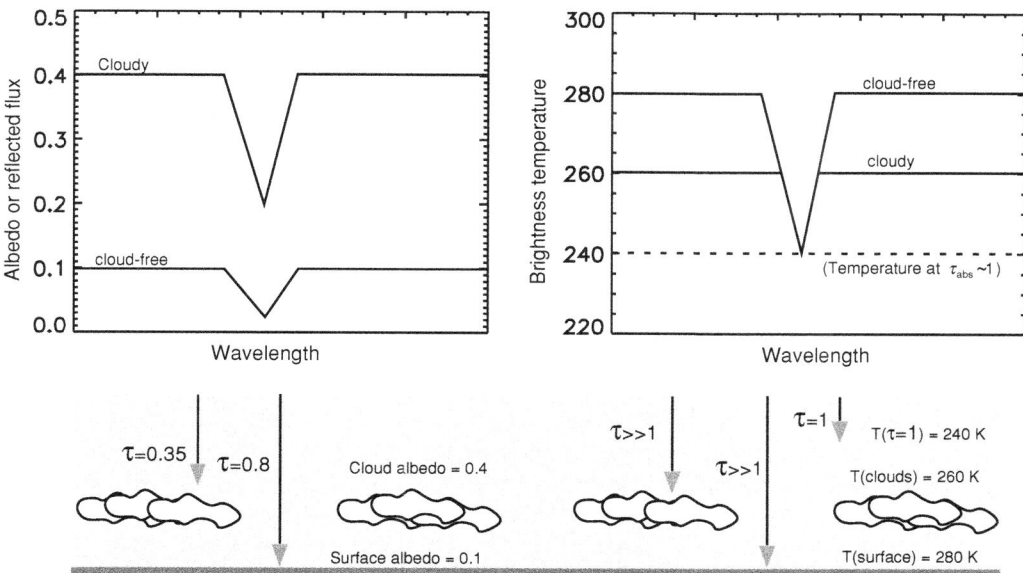

FIGURE 9.7. Effect of clouds on the depth of a molecular line in the visible reflection spectrum (left) and on a thermal emission spectrum (right). The left panel is an illustration of the fact that the effective depth of a line in a spectrum (measured in missing photons compared to the continuum) can be much greater when clouds (with a high albedo) are present, even though most of the absorbing spectrum is hidden beneath the clouds. It is thus impossible for the remote observer to establish a relation between line depth and column density. This is important because such a relation is often claimed as an advantage of the visible over the IR. The right panel illustrates that in the IR, where most of the bands are saturated, clouds that are colder than the surface lower the depth of a line significantly.

Hence, clouds help in the detection of lines by increasing the signal and the contrast between the lines and the continuum, as they increase the albedo by a factor of 2 to 10 (depending on the nature of the surface). On the other hand, they prevent us from estimating a column density. Neither the visible or the IR are able to provide a robust estimate of the column density, except for lower limits in some circumstances. However, it is not the aim of *Darwin* or *TPF* to measure the amount of species. Detection (and thus lower limits) will be the first step undertaken with these missions.

The relation between the CO_2 level and the orbital distance predicted by Kasting *et al.* (1993) requires observations in the infrared, since CO_2 cannot be detected in the visible (weak transitions start at $\lambda > 1.05$ μm). However, in the outer part of the HZ, and perhaps in the whole HZ around M-stars, CO_2 reaches levels that should prevent us from detecting anything else in the IR range. For planets in these regions, the reflected spectrum will be necessary to search for other species and in particular O_2-rich atmospheres. Also, dusty planetary systems, emitting a strong zodiacal light, could be difficult to study in the IR and less affected in the visible. All these reasons show how promising the achievement of both *Darwin* and *TPF-C* would be within a similar time frame.

9.9. Conclusion and other perspectives

As exciting as the search for life signatures can be, it remains an uncertain objective, which should be neither the single nor the main driver of missions such as *Darwin* and *TPF*. These observatories will be wonderful tools for the study of terrestrial planets, their atmosphere, their formation and their evolution. Doing planetary science with *Darwin/TPF* is essential for astrobiology: it will considerably improve our knowledge of terrestrial planets as possible sites for life. We should in particular get clues towards answering the following question of whether our planetary system and our planet are common objects in the Universe or are on the contrary extremely marginal.

First of all, the existence itself of terrestrial planets in the habitable zone of stars is considered by planetary formation models but cannot be confirmed with our current observational tools. The first major element will thus be statistical information on the distribution of these small planets. On this point, other space observatories should give us information before *Darwin*; first *COROT*, then the *Kepler* mission, based on planetary transit observations, should have the sensitivity to detect Earth-sized planets. In contrast to these transit missions, *Darwin/TPF* will give us this information for nearby systems (<20 pc approximately). For some of the closest targets, *SIM* (NASA) may also determine the mass of the detected habitable planets, down to 1–2 $M_\oplus$ (Unwin and Turyshev 2004). Also, for the *Darwin/TPF* targets, complementary data concerning the presence or otherwise of giant planets will be available in time from radial velocity and astrometry surveys. The influence of giant planets on the habitability of terrestrial planets is not yet well understood but is probably important, regarding the formation of inner telluric planets itself, the origin of water and the volatile compounds on these planets, or the bombardment of these planets by asteroids or comets (Levison and Agnor 2003; Raymond *et al.* 2004).

For each system studied, the observed properties of terrestrial planets (such as their distribution, mass, orbit, the presence or otherwise of an atmosphere, chemical composition), in combination with the properties of giant planets of the system, will give us vital clues about the origin, formation and evolution of planetary systems, their atmospheres and perhaps life. Instruments such as *Darwin* and *TPF* would also inform us considerably about the concept of habitable zone. Are these zones filled with planets as in the Solar System? Do they have atmospheres? If so, what is their composition? Is CO_2 the only greenhouse gas that provides habitability? How do the characteristics of these planets vary with the spectral type and the metallicity of their star, the distribution of giant planets, or the age of the system? The prospect of being able to observe very young planetary systems (less than 0.5–1 Gyr) is also very promising. The atmospheres of Solar System planets have evolved and differ from their primitive composition. The nature of the prebiotic environment on Earth is probably one of the keys to understanding the origins of life. However, it is possible that only the observation of young exoplanets could provide part of the answer.

For instance, χ^1 Ori is a G1 V-star located at 8.7 pc from the Earth, comparable to, or at least very similar to, the Sun (Ribas *et al.* 2005); its mass is 0.99 $M_\odot$ and its age is estimated at 300 Myr. At the same age, the Earth already had oceans (Wilde *et al.* 2001) and an atmosphere, of unknown composition. Does χ^1 Ori have planets? If so, what are they made of? An instrument such as *Darwin/TPF* might perhaps highlight these points, giving us a detailed sequence of the evolution of the terrestrial planets and their atmosphere.

REFERENCES

ANGEL, J. R. P., CHENG, A. Y. S. & WOOLF, N. J., 1986, *Nature* **322**, 341.
ARNOLD, L., GILLET, S., LARDIÈRE, O., RIAUD, P. & SCHNEIDER, J., 2002, *A&A* **392**, 231.

BARAFFE, I., CHABRIER, G., ALLARD, F. & HAUSCHILDT, P., 1998, A&A **337**, 403.

BARKER, E. S., 1972, Nature **238**, 447.

BEICHMAN, C. A., WOOLF, N. J. & LINDENSMITH, C. A. (eds.), 1999, *The Terrestrial Planet Finder (TPF): a NASA Origins Program to Search for Habitable Planets*, Pasadena: JPL Publications.

BORUCKI, W. J., KOCH, D. G., DUNHAM, E. W. & JENKINS, J. M., 1997, in *Planets Beyond the Solar System and the Next Generation of Space Missions*, ed. D. Soderblom, ASP Conference Series 119, p. 153.

BRACEWELL, R. N., 1978, Nature **274**, 780.

CARLETON, N. P. & TRAUB, W. A., 1972, Science **177**, 988.

CHARBONNEAU, D., BROWN, T. M., NOYES, R. W. & GILLILAND, R. L., 2002, ApJ **568**, 377.

CHARBONNEAU, D., ALLEN, L. A., MEGEATH, S. T., et al., 2005, ApJ **626**, 523.

CLANCY, R. T., WOLFF, M. J. & JAMES, P. B., 1999, Icarus **138**, 49.

CROVISIER, J., 1994, in IAU Symp. 160: *Asteroids, Comets, Meteors*, eds. A. Milani, M. Di Martian & A. Cellino, International Astronomical Union, Symposium no. 160. Dordrecht: Kluwer, p. 313.

DEMING, D., SEAGER, S., RICHARDSON, L. J. & HARRINGTON, J., 2005, Nature **434**, 740.

DES MARAIS, D. J., HARWIT, M. O., JUCKS, K. W., et al., 2002, Astrobiology **2**, 153.

ESPENAK, F., MUMMA, M. J., KOSTIUK, T. & ZIPOY, D., 1991, Icarus **92**, 252.

FORMISANO, V., ATREYA, S., ENCRENAZ, T., IGNATIEV, N. & GIURANNA, M., 2004, Science **306**, 1758.

GAIDOS, E. & WILLIAMS, D. M., 2004, New Astronomy **10**, 67.

GEIA, 2002, www.geiacenter.org.

HALL, D. T., STROBEL, D. F., FELDMAN, P. D., MCGRATH, M. A. & WEAVER, H. A., 1995, Nature **373**, 677.

HALL, D. T., FELDMAN, P. D., MCGRATH, M. A. & STROBEL, D. F., 1998, ApJ **499**, 475.

HOLM, N. G. & ANDERSSON, E. M., 1998, *Organic Molecules on the Primitive Earth: Hydrothermal Systems*, Cambridge: Cambridge University Press, p. 86.

JOHNSON, R. E. & JESSER, W. A., 1997, ApJ Lett. **480**, L79.

KASTING, J. F., 1988, Icarus **74**, 472.

KASTING, J. F., 1995, Planet. Space Sci. **43**, 11.

KASTING, J. F. & CATLING, D., 2003, ARA&A **41**, 429.

KASTING, J. F., WHITMIRE, D. P. & REYNOLDS, R. T., 1993, Icarus **101**, 108.

KRASNOPOLSKY, V. A., MAILLARD, J. P. & OWEN, T. C., 2004, Icarus **172**, 537.

KRESS, M. E. & MCKAY, C. P., 2004, Icarus **168**, 475.

LAWSON, S. L., JAKOSKY, B. M., PARK, H. & MELLON, M. T., 2000, J. Geophys. Res. **105**, 4273.

LEFÈVRE, F., LEBONNOIS, S., MONTMESSIN, F. & FORGET, F., 2004, J. Geophys. Res. **109(E18)**, 7004.

LÉGER, A., PIRRE, M. & MARCEAU, F. J., 1993, A&A **277**, 309.

LÉGER, A., MARIOTTI, J. M., MENNESSON, B., OLLIVIER, M., PUGET, J. L., ROUAN, D. & SCHNEIDER, J., 1996, Icarus **123**, 249.

LÉGER, A., OLLIVIER, M., ALTWEGG, K. & WOOLF, N. J., 1999, A&A **341**, 304.

LÉGER, A., SELSIS, F., SOTIN, C., et al., 2004, Icarus **169**, 499.

LEVISON, H. F. & AGNOR, C., 2003, AJ **125**, 2692.

LOVELOCK, J. E., 1975, Proc. Roy. Soc. London **B189**, 167.

MILLS, F. P., 1999, J. Geophys. Res. **104**, 30757.

MUMMA, M. J., NOVAK, R. E., DISANTI, M. A., BONEV, B. P. & DELLO RUSSO, N., 2004, Bull. AAS **36**, 1127.

NAIR, H., ALLEN, M., ANBAR, A. D., YUNG, Y. L. & CLANCY, R. T., 1994, *Icarus* **111**, 124.

NAVARRO-GONZALEZ, R., MOLINA, M. J. & MOLINA, L. T., 1998, *Geophys. Res. Lett.* **25**, 3123.

NOLL, K. S., JOHNSON, R. E., LANE, A. L., DOMINGUE, D. L. & WEAVER, H. A., 1996, *Science* **273**, 341.

NOLL, K. S., ROUSH, T. L., CRUIKSHANK, D. P., JOHNSON, R. E. & PENDLETON, Y. J., 1997, *Nature* **388**, 45.

OLLIVIER, M., 1999, Ph.D. thesis, Université de Paris XI.

OLLIVIER, M., 2004, in *Extrasolar Planets: Today and Tomorrow*, eds. J.-P. Beaulieu, A. Lecavelier des Etangs and C. Terquem, ASP Conference Proceedings, vol. 321, p. 183.

OWEN, T., 1980, in *Strategies for the Search for Life in the Universe*, ed. M. Papagiannis, Dordrecht: Reidel.

OWEN, T., BIEMANN, K., BILLER, J. E., LAFLEUR, A. L., RUSHNECK, D. R. & HOWARTH, D. W., 1977, *J. Geophys. Res.* **82**, 4635.

PAVLOV, A. A., KASTING, J. F., BROWN, L. L., RAGES, K. A. & FREEDMAN, R., 2000, *J. Geophys. Res.* **105**, 11981.

PRINN, R. G., 1993, in *Protostars and Planets III*, Tucson: University of Arizona Press, p. 1005.

PRINN, R. G. & FEGLEY, B., 1987, *Earth Planet. Sci. Lett.* **83**, 1.

RAYMOND, S. N., QUINN, T. R. & LUNINE, J. I., 2004, *Icarus* **168**, 1.

RIBAS, I., GUINAN, E. F., GÜDEL, M. & AUDARD, M., 2005, *ApJ* **622**, 680.

ROSENQVIST, J. & CHASSEFIERE, E., 1995, *Planet. Space Sci.* **43**, 3.

ROUAN, D., BAGLIN, A., COPET, E., *et al.*, 1998, *Earth, Moon, Planets* **81**, 79.

SAGAN, C., THOMPSON, W. R., CARLSON, R., GURNETT, D. & HORD, C., 1993, *Nature* **365**, 715.

SCHINDLER, T. L. & KASTING, J. F., 2000, *Icarus* **145**, 262.

SEAGER, S., TURNER, E. L., SCHAFER, J. & FORD, E. B., 2005, *Astrobiology* **5**, 372.

SEGURA, A., KRELOVE, K., KASTING, J. F., *et al.*, 2003, *Astrobiology* **3**, 689.

SELSIS, F., 2000, Ph.D. thesis, Université de Bordeaux I.

SELSIS, F., DESPOIS, D. & PARISOT, J.-P., 2002, *A&A* **388**, 985.

SPENCER JONES, H., 1959, *Life on Other Worlds*, London: Hodder and Stoughton.

TRAUGER, J. T. & LUNINE, J. I., 1983, *Icarus* **55**, 272.

UNWIN, S. & TURYSHEV, S. (eds.), 2004, *Science with the Space Interferometry Mission*, Pasadena: JPL Publications.

VIDAL-MADJAR, A., DÉSERT, J.-M., LECAVELIER DES ETANGS, A., *et al.*, 2004a, *ApJ Lett.* **604**, L69.

VIDAL-MADJAR, A., LECAVELIER DES ETANGS, A., DÉSERT, J.-M., *et al.*, 2004b, *Nature* **422**, 143.

VOLONTE, S., LAURANCE, R., WHITCOMB, G., *et al.*, 2000, *Darwin: the Infrared Space Interferometer*, Technical report, ESA.

WALKER, J. C. G., 1977, *Evolution of the Atmosphere*, New York: Macmillan.

WILDE, S. A., VALLEY, J. W., PECK, W. H. & GRAHAM, C. M., 2001, *Nature* **409**, 175.

WOOLF, N. J., SMITH, P. S., TRAUB, W. A. & JUCKS, K. W., 2002, *ApJ* **774**, 430.

Printed by Printforce, United Kingdom